Lecture Notes in Physics

Edited by J. Ehlers, München, K. Hepp, Zürich
R. Kippenhahn, München, H. A. Weidenmüller, Heidelberg
and J. Zittartz, Köln
Managing Editor: W. Beiglböck, Heidelberg

118

...um Chromodynamics

...s of the
...ernational Seminar on Theoretical Physics
..., Huesca (Spain) June 1979

Edited by J. L. Alonso and R. Tarrach

Springer-Verlag
Berlin Heidelberg New York 1980

Editors

J. L. Alonso
Facultad de Ciencias
Universidad de Zaragoza
Zaragoza
Spain

R. Tarrach
Facultad de Física
Universidad de Barcelona
Barcelona
Spain

ISBN 3-540-09969-7 Springer-Verlag Berlin Heidelberg New York
ISBN 0-387-09969-7 Springer-Verlag New York Heidelberg Berlin

Library of Congress Cataloging in Publication Data. International Seminar on Theoretical Physics, 10th, Jaca, Spain, 1979. Quantum chromodynamics. (Lecture notes in physics ; 118) Bibliography: p. Includes index. 1. Quantum chromodynamics--Congresses. I. Alonso, José L., 1942- II. Tarrach, R., 1948- III. Grupo Interuniversitario de Física Teórica. IV. Title. V. Series. QC793.3.Q35I55. 1979. 539.7'2. 80-13256

Printed in Germany

Printing and binding: Beltz Offsetdruck, Hemsbach/Bergstr.
2153/3140-543210

UNIVERSIDAD
DE
ZARAGOZA

TABLE OF CONTENTS

QUANTUM CHROMODYNAMICS

X INTERNATIONAL SEMINAR ON THEORETICAL PHYSICS

Jaca, June 4 - 9, 1979

List of Participants

- Abad, J.	(Zaragoza)
- Adeva, B.	(J.E.N.)
- del Aguila F.	(Barcelona Autonoma)
- Alonso JL.	(Zaragoza)
- Alonso F.	(C.S.I.C.)
- Antolin J.	(Zaragoza)
- Asorey M.	(Zaragoza)
- Azcárraga J.	(Valencia)
- Azcoiti V.	(Zaragoza)
- Baig M.	(Barcelona Autonoma)
- Bartels J. (Lecturer)	(Hamburg)
- Bernabéu J. (Organizer)	(Valencia)
- Berdugo FJ.	(J.E.N.)
- Botella JF.	(Valencia)
- Bramon A.	(Barcelona Autonoma)
- Bravo J.R.	(Zaragoza)
- Carbajo F.	(Madrid Autonoma)
- Cariñena JF.	(Zaragoza)
- Castillejo M.	(Madrid Autonoma)
- Cerveró JM.	(Salamanca)
- Cortes JL.	(Zaragoza)
- Cornet F.	(Barcelona Auotnoma)
- Cruz A.	(Zaragoza)
- Delgado V.	(J.E.N.)
- Dominguez R.	(Madrid Autonoma)
- Eichler R.	(DESY)
- Elizalde E.	(Barcelona Central)
- Espriu D.	(Barcelona Central)
- Fernandez JC.	(J.E.N.)
- Fernandez AE?R.	(Madrid Complutense)
- Fernandez R.A. (Director GIFT)	(Madrid Complutense)
- Fritzsch H. (Lecturer)	(Bern)
- Garcia Esteve JL.	(Zaragoza)
- Garcia F.P.	(Madrid Complutense)
- Garzon J.	(J.E.N.)
- Goñi MA.	(Madrid Complutense)
- Gerhold H.	(Wien)
- Gomez C.	(Salamanca)
- Grifols JA.	(Barcelona Autonoma)
- Hernandez M.A.	(Madrid Autonoma)
- Jacobs L.	(U.N.A.M. Mexico)
- Jordan de Urries	(C.S.I.C.)
- Julve J.	(C.S.I.C.)
- Leon J.	(C.S.I.C.)
- Leutwyler H. (Lecturer)	(Bern)

- Lopez C. (Madrid Autonoma)
- Lorente M. (Madrid COmplutense)
- Lynch H. (DESY) (Lecturer)
- Llanta E. (Barcelona Central)
- Llosá R. (J.E.N.)
- Muñoz A. (Madrid Complutense)
- Mur H. (Barcelona Autonoma)
- Narisson S. (Marseille)
- Noguera S. (Valencia)
- Nuñez Lagos R. (Zaragoza)
- del Olmo J. (Madrid Complutense)
- Palanques A. (Barcelona Central)
- Pajares C. (Organizer) (Barcelona Autonoma)
- Pascual P. (Barcelona Central)
- Pascual R. (Barcelona Autonoma)
- Peñarrocha J. (Valencia)
- Pire B. (Ecole Polytechnique Palaiseau)
- Poch A. (Barcelona Central)
- Pons JM. (Barcelona Central)
- Quiros M. (C.S.I.C)
- de Rafael E. (Lecturer) (Marseille)
- Ringland G. (Rutherford Lab)
- Roy L.J. (Madrid Autonoma)
- Sachrajda C. (Lecturer) (CERN)
- Sanchez-Guillem (Zaragoza)
- Sanchez-Velasco (Madrid Autonoma)
- Santander M. (Valladolid)
- Seguí A. (Zaragoza)
- Sierra G. (Madrid Autonoma)
- Socolovsky M. (Mexico)
- Tarrach R. (Organizer) (Barcelona Central)
- Tiemblo A. (C.S.I.C.)
- Yndurain FJ. (Organizer) (Madrid Autonoma)

FOREWORD

This volume contains the Lectures delivered at the X G.I.F.T.* International Seminar on Theoretical Physics on the subject "Quantum Chromodynamics" which was held at Jaca, Huesca, (Spain) in June 1979. The lecturers were J. Bartels, H. Fritzsch, H. Leutwyler, H. Lynch, E. de Rafael, and C. Sachrajda, who covered both theoretical and phenomenological aspects of Q.C.D. Around 80 physicists attended the Lectures at the Residence of the University of Zaragoza in Jaca. The members of the Organizing Committee of the Seminar were J.L. Alonso (Zaragoza University), J. Bernabéu (Valencia University), C. Pajares (Barcelona Autonomous University), R. Tarrach (Barcelona University) and F.J. Yndurâin (Madrid Autonomous University).

The Seminar was supported financially by the Instituto de Estudios Nucleares, Madrid and I.C.E., Zaragoza. We wish to express our thanks to the J.E.N. (Junta de Energía Nuclear) and the University of Zaragoza for their support. The efficient help of Ms. Maribel Ramoneda, Secretary of the course, is gratefully acknowledged.

J.L. Alonso

R. Tarrach

---

* The Spanish Interuniversity Group of Theoretical Physics (G.I.F.T.) associates Physicists working in Theoretical Physics all over Spain. Its aim is stimulating and coordinating research as well as the training of physicists devoted to research and teaching.

QUANTUM CHROMODYNAMICS
AS A THEORETICAL FRAMEWORK OF THE HADRONIC INTERACTIONS

E. de RAFAEL

Centre de Physique Théorique, Section 2, CNRS, Luminy

PREFACE

These notes are the written version of the lectures I gave at the Gif Summer School in September 1978 and at the GIFT Seminar in June 1979. The lectures were supposed to provide an elementary theoretical background to the topics covered by the other lectures at the Gif and Gift Seminars . Some of the topics I talked about have been considerably developped in this written version; in particular the relation between different renormalization schemes, and questions related to quark masses. On the other hand I have not included here other topics which were discussed in the lectures, like the infrared behaviour of perturbative QCD; and large p_T -behaviour of perturbative gauge theories. A short review of these two topics with earlier references can be found in the talk I gave at the France-Japan joint seminar (see ref.[R.31]). On the latter subject, large p_T -behaviour, there has been a lot of progress since then, specially in connection with applications to hard scattering hadronic processes. I recommend for a review Sachrajda's lecture at the XIIIth rencontre de Moriond (see ref.[R.32]).

Another topic not covered in these lectures is the application of QCD to deep inelastic scattering of leptons on nucleons. There is a recent review of the subject by Peterman (see ref.[R.33]).

There is another interesting development not included in these lectures, the subject of current algebra spectral function sum rules viewed from QCD. Two detailed references on this topic are [R.34] and (6.2).

References and footnotes are collected at the end of each chapter. References to textbooks, lecture notes and review articles [R] are collected at the end.

In writing these lectures I have benefited very much from the comments and questions of the stimulating audience at Gif and from my colleagues at the CPT in Marseille; in particular, Robert Coquereaux.

1. INTRODUCTION

The most succesful quantum field theory we have at present is Quantum Electrodynamics (QED). It describes the interaction of photons with matter. The quantitative success of the theory lies on the empirical fact that there exist particles in Nature, the charged leptons (electrons, muons, ...), whose dominant interaction is electromagnetic. Precise measurements of various electromagnetic observables have been confronted with perturbative approximations to the equations of motion of QED to the remarkable accuracy of a few parts per million[1]. The Lagrangean of QED reads

$$\mathcal{L}_{QED}(x) = -\frac{1}{4} F_{\mu\nu}(x) F^{\mu\nu}(x) + i\bar{\Psi}(x)\gamma^\mu\partial_\mu\Psi(x) - m\bar{\Psi}(x)\Psi(x) - e\bar{\Psi}(x)\gamma^\mu\Psi(x)A_\mu(x) \tag{1.1}$$

where

$$F_{\mu\nu}(x) = \partial_\mu A_\nu(x) - \partial_\nu A_\mu(x) \tag{1.2}$$

is the electromagnetic field tensor and $A_\mu(x)$ the vector potential describing the photon field. The first term in eq. (1.1) describes free radiation. The second and third terms are associated to free matter, spin 1/2 particles with mass m. The interaction matter-radiation is governed by the last terms where e denotes the electric charge of the matter field ($e^2/4\pi = \alpha$ the fine structure constant, $\hbar = c = 1$). When the fields $A_\mu(x)$ and $\Psi(x)$ are quantized and the interaction term is treated as a perturbation of the free Lagrangean, a well defined picture emerges with which to calculate observables to the required degree of accuracy. The theory is then a relativistic quantum field theory treated perturbatively and in that respect it is a triumph of the quantization concept when extended to relativistic systems with an infinite number of degrees of freedom.

Since the advent of QED in the late 40's the challenge of theoretical physicists has been to find dynamical theories which could describe the other interactions observed in Nature in the way that QED describes the electromagnetic interactions. During the last decade there has been a tremendous progress in developing quantum field theories potentially useful to incorporate both the electromagnetic and weak interactions of the fundamental constituents of matter: the leptons and the quarks. These are non-abelian gauge quantum field theories with spontaneous symmetry breakdown[2]. On the other hand, there are good indications that the observed strong interactions are governed by a gauge theory as well: a non abelian gauge field theory with unbroken symmetry which describes the interaction of the colour degree of freedom of the constituents of hadronic matter, the quarks, with massless gauge fields, the gluons. This theory is called Quantum Chromodynamics (QCD)[3].

The quark fields $\psi_j^\alpha(x)$ carry two types of indices: α colour index

$$\alpha = \text{blue, red and yellow} \tag{1.3}$$

and a flavour index

$$j = \text{up, down, strange, charm, bottom, ...} \tag{1.4}$$

The three colours are associated to the fundamental representation $\underline{3}$ of the gauge group

$$SU(3)_{colour} \tag{1.5}$$

The flavour index is introduced so as to incorporate in a minimal group theoretical structure all the known quantum numbers which govern the strong interaction reactions. Flavours are distributed in two axes of electric charge $Q = \frac{2}{3}$ and $Q = \frac{1}{3}$ as shown in Fig. (1.1). The doublet (u, d) defines the fundamental representation of the isospin group SU(2) . The triplet (u, d, s) defines the fundamental representation of the internal symmetry group SU(3) which incorporates isospin and strangeness. The quadruplet (u, d, s, c) the fundamental representation of a symmetry group SU(4) which incorporates isospin, strangeness and charm, etc. The observed hadrons are members of irreducible representations of flavour -SU(n) obtained from tensor products of the constituent quarks:

$$q\bar{q} \quad \text{for mesons ;} \quad qqq \quad \text{for baryons} \tag{1.6}$$

The original motivation for the quark "colour" degree of freedom is quark statistics [see refs. (1.6) and (1.7)] . Consider e.g., the Δ^{++} resonance in the $J_3 = \frac{3}{2}$ state. In terms of quarks it is described by the state

$$| \Delta^{++}, J_3 = \tfrac{3}{2} \rangle = | u^{\uparrow} u^{\uparrow} u^{\uparrow} \rangle \tag{1.7a}$$

where the arrow denotes the quark spin $(+ \frac{1}{2})$. If quarks are fermions then Fermi-Dirac statistics requires the wave function of the Δ^{++} to be antisymmetric under exchange of the space coordinates of each quark pair. On the other hand the Δ^{++} being described by the ground state of the uuu system one expects it to be in an s-wave state and hence symmetric under space coordinate exchanges. One way out of this paradox is to assume that each of the u quarks comes in three colours and that the baryon wave function is indeed symmetric in space and spin but antisymmetric in colour:

$$| uuu \rangle \rightarrow \frac{1}{\sqrt{6}} \epsilon^{\alpha\beta\gamma} | u_\alpha u_\beta u_\gamma \rangle \tag{1.7b}$$

α, β, γ = blue, red and yellow.

So far we have spoken of two different groups for the quarks:

$$\text{flavour} - SU(n) \equiv SU(n)_F \text{ and colour} - SU(3) \equiv SU(3)_c \tag{1.8a,b}$$

The first one is a generalization of the old Eightfold way SU(3) (see ref. [R.8] for a review). The generators do not correspond to exact conservation laws. By contrast colour -SU(3) is assumed to be an exact symmetry: quarks of the same flavour and different colours are otherwise indistinguishable. Furthermore, all observables in Nature are assumed to be color singlets. This is the so-called confinement hypothesis, which is expected to be implemented by the dynamical content of QCD itself.

With quarks assigned to the 3 representation of $SU(3)_c$, the simplest colour singlet configurations we can make out of colour-triplets are

$$3 \times \bar{3} = 1 \text{ (MESONS)} + 8 \quad ; \tag{1.9}$$

and

$$3 \times 3 \times 3 = 1 \text{ (BARYONS)} + 8 + 8 + 10 \tag{1.10}$$

Notice that we cannot make colour singlets out of two-quarks since

$$3 \times 3 = \bar{3} + 6 \not\ni 1 \tag{1.11}$$

neither out of 4-quarks . There are of course other configurations than (1.9) and (1.10) which can exist as colour singlets, but they are more complicated e.g.,

$$\bar{3} \times 3 \times \bar{3} \times 3 \rightarrow \text{DIMESONS} \qquad (1.12)$$

$$3 \times 3 \times 3 \times \bar{3} \times 3 \rightarrow \text{MESOBARYONS} \qquad (1.13)$$

In fact, candidates for bayonium states of the type $\bar{q}q\bar{q}q$ have already been observed[4)].

The colour forces between quarks are mediated by massless vector bosons -the so called gluons much the same as photons mediate the electromagnetic forces between charged leptons. The gluons are the gauge fields belonging to the adjoint representation of $SU(3)_c$. There are altogether eight gluons, one associated to each generator of the group $SU(3)_c$. In the case of QED, the photon field is the gauge field associated to the generator of the abelian group U(1) : the group of gauge transformations. The crucial difference between QED and QCD is that, because of the non-abelian structure of SU(3) the gluons have self-interactions described by a Yang-Mills type Lagrangean [see ref. (1.8)] . The full Lagrangean of QED is given in eq. (2.1). Before we enter into a technical description of this Lagrangean, which is the subject of section 2, I shall spend the rest of this introduction giving a brief review of the evolution of ideas which has lead to our present understanding of particle physics. This is by no means necessary to follow the next lectures. I think, however, that it may help to have a certain perspective of how many of the ideas developed in the 50's and the 60's are incorporated in the present scheme. Also you will see that in spite of the progress made there remain many old problems unsolved.

Wisdom of the 50's:

1. Largely motivated by the challenge of giving a field theoretical framework to the concept of isospin invariance, Yang and Mills (1.8) extend the concept of local gauge invariance from abelian to non-abelian groups. They show this explicitly in the case of SU(2), and construct a gauge invariant Lagrangean out of the three gauge fields $\vec{W}_\mu(x)$ associated to the three generators of isospin $\vec{T}$

$$\mathcal{L}_{\text{Yang-Mills}}(x) = -\frac{1}{4} \vec{F}_{\mu\nu}(x) \vec{F}^{\mu\nu}(x) \tag{1.14a}$$

where $\vec{F}_{\mu\nu}$ denotes the three field-strength tensors

$$\vec{F}_{\mu\nu}(x) = \partial_\mu \vec{W}_\nu(x) - \partial_\nu \vec{W}_\mu(x) + g\, \vec{W}_\mu(x) \times \vec{W}_\nu(x) \tag{1.14b}$$

with g a coupling constant describing the self interaction of the gauge fields.

2. In an attempt to understand the short-distance behaviour of QED Gell-Mann and Low (1.9) and independently Stueckelberg and Peterman (1.10) develop the concept of Renormalization group Invariance. The importance of the renormalization group as a potential tool to understand the hadronic interactions at short distances will be repeatedly stressed later on by K.G. Wilson (see e.g. ref. (1.11)).

3. The formulation of an effective Lagrangean of the current x current form[5)] for the description of Weak interaction phenomenology

$$\mathcal{L}(x) = \frac{G}{\sqrt{2}} J_\mu(x) J^\mu(x)^+ \tag{1.15}$$

with

$$J_\mu(x) = \bar{e}(x) \gamma_\mu (1-\gamma_5) \nu_e(x) + e \leftrightarrow \mu + \text{hadronic currents} \tag{1.16}$$

and G a universal coupling constant, the Fermi constant, fixed from μ decay,

$$G = 1.03\ 10^{-5}\ M_{\text{proton}}^{-2} \tag{1.17}$$

An important concept introduced in the late 50's by Feynman and Gell-Mann (1.12) is that of the conserved vector current CVC. For the first time the abstraction of currents from their probe interactions is made: the weak hadronic current and the electromagnetic current are considered as components of the same entity. This will be the basis of the Current Algebra development in the 60's.

Wisdom of the 60's :

1. The Eightfold Way [R.8] . A successful description of the spectroscopy of particles is obtained when the SU(2) group of isospin is enlarged to SU(3) . Hadrons are classified in families of irreducible representations of SU(3) : octets and decuplets.

2. Current Algebra (See refs. [R.8] and [R.11] for a review). The CVC concept is extended to an octet of vector currents and an octet of axial vector currents.

3. The Quark Model (refs. (1.13), (1.14)). Hadrons are considered as being built up out of quarks: fundamental entities associated to the representations 3 and $\bar{3}$ (antiquarks) of the eightfold way SU(3)). The quantum numbers of the quarks are:

	UP	DOWN	STRANGE
Baryon Number	1/3	1/3	1/3
Electric Charge	2/3	-1/3	-1/3
Isospin (I, I_3)	1/2, 1/2	1/2, -1/2	0
Strangeness	0	0	1

In the SU(3) quark model, the weak hadronic current takes the specific form, as proposed by Cabibbo

$$J_\mu^+(x) = \bar{u}(x)\gamma_\mu(1-\gamma_5)\left[d(x)\cos\theta + s(x)\sin\theta\right] \tag{1.18}$$

with θ the Cabibbo angle, a phenomenological parameter fixed from hadronic weak decays (θ = 0.23 radians) .
The electromagnetic hadronic current is then (in units of the electric charge e)

$$J_\mu^{EM}(x) = \frac{2}{3}\bar{u}(x)\gamma_\mu u(x) - \frac{1}{3}\bar{d}(x)\gamma_\mu d(x) - \frac{1}{3}\bar{s}(x)\gamma_\mu s(x) \tag{1.19}$$

These currents (1.18) and (1.19) are two particular combinations of an octet of vector currents and an octet of axial-vector currents one can construct with the quark fields i.e., the quark model gives a precise construction of the algebra of hadronic currents. The basic hypothesis of the algebra of currents is that the equal-time commutators of the time components of the hadronic currents are precisely those of the quark model.

The Lagrangean in eq. (1.15) with currents defined by eqs. (1.16) and (1.18) is a non-renormalizable Lagrangean. This is more than a technical difficulty in the sense that attempts to formulate a phenomenological theory by introducing an arbitrary regulator, a mass scale Λ, to give a meaning to a perturbation theory in powers of the Fermi coupling constant fail. Inconsistent limits for Λ are obtained from different processes.

The idea that $\mathcal{L}(x)$ in eq. (1.15) must be some effective limit of a Yang-Mills type theory has been suspected for a long time. Two stumbling blocks, however, had to be overcomed to pursue that line of thought:

1. the empirical fact that neutral currents were not observed in the easiest processes where they could be detected: strangeness changing decays like $K \to \mu^+\mu^-$ and $K \to \pi e^+e^-$. In a gauge theory, neutral currents appear in a natural way because the commutator of two charged currents gives a neutral current

$$[J^+, J^-] = J^0$$

2. the theoretical difficulty to formulate a renormalizable non-abelian gauge field theory with massive gauge fields (the intermediate vector bosons). The fact that intermediate vector bosons have to be massive is an inevitable constraint dictated by the short range character of the weak forces.

The way out of the first difficulty was found by Glashow-Iliopoulos and Maiani (1.16): there are four flavoured quarks instead of three which, as regards the weak interactions, combine in two fundamental doublets

$$\begin{pmatrix} u \\ d\cos\theta + s\sin\theta \end{pmatrix} \quad \text{and} \quad \begin{pmatrix} c \\ -d\sin\theta + s\cos\theta \end{pmatrix} \tag{1.20}$$

much the same as their leptonic partners

$$\begin{pmatrix} \nu_e \\ e \end{pmatrix} \quad \text{and} \quad \begin{pmatrix} \nu_\mu \\ \mu \end{pmatrix} \tag{1.21}$$

The new charm quark c chooses as a partner precisely the orthogonal combination of the d and s quarks introduced by Cabibbo. Neutral currents in the strangeness changing sector are then avoided. This mechanism, when incorporated into the minimal $SU(2)_L \times U(1)$ model previously suggested[6)] by Weinberg (1.17) and Salam (1.18), leads to three specific predictions:

i) there should be sizeable neutral leptonic currents e.g. of the type

$$\nu_\mu + e^- \rightarrow \nu_\mu + e^- \tag{1.22}$$

ii) there should be sizeable neutral hadronic strangeness conserving currents of the type

$$\nu_\mu + N \longrightarrow \nu_\mu + \text{HADRONS} \tag{1.23}$$

iii) hadrons spectroscopy must require a new quantum number associated to the postulated charm quark.

As you know, the three predictions have been confirmed and the experimental evidence of the three of them is now well established.

The second difficulty has been solved thanks to the work of 't Hooft (1.20) which proved the renormalizability of non-abelian Lagrangean field theories[7)].

The introduction of arbitrary mass terms in the symmetric $SU(2)_L \times U(1)$ model violates the local gauge invariance of the Lagrangean and leads in general to a non-renormalizable theory. Up to now, only one mechanism of generating masses has been proved to be tolerable: the spontaneous symmetry breakdown mechanism [see ref. [R.2] for a review and references].

An intersting interplay between the weak and electromagnetic interactions as described by the $SU(2)_L \times U(1)$ model on the one hand, and the colour degree of freedom appears when the perturbation theory of the Weinberg-Salam Lagrangean is examined at higher orders. Because of the presence of both vector and axial-vector currents there appear anomalies of the type encountered in the soft pion analyses of $\pi^0 \to \gamma\gamma$ decay (see ref. [R.13] for a review and references). These anomalies which a priori would spoil the renormalizability of the theory, cancel if as pointed out by Bouchiat Iliopoulos and Meyer (1.2), the hadronic doublets in eq. (1.20) are taken in three colours.

We have mentioned two examples where the hidden colour quantum number plays a crucial role: the Δ-resonance; and the cancellation of anomalies in the $SU(2)_L \times U(1)$ model of the weak and electromagnetic interactions. There are in fact two other observables where the colour degree of freedom plays a dramatic rolé:

i) the ratio of cross-sections

$$R = \frac{\sigma(e^+e^- \to \text{hadrons})}{\sigma(e^+e^- \to \mu^+\mu^-)} \tag{1.24}$$

ii) the decay $\pi^0 \to \gamma\gamma$

As we shall see in detail in section 7, the ratio R at large C.M. energy is given by the sum of all the squared quark charges (in units of the electric charged) seen by the electromagnetic field at that energy i.e.,

$$\lim_{s \to \infty} R = 3 \sum_{i=\text{flavour}} Q_i^2 \tag{1.25}$$

where the factor 3 stands for the colour degree of freedom of each quark flavour. Notice that

$$R(u,d,s) = 3\left\{\left(\tfrac{2}{3}\right)^2 + \left(\tfrac{1}{3}\right)^2 + \left(\tfrac{1}{3}\right)^2\right\} = 2 \tag{1.26a}$$

$$R(u,d,s,c) = \frac{10}{3} \quad ; \quad R(u,d,s,c,b) = \frac{11}{3} \tag{1.26b,c}$$

The experimental value of R as a function of the total e^+e^- C.M. energy is shown in Fig. 1.2. One can distinctly see a first flat region below 3 GeV. and another one after 5 GeV. The first region can be interpreted as asymptotic with respect to the u, d and s quarks;

the other settles after the charm threshold has been passed.

The decay amplitude for the process $\pi^0 \to \gamma\gamma$ is governed via PCAC[8], by the coupling of the axial vector current to two electromagnetic currents. In the low energy limit his coupling is given, to all orders of QCD [see ref. (1.22)] , by the triangle diagram shown in fig. (1.3). Thus one obtains

$$\Gamma(\pi^0 \to \gamma\gamma) = \frac{m_\pi^3}{32\pi} \frac{1}{f_\pi^2} \left(\frac{\alpha}{\pi}\right)^2 \left(\frac{N}{3}\right)^2 \tag{1.27}$$

where f_π is the $\pi \to \mu\nu$ coupling constant $(f_\pi = .96\ m_\pi)$ and N the number of colours. The observed width is

$$\Gamma(\pi^0 \to \gamma\gamma) = 7.95 \pm 0.55\ eV \tag{1.28}$$

in excellent agreement with eq. (1.27) with N=3.

The remarkable property of QCD which makes of the QCD Lagrangean such a good candidate for a theoretical framework of the hadronic interaction is asymptotic freedom [ref. (1.3) and (1.4)] . This means that the short-distance behaviour of the Green's function of QCD is governed the free field theory (the old quark model). Corrections to the free field behaviour can be computed perturbatively and the specific theoretical predictions of QCD can thus be tested. This property of asymptotic freedom has a tremendous phenomenological impact. It implies specific predictions for all those observables governed by the short distance behaviour of products of field operators. The cleanest examples of such observables are the ratio R defined in eq. (1.24) and the nucleon structure functions measured in the inclusive reactions

$$e N \to e + \Gamma \ , \quad \nu N \to \mu + \Gamma \tag{1.29a,b}$$

at large momentum transfer and large inelasticity.

In spite of the successes of QCD and the weak and electromagnetic gauge model $SU(2)_L \times U(1)$ there remain many unsolved problems. I shall finish this introduction with a list of what it seems to me many theoreticians would agree to be the fundamental ones. In increasing order of complexity (in my opinion):

i) the origin of quark masses; Cabibbo angle and CP-violation.

ii) the confinement of quarks. In particular, we would like to extent the predictive power of QCD to diffraction phenomena in Hadron physics.

iii) the origin of flavour and leptonic quantum numbers.

I hope these will be the subjects of the future GIFT summer schools.

FOOTNOTES OF SECTION I

1) For a recent review see e.g. Kinoshita's talk in Ref. [R.1]

2) They have been extensively discussed in previous Gif summer schools. For other reviews see e.g. ref. [R.2,3,4,5]

3) A list of references which accounts for the major steps in the development of QCD as a theoretical framework of the hadronic interactions is given en refs. (1.1) to (1.5). See also the review articles quoted in refs. [R.6] and [R.7]

4) For a recent review see e.g. ref. [R.9]

5) See e.g. ref. [R.10] for an account of the earlier development of weak interaction theory.

6) See also ref. (1.19)

7) The reader interested in the historical steps which have led to this remarkable theoretical achievement should consult ref. [R.12]

8) PCAC stands for partial conservation of the axial current. It is another piece of wisdom inherited from the 50's and 60's. See ref. [R.11] for a review and references.

REFERENCES OF SECTION 1

1) M. Gell-Mann, Acta Phys. Austriaca Suppl. IX (1972) 733.
2) H. Fritzsch and M. Gell-Mann, Proc. XVI International Conference on High-Energy Physics (1972), Vol. 2 p. 135.
3) H.D. Politzer, Phys. Rev. Letters 30 (1973) 1346.
4) D.J. Gross and F. Wilczek, Phys. Rev. Letters 30 (1973) 1343.
5) H. Fritzsch, M. Gell-Mann and H. Leutwyler, Phys. Letters B, 47 (1973) 365.
6) O.W. Greenberg, Phys. Rev. Letters 13 (1964) 598.
7) M. Han and Y. Nambu, Phys. Rev. 139 (1965) 1006.
8) C.N. Yang and R.L. Mills, Phys. Rev. 96 (1954) 191.
9) M. Gell-Mann and F.E. Low, Phys. Rev. 95 (1954) 1300.
10) E.C.G. Stueckelberg and A. Peterman, Helv. Phys. Acta 26 (1953) 499.
11) K.G. Wilson, Phys. Rev. 3D (1971) 1818.
12) R.P. Feynman and M. Gell-Mann, Phys. Rev. 109 (1958) 193.
13) M. Gell-Mann, Physics Letters 8 (1964) 214.
14) G. Zweig, CERN Preprint, unpublished (1964).
15) N. Cabibbo, Phys. Rev. Letters 10 (1963) 531.
16) S.L. Glashow, J. Iliopoulos and L. Maiani, Phys. Rev. D2 (1970)1285.
17) S. Weinberg, Phys. Rev. Letters 19 (1967) 1264.
18) A. Salam, Proc. 8th Nobel Sympos., Stockholm, ed. N. Svartholm (Almquist and Wiksells, Stockholm, 1968) p. 367.
19) S.L. Glashow, Nucl. Phys. 22 (1961) 579
20) G.'tHooft, Nucl. Phys. B33 (1971) 173 ; ibid B35 (1971) 167.
21) C. Bouchiat, J. Iliopoulos and Ph. Meyer, Phys. Letters 38B (1972) 519.
22) S.L. Adler and W.A. Bardeen, Phys. Rev. 182 (1969) 1517.

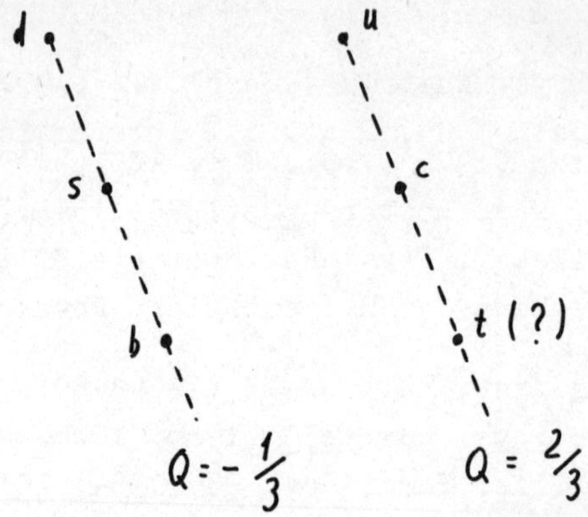

Fig. 1.1 The fundamental constituents of hadronic matter, the quarks, come in various flavours: *u* (up), *d* (down), *c* (charm), *b* (bottom), ... distributed in two axes of electric charge Q = 2/3 and Q = -1/3 . Each flavoured quark comes in three colours: blue, red and yellow.

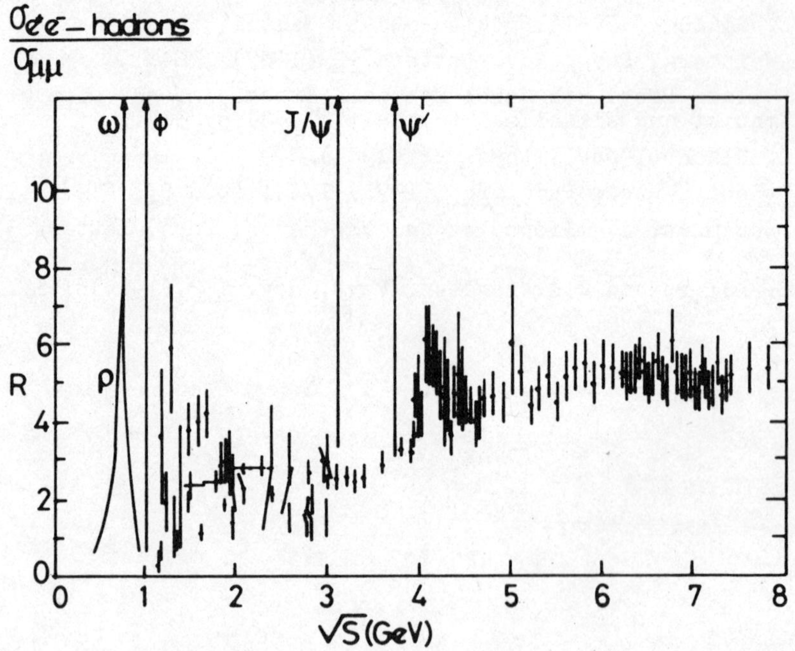

Fig. 1.2 The ratio R in eq. (1.24) as a function of the total e^+e^- C.M. energy in GEV.

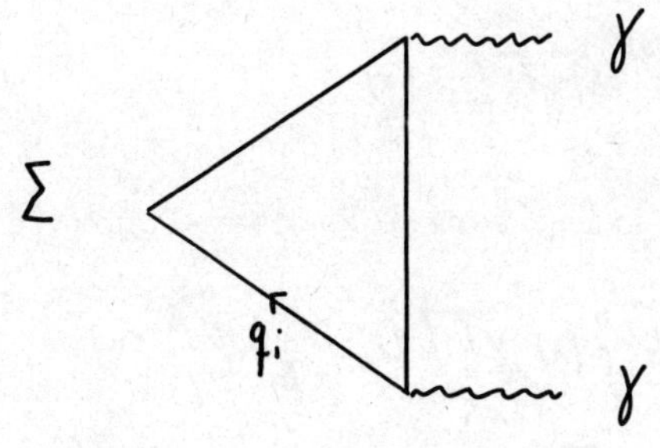

Fig. 1.3 The triangle diagram which governs $\pi^0 \to \gamma\gamma$ decay.

2. THE QCD LAGRANGEAN

The Lagrangean of QCD is

$$\mathcal{L}_{QCD}(x) = -\frac{1}{4} F^{(a)}_{\mu\nu}(x)\, F^{\mu\nu}_{(a)}(x)$$

$$+ \text{ gauge fixing term } + \text{ Faddeev-Popov term} \tag{2.1}$$

$$+ i \sum_{j=1}^{n} \bar{\Psi}^{\alpha}_{j}(x)\, \gamma^{\mu} (D_{\mu})_{\alpha\beta}\, \Psi^{\beta}_{j}(x)$$

$$- \sum_{j=1}^{n} m_j\, \bar{\Psi}^{\alpha}_{j}(x)\, \Psi_{\alpha j}(x)$$

Let us first explain what is the formal meaning of each term and then we shall discuss the rationale which is behind the justification of this particular Lagrangean.

i) $F^{(a)}_{\mu\nu}(x)$ with $a = 1,2, \ldots\ldots 8$

are the Yang-Mills (2.1) field strengths constructed from the gluon fields $W^{(a)}_{\mu}(x)$ and μ,ν are space-time indices:

$$F^{(a)}_{\mu\nu}(x) = \partial_{\mu} W^{(a)}_{\nu}(x) - \partial_{\nu} W^{(a)}_{\mu}(x) + g\, f_{abc}\, W^{(b)}_{\mu}(x)\, W^{(c)}_{\nu}(x) \tag{2.2}$$

g is the QCD coupling constant and f_{abc} the structure constants of the SU(3) algebra.

$$[T^{(a)}, T^{(b)}] = i\, f_{abc}\, T^{(c)} \tag{2.3}$$

where $T^{(a)}$ are the eight generators of the SU(3) algebra.

ii) $\Psi^{\alpha}_{j}(x)$ and $\Psi^{\beta}_{j}(x)$ with α,β color indices

α,β = blue, red and yellow (2.4a)

and j a flavor index:

$$j = \text{up, down, strange, charm, bottom,}\dots\ n, \qquad (2.4b)$$

are 4-component Dirac spinors associated to each quark field of color α and flavor j .

iii) $(D_\mu)_{\alpha\beta}$ is the covariant derivative acting on the quark color components

$$(D_\mu)_{\alpha\beta} = \delta_{\alpha\beta}\,\partial_\mu - i g \sum_a \frac{1}{2}\,\lambda^{(a)}_{\alpha\beta}\; W^{(a)}_\mu(x) \qquad (2.5)$$

with $\lambda^{(a)}_{\alpha\beta}$ the eight 3 x 3 Gell-Mann matrices. They are the representation of the SU(3) generators in the fundamental $\underline{3}$ dimensional representation i.e., the color basis of the quark fields

$$(T^{(a)})_{\alpha\beta} = \frac{1}{2}\,\lambda^{(a)}_{\alpha\beta} \qquad (2.6)$$

much the same as the Pauli matrices $\tau^{(a)}$, a = 1,2,3, are the representation of the three generators in a two component basis[1]. The structure constants f_{abc} can also be viewed as the matrix representation of the eight SU(3) generators $T^{(a)}$ in the regular (adjoint) $\underline{8}$ dimensional representation i.e., the gluon basis

$$(T^{(a)})_{bc} = -i\,f_{abc} \qquad (2.7)$$

iv) The arbitrary parameters of the theory are the coupling g and the quark masses m_j of the different quark flavors, n in total.

We shall specify later on what is meant by the heading: "gauge fixing term" + "Faddeev-Popov term".

In order to gain some familiarity with this Lagrangean let us discuss some of the fundamental interactions that it describes, as well as some simple group theoretical relations which will be useful to have in hand later on.

Quarks of a fixed flavor and gluons interact via the term

$$g \sum_a \frac{1}{2} \lambda^{(a)}_{\alpha\beta} \bar{\psi}_\alpha(x) \gamma^\mu \psi_\beta(x) W_\mu^{(a)}(x) \tag{2.8a}$$

This is much the same as the interaction of electrons with photons in QED

$$- e \bar{\psi}(x) \gamma^\mu \psi(x) A_\mu(x) \tag{2.8b}$$

but for the group theoretical matrix element factor $\lambda^{(a)}_{\alpha\beta}$. The graphical representation of these interactions is shown in Fig. (2.1).

The new feature of the QCD Lagrangean in eq. (2.1) as compared to the QED Lagrangean in eq. (1.1), is the presence of fundamental interactions among the gluon fields themselves. (In QED photon photon interactions are always induced by their coupling to electrons, i.e., via electron loops, like e.g. in Fig. (2.2).)

In QCD these self interactions are due to the presence of the coupling constant g already at the level of the field strengths, (see eq. (2.2)). There are two types of fundamental self-gluon interactions: those involving three gluon fields,

$$-\frac{1}{2} g f_{abc} \left(\partial_\mu W_\nu^{(a)}(x) - \partial_\nu W_\mu^{(a)}(x)\right) W_{(b)}^\mu(x) W_{(c)}^\nu(x) \tag{2.9}$$

representated graphically in Fig. (2.3); and those involving four gluon fields,

$$-\frac{1}{4} g^2 f_{abc} f_{ade} W_\mu^{(b)}(x) W_\nu^{(c)}(x) W_{(d)}^\mu(x) W_{(e)}^\nu(x) \tag{2.10}$$

represented in Fig. (2.4). This completes the set of fundamental interactions described by the Lagrangean in eq. (2.1). Notice that they are all governed by the <u>same</u> coupling constant g .

In practical calculations we shall often encounter the quadratic combination of matrices

$$\sum_a \sum_\beta (T^{(a)})_{\alpha\beta} (T^{(a)})_{\beta\gamma} = C_2(R)\, \delta_{\alpha\beta} \tag{2.11a}$$

It appears for example in the evaluation of the one loop diagram associa ted to the selfenergy of a quark (see Fig. (2.5)). In group theory language $C_2(R)$ is the eigenvalue of the quadratic Casimir operator in the fundamental representation. For SU(N) , in general,

$$C_2(R) = \frac{N^2-1}{2N} = \frac{4}{3}, \quad N=3 \tag{2.11b}$$

Another useful combination is

$$\sum_a \sum_c (T^{(a)})_{bc} (T^{(a)})_{cd} = C_2(G)\,\delta_{bd} \tag{2.12a}$$

$C_2(G)$ is the eigenvalue of the quadratic Casimir operator in the adjoint representation,

$$C_2(G) = N = 3 \quad , \quad N = 3 \tag{2.12b}$$

It appears for example in the evaluation of the gluon self-energy diagram shown in Fig. (2.6).

Another group theoretical quantity which we shall need in one-loop calculations is

$$\mathrm{Tr}\,(T^{(a)} T^{(b)}) = T(R)\,\delta_{ab} \tag{2.13a}$$

$T(R)$ is representation dependent:

$$T\,(\text{adjoint}) = C_2\,(G) = 3 \tag{2.13b}$$

$$T\,(\text{fundamental}) = \frac{1}{2} \tag{2.13c}$$

The last one appears in the evaluation of the quark loop diagram shown in Fig. (2.7)

$$\sum_{\alpha,\beta} (T^{(b)})_{\beta\alpha} (T^{(a)})_{\alpha\beta} = \frac{1}{2}\,\delta_{ab} \tag{2.14}$$

This is all the group theory we shall need in these lectures[2].

2a. Gauge Transformations

Here we shall generalize the local gauge invariance argument which leads to the QED Lagrangean in Section 1, to the non-abelian case[3]. As we shall see, it is the local gauge invariance concept which provides the rationale for the specific form of the QCD Lagrangean in eq. (2.1).

We start with the free quark field Lagrangean,

$$i \; \bar{\psi}^\alpha(x) \gamma^\mu \partial_\mu \psi_\alpha(x) - m \; \bar{\psi}^\alpha(x) \psi_\alpha(x) \tag{2.15}$$

where for convenience we have dropped the flavour indices.
The generalization of the local gauge transformation

$$\psi(x) \longrightarrow \exp\left(ie\,\theta(x)\right) \psi(x) \tag{2.16}$$

in the abelian case to the non-abelian case is the following

$$\psi(x) \longrightarrow \exp\left(-ig\,\vec{T}\cdot\vec{\theta}(x)\right) \psi(x) \tag{2.17}$$

In the abelian case $U(1)$ we only had one generator. Now we have eight generators $T^{(a)}$ $a = 1,2,\ldots 8$ and to each one we associate an arbitrary function $\theta^{(a)}(x)$ of space-time.
The operator:

$$G(x) \equiv \exp\left(-ig\,\vec{T}\cdot\vec{\theta}(x)\right) \tag{2.18}$$

defines a gauge transformation. It is a function from space-time to the gauge group.

Clearly $\psi(x)$ and $\partial_\mu \psi(x)$ do not transform alike under gauge transformations. (This is due precisely to the x-dependence of the $\vec{\theta}(x)$ functions). Therefore the free quark field Lagrangean in eq. (2.15) is not a local gauge invariant. Following the analogy with QED we seek for a covariant derivative D_μ such that $\psi(x)$ and $D_\mu\psi(x)$ transform alike under gauge transformation . In QED, this was done at the expense of introducing a gauge field $A_\mu(x)$.

$$\partial_\mu \rightarrow D_\mu = \partial_\mu + ieA_\mu \tag{2.19}$$

and the requirement that $\psi(x)$ and $D_\mu\psi(x)$ transform alike under the gauge transformation in eq. (2.16) fixed the transformation properties of A_μ itself

$$A_\mu(x) \rightarrow A'_\mu(x) = A_\mu(x) - \partial_\mu\theta(x) \tag{2.20}$$

The generalization of the covariant derivative to the non-abelian case requires the introduction of eight gauge fields $\vec{W}_\mu(x)$(one associated to each generator)

$$\partial_\mu \rightarrow D_\mu = \partial_\mu - ig\,\vec{T}\cdot\vec{W}_\mu(x) \equiv \partial_\mu + g\,W_\mu(x) \tag{2.21}$$

where we have introduced the short-hand notation

$$W_\mu(x) \equiv \frac{1}{i}\,\vec{T}\cdot\vec{W}_\mu(x) \tag{2.22}$$

The requirement that $\psi(x)$ and $D_\mu\psi(x)$ transform alike under the gauge transformation in eq. (2.18) implies

$$\psi(x) \rightarrow \psi'(x) = G(x)\,\psi(x) \tag{2.23a}$$

$$D_\mu(W)\,\psi(x) \rightarrow D_\mu(W')\,\psi'(x) = G(x)\,D_\mu(W)\,\psi(x) \tag{2.23b}$$

i.e.,

$$(\partial_\mu + g\,W'_\mu(x))\,G(x)\,\psi(x) = G(x)(\partial_\mu + g\,W_\mu(x))\,\psi(x) \tag{2.24}$$

This fixes the transformation properties of $W_\mu(x)$:

$$W_\mu(x) \rightarrow W_\mu'(x) = G(x)\, W_\mu(x)\, G^{-1}(x) - g^{-1}(\partial_\mu G(x))\, G^{-1}(x) \qquad (2.25)$$

The free quark field Lagrangean in eq. (2.15) is then replaced via the local gauge invariance requirement, by the Lagrangean

$$i\,\bar{\psi}^\alpha(x)\gamma^\mu D_\mu \psi_\alpha(x) - m\,\bar{\psi}^\alpha(x)\psi_\alpha(x) \qquad (2.26)$$

which is the sum of the free quark field Lagrangean, eq. (2.15) and the quark-quark-gluon interaction term in eq. (2.8a). This Lagrangean, however, does not fix the equations of motion of the gluon fields. For that we need the generalization of the pure radiation term in QED

$$-\frac{1}{4} F_{\mu\nu}(x)\, F^{\mu\nu}(x) \quad , \quad F_{\mu\nu}(x) = \partial_\mu A_\nu(x) - \partial_\nu A_\mu(x) \qquad (2.27a,b)$$

to the non-abelian case. This we do via a generalization of the curl to a gauge covariant curl

$$g\, F_{\mu\nu} \equiv g\,\frac{1}{i}\,\vec{T}\cdot\vec{F}_{\mu\nu}(x) = [\,D_\mu(W), D_\nu(W)\,] \qquad (2.28)$$

with $D_\mu(W)$ the covariant derivative introduced in eq. (2.21); i.e.,

$$\begin{aligned} g\,\frac{1}{i}\,\vec{T}\cdot\vec{F}_{\mu\nu}(x) &= [\,\partial_\mu - ig\,\vec{T}\cdot\vec{W}_\mu(x),\ \partial_\nu - ig\,\vec{T}\cdot\vec{W}_\nu(x)\,] \\ &= -ig\,(\partial_\mu \vec{W}_\nu(x) - \partial_\nu \vec{W}_\mu(x))\,\vec{T} + g^2[\,W_\mu(x), W_\nu(x)\,] \end{aligned} \qquad (2.29)$$

In the abelian case there is only the unit generator and the last term in the r.h.s. vanishes. Then we get the field strength tensor in eq. (2.27b). In the non-abelian case see eqs. (2.22) and (2.3)

$$g^2\,[\,W_\mu, W_\nu\,] = -ig^2 f_{abc}\, T^{(c)}\, W_\mu^{(a)}\, W_\nu^{(b)} \qquad (2.30)$$

and the eight (a= 1,...8) non-abelian field strength tensors are:

$$F_{\mu\nu}^{(a)}(x) = \partial_\mu W_\nu^{(a)}(x) - \partial_\nu W_\mu^{(a)}(x) + g\, f_{abc}\, W_\mu^{(b)}(x)\, W_\nu^{(c)}(x) \qquad (2.31)$$

The final step is to construct a gauge invariant object out of the field strengths $F_{\mu\nu}^{(a)}$. The gauge transformation properties of $F_{\mu\nu}^{(a)}$ can be easily read from eq. (2.28). It follows from eq. (2.23) that

$$D_\mu(W') = G(x)\, D_\mu(W)\, G^{-1}(x) \qquad (2.32)$$

Therefore (see eq. (2.28))

$$F_{\mu\nu}(x) \rightarrow G(x)\, F_{\mu\nu}(x)\, G^{-1}(x) \qquad (2.33)$$

and we see that the combination

$$-\frac{1}{4}\, \vec{F}_{\mu\nu}(x) \cdot \vec{F}^{\mu\nu}(x) \qquad (2.34)$$

is the local gauge invariant quantity we were looking for. This completes the rationale for the specific form of the QCD Lagrangean in eq. (2.1). It is the generalization of the QED Lagrangean to the case where the local gauge froup is color-SU(3) instead of the U(1) associated to the electric charge.

2b. Global Symmetry properties: $SU_L(n) \otimes SU_R(n) \otimes U_B(1) \otimes U_A(1)$

The QCD Lagrangean in eq. (2.1) is invariant with respect to the set of one-parameter transformations

$$\psi(x) \rightarrow \exp(-i\theta \mathbb{1})\, \psi(x) \qquad (2.35)$$

acting on the quark flavor components. $\mathbb{1}$ is the n x n dimensional unit matrix. To this gauge invariance of the 1st kind (or global gauge invariance, θ in eq. (2.35) is a constant) there is an associated current, via Noether's theorem[3], the baryonic current

$$J_\mu(x) = \sum_{i=1}^{n} \bar{\psi}_i(x) \gamma_\mu \psi_i(x) \tag{2.36}$$

with a trace over the colour indices understood. This current is conserved

$$\partial^\mu J_\mu(x) = 0 \tag{2.37}$$

and the associated charge

$$B = \int d^3x \; J^0(\vec{x},t) \tag{2.38}$$

is the <u>baryonic charge</u>, the generator of a $U_B(1)$ group, which is a gloval symmetry of $\mathcal{L}_{QCD}(x)$.

$\mathcal{L}_{QCD}(x)$ in eq. (2.1) is also invariant with respect to <u>each</u> set of one-parameter transformations

$$\psi_j(x) \rightarrow \exp(-i\,\theta_j)\,\psi_j(x) \qquad j = 1,\cdots n \tag{2.39}$$

acting on a fixed flavor component $j = 1,2,\ldots, n$. To each flavor there corresponds an associated $U_j(1)$ group which is a global symmetry of the Lagrangean, i.e., $\mathcal{L}_{QCD}$ has also a

$$U(1)_{up} \otimes U(1)_{down} \otimes \cdots \otimes U(1)_n \tag{2.40}$$

global symmetry. Physically this symmetry corresponds to the separate conservation of flavor in the strong interactions. Clearly this symmetry is intimately related to the specific form of the mass terms in the QCD Lagrangean

$$\sum_{j=1}^{n} m_j \; \bar{\psi}_j(x)\, \psi_j(x) \tag{2.41}$$

and it would be broken by non-diagonal mass terms of the type

$$m_{ij} \quad \bar{\psi}_j(x) \; \psi_i(x) \tag{2.42}$$

This brings up the question of generality of the mass term in eq. (2.41). In answer to that question there is a relevant theorem which we next discuss.

<u>THEOREM on the generality of the mass terms in</u> $\mathcal{L}_{QCD}$:

with

$$q_{R,L}(x) \equiv \frac{1}{2}(1 \pm \gamma_5) \begin{pmatrix} \psi_u \\ \psi_d \\ \vdots \\ \psi_n \end{pmatrix} \tag{2.43}$$

an interaction term of the general form

$$q_L^* \gamma^0 \mathcal{M} q_R + q_R^* \gamma^0 \mathcal{M}^+ q_L \tag{2.44}$$

in the QCD Lagrangean, where $\mathcal{M}$ is an arbitrary n x n matrix which commutes with the colour generators, can be transformed without loss of generality into the diagonal form of eq. (2.41).

The proof os this theorem is rather simple, $\mathcal{M}$ can always be written in the form

$$\mathcal{M} = \mathcal{M}^H \mathcal{U} \tag{2.45}$$

where $\mathcal{M}^H$ is a hermitean matrix and $\mathcal{U}$ a unitary matrix. Then eq. (2.44) can also written in the form

$$\begin{aligned} & q_L^* \gamma^0 \mathcal{M}^H (\mathcal{U}) q_R + (q_R^* \gamma^0 \mathcal{U}^{-1}) \mathcal{M}^H q_L \\ = \; & q_L^* \gamma^0 \mathcal{M}^H q_R' + q_R^{*\prime} \gamma^0 \mathcal{M}^H q_L \end{aligned} \tag{2.46}$$

If we now define

$$q' \equiv q_R' + q_L \tag{2.47}$$

we can also write eq. (2.46) in the form

$$\bar{q}' m^H q' = (\bar{q}' V^{-1}) V m^H V^{-1} (V q') \tag{2.48}$$

where V is the matrix which diagonalizes m i.e.,

$$V m^H V^{-1} \equiv \begin{pmatrix} m_u & & & \\ & m_d & & \\ & & \ddots & \\ & & & m_n \end{pmatrix} \tag{2.49a}$$

with the following redefinition of the ψ_j fields

$$V q' \equiv \begin{pmatrix} \psi_u \\ \psi_d \\ \vdots \\ \psi_n \end{pmatrix} \tag{2.49b}$$

we arrive at a final expression of the form shown in eq. (2.41). Since all the other terms in the Lagrangean involving quarks fields are bilinear in flavor the redefinition of the fermion fields in eqs. (2.47) and (2.49b) which only involves unitary matrices does not change the form of the Lagrangean.

If two masses are equal

$$m_i = m_j$$

the Lagrangean $\mathcal{L}_{QCD}$ has a larger global symmetry than $U_i(1) \otimes U_j(1)$. It is now invariant under the set of SU(2) transformation acting on the two dimensional subspace

$$\begin{pmatrix} \psi_i \\ \psi_j \end{pmatrix} \longrightarrow \mathcal{U} \begin{pmatrix} \psi_i \\ \psi_j \end{pmatrix}$$

with $\mathcal{U}$ an arbitrary SU(2) matrix.

In general, if $m_1 = m_2 = \ldots\ldots = m_n$

the $U_1(1) \otimes \ldots\ldots \ \ldots\ldots \otimes U_n(1)$ invariance of $\mathcal{L}_{QCD}$ is enlarged to a global SU(n) symmetry.

In the absence of mass terms i.e.,

$$m_j = 0 \quad , \quad j = u, d, \cdots n$$

$\mathcal{L}_{QCD}$ is invariant under the set of global gauge transformations

$$\psi \rightarrow \exp\left(-i\,\theta^{(A)}\,T^{(A)}\right)\psi \qquad (2.50a)$$

and

$$\psi \rightarrow \exp\left(-i\,\theta_5^{(A)}\,T^{(A)}\gamma_5\right)\psi \qquad (2.50b)$$

where $\theta^{(A)}$ are constant parameters and there is no summation over the index $A = 1,2,\ldots n^2-1$, n is the total number of flavors, Here $T^{(A)}$ are the infinitesimal generators of the group $SU(n)$ acting on the quark flavor components

$$\psi \equiv \begin{pmatrix} \psi_u \\ \psi_d \\ \vdots \\ \psi_n \end{pmatrix} \qquad (2.50c)$$

The corresponding Noether currents (where a trace over the color indices is understood) :

$$V_\mu^{(A)}(x) \equiv \bar{\psi}^i(x)\,\gamma_\mu\,T_{ij}^{(A)}\,\psi^j(x) \qquad (2.51a)$$

and

$$A_\mu^{(A)}(x) \equiv \bar{\psi}^i(x)\,\gamma_\mu\gamma_5\,T_{ij}^{(A)}\,\psi^j(x) \qquad (2.51b)$$

are respectively the vector and axial-vector currents of the algebra of currents of Gell-Mann (2.3). They are conserved, and their associated charges

$$Q^{(A)} = \int d^3x\; V_0^{(A)}(\vec{x},t) \qquad (2.52a)$$

and

$$Q_5^{(A)} = \int d^3x\; A_0^{(A)}(\vec{x},t) \qquad (2.52b)$$

satisfy the commutation relations

$$[Q^{(A)}, Q^{(B)}] = i f^{ABC} Q^{(C)} \tag{2.53a}$$

$$[Q^{(A)}, Q_5^{(B)}] = i f^{ABC} Q_5^{(C)} \tag{2.53b}$$

$$[Q_5^{(A)}, Q_5^{(B)}] = i f^{ABC} Q^{(C)} \tag{2.53c}$$

The combination of charges

$$Q_L^{(A)} = Q^{(A)} - Q_5^{(A)} \tag{2.54a}$$

and

$$Q_R^{(A)} = Q^{(A)} + Q_5^{(A)} \tag{2.54b}$$

are the generators of Chiral $SU_L(n) \otimes SU_R(n)$, which is a global symmetry of $\mathcal{L}_{QCD}$ at the limit of massless quarks.

With all the masses $m_j = 0$, $\mathcal{L}_{QCD}$ has yet an additional $U_A(1)$ global symmetry implemented by the set of one-parameter transformations

$$\psi \rightarrow \exp(-i\theta \mathbb{1} \gamma_5) \psi \tag{2.55}$$

acting on the quark flavor components. The corresponding Noether current (as always, trace over color indices in understood) is the so called axial baryonic current

$$J_5^\mu(x) = \sum_{i=1}^{n} \bar{\psi}_i(x) \gamma^\mu \gamma_5 \psi_i(x) \tag{2.56}$$

2c. Comment on the realization of Global Symmetry

In this section we shall review very briefly our present understanding on how chiral symmetry and the $U_A(1)$ symmetry are realized in Nature. These questions fall outside the scope of perturbative QCD and therefore outside the limits of the subject of these lectures. They are however so fundamental that I cannot refrain from making a few introductory remarks[4)].

There are two options for the chiral symmetry of the QCD Lagrangean to be realizaed on the physical states:

i) the WIGNER-WEYL realization, where the chiral charges annihilate the vacuum

$$Q^{(A)}|0\rangle = 0 \quad , \quad Q_5^{(A)}|0\rangle = 0 \qquad (2.57a,b)$$

ii) the NAMBU-GOLDSTONE realization, where

$$Q^{(A)}|0\rangle \neq 0 \quad , \quad Q_5^{(A)}|0\rangle \neq 0 \qquad (2.58a,b)$$

The Wigner-Weyl realization is familiar to us. It is the way that Poincaré group invariance is realizaed on the physical states. Examples of the Namb-Goldstone realization are known from the study of superconductivity phenomena[5)].

To each of these options there is an associated theorem which applies to charges defined as spatial integrals of local current densities, hence to $Q^{(A)}$ and $Q_5^{(A)}$. These theorems say the following:

i) COLEMAN'S Theorem (2.4) states that the realization à la Wigner-Weyl implies that the physical states can be classified according to the irreducible unitary representations of the group generated by the charges which annihilate the vacuum. This means that particles should appear in parity doublets

$$(0^-,0^+) \; ; \; (1^-,1^+) \; ; \; \ldots$$

with mass degeneracy at the limit of chiral symmetry.

ii) GOLDSTONE'S theorem (2.5) states that to each generator which does not annihilate the vacuum there is an associated spin zero massless particle.

Neither of these two possibilities seems to be the one realized in Nature. On the one hand parity doublets are not observed. We see an octet of pseudoscalars and an octet of vector particles, but there does not seem to correspond any octet of scalars nor axial vector particles. On the other hand, if we identify the octet of pseudoscalars with the Goldstone particles there does not seem to be any corresponding octet of Golstone like scalars.

The picture which has emerged from the study of chiral symmetry problems is a mixed one:

$$Q^{(A)}|0\rangle = 0 \quad ; \quad Q_5^{(A)}|0\rangle \neq 0 \qquad A = 1,2 \cdots n^2-1 \tag{2.59a,b}$$

at the limit of chiral symmetry. This implies the existence of a n^2-1 multiplet of zero-mass pseudoscalars, and a set of massive multiplets with degenerate masses within each multiplet. In the case of $SU_L(3) \otimes SU_R(3)$ these correspond to the observed multiplets

$$1^- \text{ octet}; \ \frac{1}{2}^+ \text{ octet}; \ \frac{3}{2}^- \text{ octet}; \ 2^+ \text{ octet}; \ ,,,$$

The symmetry is broken because neither the observed pseudoscalars are massless, nor the other multiplets are degenerated in mass. However, the fact that the pion mass is small (m^2 as compared to the next light hadron, the kaon, is given by $m_\pi^2/m_K^2 = .075$) is associated to an approximate chiral $SU_L(2) \otimes SU_R(2)$ invariance. The successes of PCAC and current algebra are rooted in this aproximate invariance. The diagonal $SU(2)$ of this chiral $SU_L(2) \otimes SU_R(2)$ is the isospin group and the fact that isospin invariance is well realized in Nature is associated to the smallness of the up and down quarks in units of the mass scale associated to perturbative QCD which as we shall see is $\Lambda \sim 500$ MeV. The qualitative successes of the eightfold way are also associated to the fact that in units of Λ the masses of the up, down and strange quarks are small numbers. To the extent that the other quark flavors have masses larger than Λ we do not expect to see much of a direct symmetry pattern when we go beyond $SU(3)$ in the spectroscopy classification of physical states.

How is the $U_A(1)$ symmetry realized in Nature? A Nambu-Goldstone realization analogous to the other axial charges would imply the existence of a flavor singlet massless pseudoscalar. In the case of two flavors, the natural candidate is the η -particle. However m_η^2 / $m_\pi^2 \sim 16$, where does this big breaking come from? First remark, relevant (a priori) to this question is that the Noether current associated to the $U_A(1)$ invariance of the QCD Lagrangean is in fact nos conserved. This is due to the presence of anomalous terms in the naive Ward identities, much the same as the Adler-Bell-Jackiw anomaly associated to the $\pi^0 \to \gamma\gamma$ amplitude (2.6). These anomalous terms are uniquely determined by the triangle diagram shown in Fig. 2.8 and they lead to the result:

$$\partial_\mu J_5^\mu = \frac{g^2}{4\pi^2} \frac{n}{8} \epsilon^{\mu\nu\rho\sigma} F_{\mu\nu}^{(a)}(x) F_{(a)\rho\sigma}(x)$$

where n is the total number of flavors. One might think that this non-zero divergence is at the origin of a different symmetry realization for the states with the quantum numbers of the axial baryonic charge than for the others. However, it turns out that the structure of this divergence is such that the rate of change of the associated axial charge

$$\dot{Q}_5(t) = \int d^3x \; \partial_0 J_{05}(\vec{x},t) \tag{2.61}$$

is still null, in the absence of instanton type solutions (2.7). The crucial role of instantons in the realization of the $U_A(1)$ symmetry has been shown by 't Hooft (2.8). An excellent review on the present status of this problem can be found in ref. [R.18].

FOOTNOTES OF SECTION 2

1) The reader unfamiliar with these concepts should consult e.g. The eightfold way quoted in ref. [R.8]

2) For the reader interested in the generalization of group weight factors to arbitrary Feynman diagrams in gauge theories we recommend a paper by Cvitanovic, ref. (2.2)

3) The argument in the QED case is described in detail in ref. [R. 14].

4) Suggestions for further reading: Heinz Pagels "Departures from Chiral Symmetry" in ref. [R. 15] provides a good review of the problem prior to QCD. See also B.W. Lee's lectures in ref. [R. 16] . Sidney Coleman's 1975 lectures in ref. [R. 17] are an excellent introduction to the $U_A(1)$ problem and contain many references. See also the more recent set of lectures of Crewther at Schladming ref. [R. 18] .

5) For a review, where the original literature can be found, see e.g. ref. [R. 19] .

REFERENCES OF SECTION 2

1) C.N. Yang and R.L. Mills, Phys. Rev. 96 (1954) 191.

2) P. Cvitanovic, Phys. Rev. 14D (1976) 1536.

3) M. Gell-Mann, Physics 1 (1964) 63.

4) S. Coleman, J. Math. Phys. 7 (1966) 787 .

5a) J. Golstone, Nuovo Cim. 19 (1961) 154.

5b) J. Golstone, A. Salam and S. Weinberg, Phys. Rev. (1962) 127.

5c) D. Kastler, D. Robinson and A. Swieca, Comm. Math. Phys. 2(1960)108.

6a) J.S. Bell and R. Jackiw, Nuovo Cimento 60A (1974) 470.

6b) S.L. Adler, Phys. Rev. 177 (1969) 2426.

7) A.A. Belavin, A.M. Polyakov, A.S. Schwartz, and Yu. S. Tyupkin, Phys. Letters, 59B (1975) 85.

8a) G't Hooft, Phys. Rev. Letters 37 (1976) 8.

8b) G't Hooft, Phys. Rev. D14 (1976) 3432.

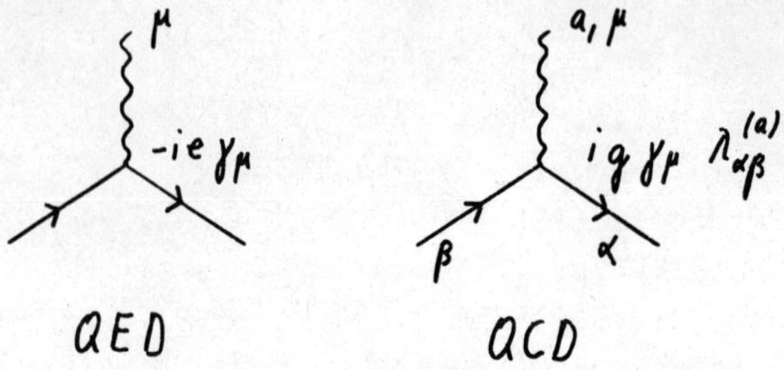

Fig. 2.1 Graphical representation of the interactions corresponding to eq. (2.8b), QED and eq. (2.8a), QCD.

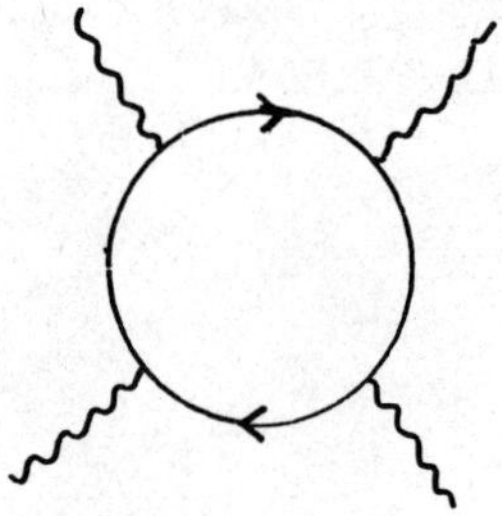

Fig. 2.2 Lowest order photon-photon interaction in QED, induced by an electron loop.

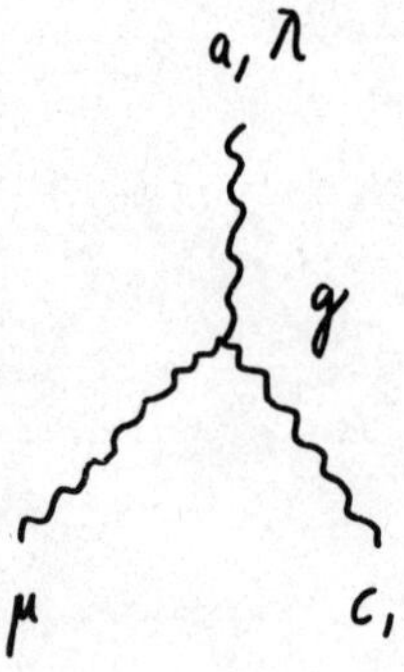

Fig. 2.3 Self gluon fundamental interaction involving 3-gluons corresponding to eq. (2.9).

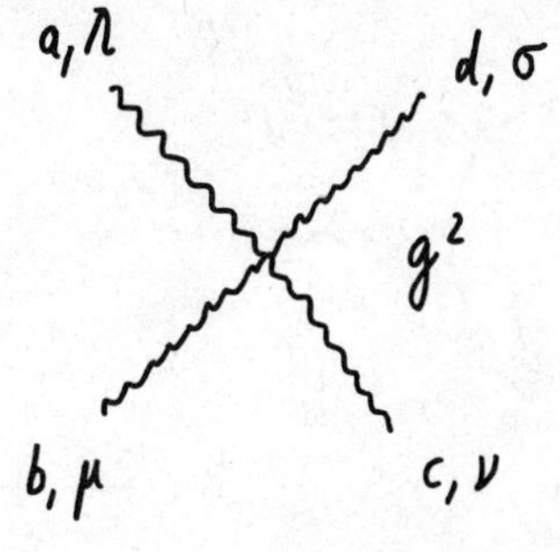

Fig. 2.4 Self gluon fundamental interaction involving 4-gluons corresponding to eq. (2.10).

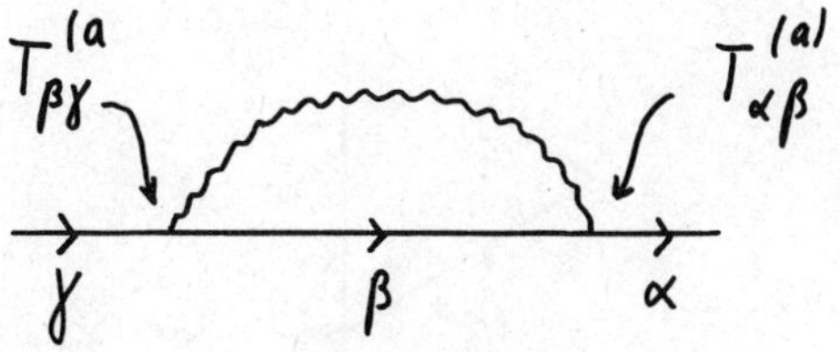

Fig. 2.5 Example of diagram where the group factor $\sum_a \sum_\beta T^{(a)}_{\alpha\beta} \; T^{(a)}_{\beta\gamma}$ appears.

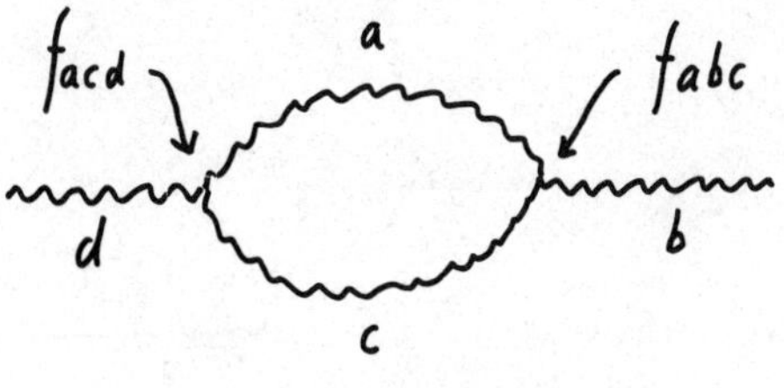

Fig. 2.6 Example of diagram where the group factor $\sum_a \sum_c T^{(a)}_{bc} \, T^{(a)}_{cd}$ appears.

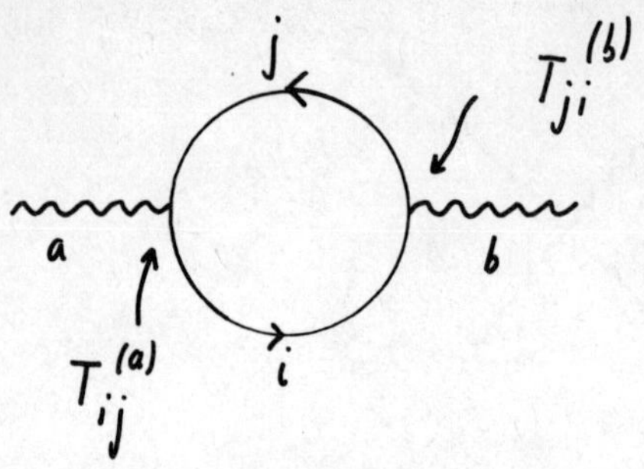

Fig. 2.7 Example of diagram where the group factor $\sum_\alpha \sum_\beta T^{(b)}_{\alpha\beta} T^{(a)}_{\alpha\beta}$ appears.

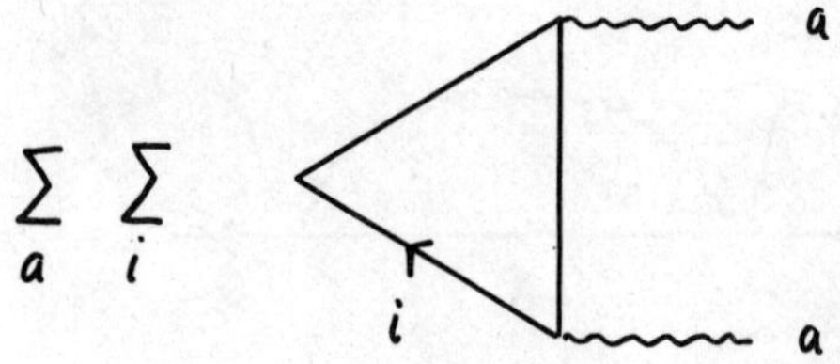

Fig. 2.8 Triangle diagram which gives the anomalous term in the Ward identity associated to the axial-baryonic current.

3. COVARIANT QUANTIZATION AND FEYNMAN RULES.

In the QCD Lagrangean exhibited in eq. (2.1) we have written

+ "gauge fixing term" + "Fadeev-Popov term"

The purpose of this section is the discussion of these terms, which are intimately related to the covariant quantization of the theory; to describe the Feynman rules in a covariant gauge, and to discuss the dynamical symmetries of $\mathcal{L}_{QCD}$.

3a. Covariant Quantization and Gauge-Fixing Term

First, I would like to remind you that these questions already appear at the level of the abelian gauge theory i.e. in QED. Remember that in QED the gauge fixing term

$$-\frac{1}{2a}\left(\partial_\mu A^\mu(x)\right)^2 \tag{3.1}$$

is added to the QED Lagrangean

$$\begin{aligned}\mathcal{L}_{QED} = -\frac{1}{4}F_{\mu\nu}(x)F^{\mu\nu}(x) + i\,\bar{\Psi}(x)\gamma^\mu(\partial_\mu + ie A_\mu(x))\Psi(x) \\ - m\,\bar{\Psi}(x)\,\Psi(x)\end{aligned} \tag{3.2}$$

so as to make covariant canonical quantization possible. Indeed, straightforward applications of the canoncial quantization procedure to the QED Lagrangean in eq. (3.2) leads to difficulties. These are due to the fact that $\partial A_0/\partial t$ does not appear in $\mathcal{L}_{QED}$, hence, the conjugate momentum

$$\Pi^\rho(x) = \frac{\partial\mathcal{L}}{\partial(\partial_0 A_\rho(x))} = F^{\rho 0}(x) \tag{3.3}$$

is such that

$$\Pi^0(x) = 0 \tag{3.4}$$

From the canonical equal-time commutation relations

$$[A_\mu(x), A_\nu(y)]\,\delta(x_0-y_0) = 0 \tag{3.5a}$$

$$[\Pi_\mu(x), \Pi_\nu(y)]\,\delta(x_0-y_0) = 0 \tag{3.5b}$$

$$[\Pi_\mu(x), A_\nu(y)]\,\delta(x_0-y_0) = -i\, g_{\mu\nu}\, \delta^{(4)}(x-y) \tag{3.5c}$$

it then follows that $A_0(x)$ commutes with all other operators i.e., it behaves like a c-number in contrast to the space components $A_i(x)$. Manifest covariance is then lost.

This covariance difficulty is related to the particle content of the field A_μ which has four components. However, it is expected that it describes photon states which are known to have two independent components. One way to solve the covariance difficulty is to apply the canonical quantization procedure to the Lagrangean in eq. (3.2) when the gauge field A_μ is restricted to a covariant subsidiary condition. The simplest way to implement this[1)] is by adding to $\mathcal{L}_{QED}$ the term in eq. (3.1). The new Lagrangean

$$\mathcal{L}_{QED} = -\frac{1}{4} F_{\mu\nu} F^{\mu\nu} + i\,\bar{\Psi}\gamma^\mu D_\mu \Psi - m\bar{\Psi}\Psi - \frac{1}{2a}(\partial_\mu A^\mu)^2 \tag{3.6}$$

is still invariant under the set of local gauge transformations

$$\Psi(x) \rightarrow \exp(ie\,\theta(x))\,\Psi(x) \tag{3.7a}$$

$$A_\mu(x) \rightarrow A_\mu(x) - \partial_\mu \theta(x) \tag{3.7b}$$

<u>provided</u>

$$\partial^\mu \partial_\mu \theta(x) = 0 \tag{3.7c}$$

The same covariance difficulties appear when the covariant canonical quantization procedure is applied to the non-abelian case.

The natural generalization of the abelian gauge fixing term, eq. (3.1), to the non-abelian case is the term

$$-\frac{1}{2a}\,\partial_\mu \vec{W}^\mu(x)\cdot\partial_\nu \vec{W}^\nu(x) \tag{3.8}$$

The difficulty in the non-abelian case lies in the impossibility of finding c-number restrictions, like eq. (3.7), for the $\vec{\theta}$ functions, so that the new Lagrangean

$$\mathcal{L}_{QCD} = -\frac{1}{4}\vec{F}_{\mu\nu}\cdot\vec{F}^{\mu\nu} - \frac{1}{2a}(\partial_\mu \vec{W}^\mu)^2 \tag{3.9}$$

$$+ \text{ fermion term}$$

remains locally gauge invariant. It is precisely this difficulty which is at the basis of the introduction of yet another term in the QCD Lagrangean, the so called Fadeev-Popov term which we discuss in the next subsection.

3b. Indefinite Metric and Fadeev-Popov Term

Again, let us first recall our experience with this problem in QED.

With the choice of the QED Lagrangean in eq. (3.6), the canonical momentum conjugate to $A_\mu(x)$ is

$$\pi_\mu(x) = F_{\mu 0}(x) - \frac{1}{a}\,g_{\mu 0}\,\partial_\nu A^\nu(x) \tag{3.10}$$

and the canoncial commutation relations in eqs. (3.5), plus the fermion anticommutation relation,

$$\{\psi(x), \psi^+(y)\}\,\delta(x_0-y_0) = \delta^{(4)}(x-y) \tag{3.11}$$

can be applied without loss of covariance. The next task is to study the structure of the Fock space generated by the quantization rules obeyed by the creation and annihilation operators[1]. Associated to A_μ there are: annihilation operators of transverse-photons

$$a_i(\vec{p})\,, \qquad i = 1, 2$$

longitudinal photons $a_3(\vec{p})$; and time-like photons $a_0(\vec{p})$. The non-zero commutation relation among these operators and their corresponding creation operators a_i^+ ; a_3^+ and a_0^+, which follow from the canonical commutation relations in eqs. (3.5), are

$$[a_i(\vec{p}), a_j^+(\vec{p}')] = 2|\vec{p}|\, \delta^{(3)}(\vec{p}-\vec{p}')\, \delta_{ij} \qquad (3.12a)$$

$$i,j = 1,2,3$$

and

$$[a_0(\vec{p}), a_0^+(\vec{p}')] = -2|\vec{p}|\, \delta^{(3)}(\vec{p}-\vec{p}') \qquad (3.12b)$$

Because of the minus sign in the r.h.s. of the commutator of time-like photons it turns out that the space of states, the Fock space, has an indefinite metric, e.g. the one particle states of time-like photons have negative probability. A priori, there is nothing alarming about this indefinite metric structure. In order to perform a covariant canonical quantization we were obliged to enlarge the dynamical content of the theory (term $-\frac{1}{2a}(\partial_\mu A^\mu)^2$ in the Lagrangean). What we are eventually interested in is the reduction of the quantized theory to the physical subspace of transverse photons. It is in this subspace that one has to check that we have a consistent theory. That this is certainly the case in QED happens thanks to a cancellation of the probabilities of observing scalar (or time-like) photons and longitudinal photons, i.e., the restriction of the theory to the subspace of physical (transverse) photons, preserves unitary.

In QCD, covariant, canonical quantization in the presence of the gauge fixing term in eq. (3.8) also leads to a Fock space with indefinite metric. The difference with QED is that now, the restriction to the subspace of transverse gluons is no longer automatically unitary, as it was in QED. One is forced to introduce supplementary non-physical fields, the so called Fadeev-Popov ghosts, to restore unitary in the subspace of physical states. The best way to understand the problem is to examine an example[2].

Let us consider the process

$$e^+e^- \to \gamma\gamma \qquad \text{in QED} \qquad (3.13)$$

versus

$$q\bar{q} \rightarrow WW \qquad \text{in QCD} \tag{3.14}$$

and describe what happens at the lowest order in perturbation theory. The corresponding Feynman diagrams are shown in Figs. (3.1) and (3.2) respectively.

In order to define the polarization degree of freedom associated to a photon (gluon) of energy-momentum k_μ let us introduce a tetrad of unit 4-vectors

$$\{\epsilon_\mu^{(0)}, \epsilon_\mu^{(i)}\} \qquad i = 1,2,3,$$

with (recall that $k^2 = 0$)

$$\epsilon_\mu^{(0)} = \eta_\mu \quad \text{unit time-like 4-vector, such that } k^0 = k\cdot\eta \tag{3.15a}$$

$$\epsilon_\mu^{(i)} \; i = 1, 2 \quad \text{two orthogonal unit 4-vectors in the 2-plane } \perp \text{ to } (k,\eta) \tag{3.15b,c}$$

$$\epsilon_\mu^{(3)} = \frac{k_\mu - (k\cdot\eta)\eta_\mu}{k\cdot\eta} \tag{3.15d}$$

With the aid of the tetrad, we can write the metric tensor as follows

$$-g_{\mu\nu} = -\epsilon_\mu^{(0)}\epsilon_\nu^{(0)} + \sum_{i=1,2} \epsilon_\mu^{(i)}\epsilon_\nu^{(i)} + \epsilon_\mu^{(3)}\epsilon_\nu^{(3)} \tag{3.16}$$

and define a transverse tensor $\tau_{\mu\nu}$ as follows

$$\tau_{\mu\nu} = \sum_{i=1,2} \epsilon_\mu^{(i)}\epsilon_\nu^{(i)} = -g_{\mu\nu} - \frac{k_\mu k_\nu}{(k\cdot\eta)^2} + \frac{1}{k\cdot\eta}(k_\mu\eta_\nu + k_\nu\eta_\mu) \tag{3.17}$$

Now let us consider the annihilation amplitude to a final state with fixed polarizations a and b (a,b = 0,1,2,3). In both cases QED and QCD we shall get an expression of the type

$$(\text{AMPLITUDE})_{ab} = \epsilon_{\mu_1}^{(a)}\epsilon_{\mu_2}^{(b)} J^{\mu_1\mu_2} \tag{3.18}$$

Of course $J^{\mu_1\mu_2}$ will be different for QED and QCD. The corresponding transition probability will be

$$P_{ab} = \epsilon_{\mu_1}^{(a)} \epsilon_{\nu_1}^{(a)} \epsilon_{\mu_2}^{(b)} \epsilon_{\nu_2}^{(b)} J^{\mu_1\mu_2} (J^{\nu_1\nu_2})^+$$
$$\equiv \epsilon_{\mu_1}^{(a)} \epsilon_{\nu_1}^{(a)} \epsilon_{\mu_2}^{(b)} \epsilon_{\nu_2}^{(b)} W^{\mu_1\nu_1\mu_2\nu_2} \tag{3.19}$$

The transition probability to transverse photons (gluons) will be according to eqs. (3.17) and (3.19)

$$P_{TT} = \frac{1}{2} \tau_{\mu_1\nu_1} \tau_{\mu_2\nu_2} W^{\mu_1\nu_1\mu_2\nu_2} \tag{3.20}$$

The factor 1/2 coming from Bose statistics (2 identical particles in the final state). Let us now define the covariant transition probability as

$$P_{CC} = \frac{1}{2} (-g_{\mu_1\nu_1})(-g_{\mu_2\nu_2}) W^{\mu_1\nu_1\mu_2\nu_2} \tag{3.21}$$

In QED we have

$$P_{CC} = P_{TT} \tag{3.22}$$

i.e., the total physical probability coincides with the covariant probability. This follows from the identity in eq. (3.17) and gauge invariance which implies

$$k_{\mu_1} J^{\mu_1\mu_2} = k_{\mu_2} J^{\mu_1\mu_2} = 0 \tag{3.23}$$

In QCD, with eq. (3.9) as the underlying Lagrangean, the covariant probability does not coincide with the total physical probability. In fact it turns out that

$$P_{CC} = P_{TT} + (P_{TL} + P_{TS} = 0) + (P_{LT} + P_{ST} = 0)$$
$$+ (P_{LL} + P_{LS} + P_{SL} + P_{SS} > 0) \tag{3.24}$$

i.e., there is still a partial cancellation of longitudinal and scalar gluons but not complete. In fact a fraction of the total covariant probability for $q\bar{q} \to WW$ corresponds to nothing observable!

The fields which are required to restore unitarity in the physical sector are a set of scalar fields

$$\varphi_{(a)}(x) \qquad a = 1, 2, \cdots 8 \tag{3.25}$$

with Fermi-Dirac statistics (hence they have negative metric in the Fock space) coupled to the gluon fields $W_\mu^{(a)}(x)$ via the coupling

$$- \partial_\mu \bar{\varphi}_{(a)}(x) (D^\mu \varphi)^{(a)}(x) \tag{3.26}$$

The fields $\varphi^{(a)}(x)$ are the so called "ghosts" or Fadeev-Popov fields. A term like (3.26) when added to the QCD Lagrangean in eq. (3.9) induces new diagrams like the one shown in Fig. (3.3). The corresponding transition probability is negative (hence the name ghost) and exactly cancells the offending term in eq. (3.24) i.e.,

$$P(\text{Ghost - Ghost}) = -(P_{LL} + P_{LS} + P_{SL} + P_{SS}) \tag{3.27}$$

and therefore

$$P_{CC} + P(\text{Ghost - Ghost}) = P_{TT} \tag{3.28}$$

What is left of the initial local gauge invariance in the presence of the gauge fixing term and the Fadeev-Popov term? This is the question we discuss in the next subsection.

3c. Becchi-Rouet-Stora Transformations and Slavnov-Taylor identities

The full QCD Lagrangean, with gauge fixing term included, eq. (3.8), and Fadeev-Popov term included, eq. (3.26), is then

$$\mathcal{L}_{QCD}(x) = -\frac{1}{4} F_{\mu\nu}^{(a)}(x) F^{\mu\nu}_{(a)}(x) - \frac{1}{2a} \partial_\mu W^\mu_{(a)}(x) \partial^\nu W_\nu^{(a)}(x)$$
$$- \partial_\mu \bar{\varphi}_{(a)}(x) [\delta_{ab} \partial^\mu - ig(-if_{cab}) W^\mu_{(c)}(x)] \varphi_{(b)}(x)$$

$$+ i \sum_{j=1}^{n} \bar{\psi}_j^{\alpha}(x) \gamma^{\mu} [\delta_{\alpha\beta} \partial_\mu - ig \sum_a \frac{1}{2} \lambda_{\alpha\beta}^{(a)} W_\mu^{(a)}(x)] \psi_j^{\beta}(x) \tag{3.29}$$
$$- \sum_{j=1}^{n} m_j \bar{\psi}_j^{\alpha}(x) \psi_{\alpha j}(x)$$

It has been pointed out by Becchi, Rouet and Stora, ref. (3.3), that there exists a set of one-parameter transformations which leave the action associated to $\mathcal{L}_{QCD}(x)$ invariant[3)]. The transformations are

$$W_\mu^{(a)}(x) \to W_\mu^{(a)}(x) + \omega (D_\mu \varphi)^{(a)}(x) \tag{3.30a}$$

$$\psi(x) \to \exp(-ig\omega \vec{T}\cdot\vec{\varphi}(x)) \psi(x) \tag{3.30b}$$

$$\bar{\varphi}^{(a)}(x) \to \bar{\varphi}^{(a)}(x) + \frac{\omega}{a} \partial^\mu W_\mu^{(a)}(x) \tag{3.30c}$$

$$\varphi^{(a)}(x) \to \varphi^{(a)}(x) - \frac{1}{2} g \omega \varphi^{(b)}(x) f_{abc} \varphi^{(c)}(x) \tag{3.30d}$$

The interesting thing about these transformations is that they generate all the identities among Green's functions which follow from the local gauge invariance of the theory, i.e., the so called Slavnov-Taylor identities (see refs. (3.5) and (3.6)), the equivalent of the Ward-Takahashi identities in QED.

In applying the B-R-S Transformations it is useful to remember that under (3.30) the changes of the fields vanish:

$$\delta((D_\mu \varphi)^{(a)}) = \delta(f_{abc} \varphi^{(b)} \varphi^{(c)}) = \delta(\partial_\mu W^{\mu(a)}) = 0 \tag{3.31a}$$

or

$$\delta^2 A_\mu = \delta^2 \varphi = \delta^2 \bar{\varphi} = 0 \tag{3.31b}$$

As an example of the usefulness of B-R-S transformations to derive Salnov-Taylor identities let us apply them to the gluon propagator and prove that order by order in perturbation theory the non-transverse part of the gluon propagator remains the same as for the free propagator. The formal expression for the gluon propagator reads

$$i\, D^{ab}_{\mu\nu}(k) = \int d^4x\, e^{ik\cdot x} \langle 0 | T(W^{(a)}_{\mu}(x)\, W^{(b)}_{\nu}(0)) | 0 \rangle \tag{3.32}$$

and the identity in question states

$$k^{\mu} k^{\nu}\, i\, D^{ab}_{\mu\nu}(k) = -i\, a\, \delta_{ab} \tag{3.33}$$

Incidences of this identity will show up several times in the course of these lectures. To derive this identity we start with the trivial identity,

$$\langle 0 | T(\partial_{\mu} W^{\mu}_{(a)}(x), \bar{\varphi}_{(b)}(y)) | 0 \rangle = 0 \tag{3.34}$$

By B-R-S invariance, we also have

$$\langle 0 | T(\partial_{\mu} W^{\mu}_{(a)}(x), \bar{\varphi}_{(b)}(y)) | 0 \rangle = \langle 0 | T(\partial_{\mu} W'^{\mu}_{(a)}(x), \bar{\varphi}'_{(b)}(y)) | 0 \rangle \tag{3.35}$$

where

$$W'^{\mu}_{(a)} = W^{\mu}_{(a)} + \omega\, (D^{\mu} \varphi)_{(a)} \tag{3.30a}$$

$$\bar{\varphi}'_{(b)} = \bar{\varphi}_{(b)} + \frac{\omega}{a}\, \partial_{\mu} W^{\mu}_{(b)} \tag{3.30b}$$

It then follows that

$$\frac{\omega}{a} \langle 0 | T(\partial_{\mu} W^{\mu}_{(a)}(x), \partial_{\nu} W^{\nu}_{(b)}(y)) | 0 \rangle = 0 \tag{3.36}$$

If we now Fourier transform and double partial integrate, we get

$$\int d^4x\, d^4y\; e^{ik'x}\, e^{-iky} \langle 0 | T(\partial_{\mu} W^{\mu}_{(a)}(x), \partial_{\nu} W^{\nu}_{(b)}(y)) | 0 \rangle$$

$$= k'_{\mu} k_{\nu} \int d^4x\, d^4y\; \exp(ik'x) \exp(iky) \langle 0 | T(W^{\mu}_{(a)}(x)\, W^{\nu}_{(b)}(y)) | 0 \rangle$$

$$+ i k'_\mu \int d^4x\, d^4y\, \exp(ik'x) \exp(-iky)\, \delta(x_0-y_0) \langle 0 | [W^\mu_{(a)}(x), W^0_{(b)}(y)] | 0 \rangle$$

$$- \int d^4x\, d^4y\, \exp(ik'x) \exp(-iky)\, \delta(x_0-y_0) \langle 0 | [W^0_{(a)}(x), \partial_\nu W^\nu_{(b)}(y)] | 0 \rangle \qquad (3.37)$$

Next we use the canonical commutation relation

$$[\Pi^\mu_{(a)}(x) , W^\nu_{(b)}(y)] \, \delta(x_0-y_0) = -i g^{\mu\nu} \delta_{ab}\, \delta^{(4)}(x-y) \qquad (3.38)$$

with

$$\Pi^\mu_{(a)}(x) = - F^{0\mu}_{(a)}(x) - \frac{1}{a} g^{0\mu} \partial_\nu W^\nu_{(a)}(x) \qquad (3.39)$$

as follows from the Lagrangean in eq. (3.29). Thus we obtain

$$[W^0_{(a)}(x) , \partial_\nu W^\nu_{(b)}(y)] \, \delta(x_0-y_0) = -i a\, \delta^{(4)}(x-y) \qquad (3.40)$$

With this result incorporated in eq. (3.37), and the definition in eq. (3.32) we inmediatly obtain the identity in eq. (3.33).

3d. <u>Feynman Rules</u>

Quantization in a covariant gauge leads to the following Feynman rules for tree vertices and propagators:

<u>quark-gluon-quark</u>

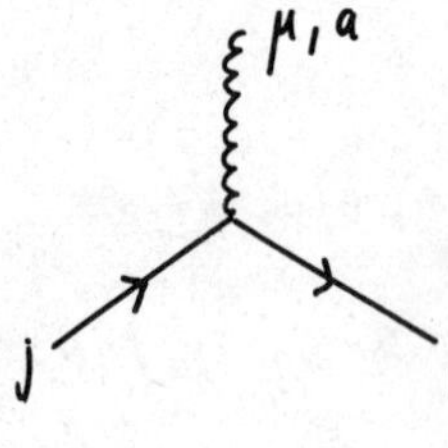

$$i g \gamma^\mu T^{(a)}_{ij} \qquad (3.41)$$

3-gluon vertex

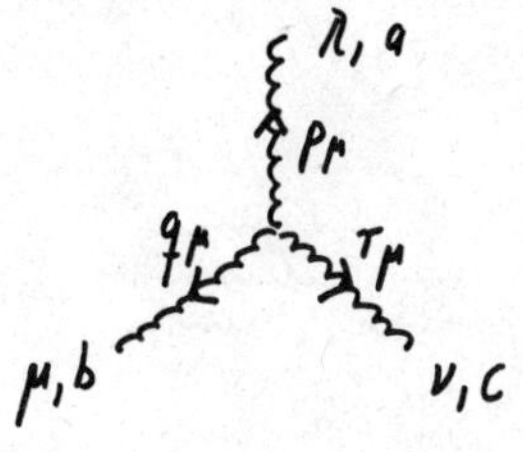

$$-g f^{abc}[(p-q)_\nu g_{\lambda\mu} + (q-r)_\lambda g_{\mu\nu} + (r-p)_\mu g_{\nu\lambda}] \tag{3.42}$$

4-gluon vertex

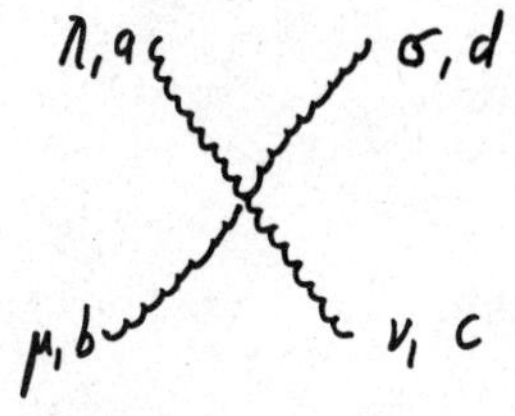

$$\begin{aligned}&-ig^2 f^{abe} f^{cde} (g_{\lambda\nu} g_{\mu\sigma} - g_{\lambda\sigma} g_{\mu\nu})\\ &-ig^2 f^{ace} f^{bde} (g_{\lambda\mu} g_{\nu\sigma} - g_{\lambda\sigma} g_{\mu\nu})\\ &-ig^2 f^{abe} f^{cbe} (g_{\lambda\nu} g_{\mu\sigma} - g_{\lambda\mu} g_{\sigma\nu})\end{aligned} \tag{3.43}$$

ghost-gluon-ghost

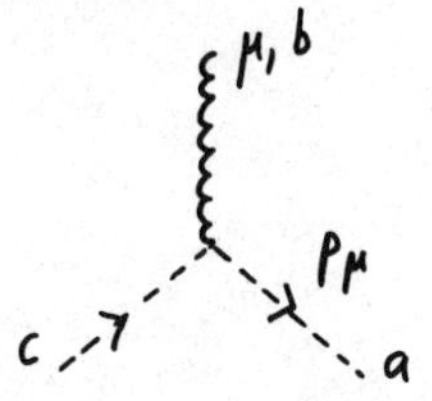

$$g f^{abc} p_\mu \tag{3.44}$$

Propagators:

quark

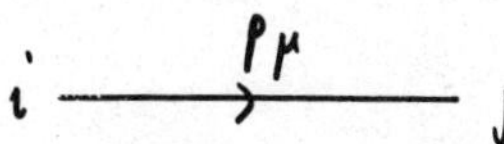

$$\frac{i}{\not p - m + i\epsilon} \delta_{ij} \tag{3.45}$$

gluon

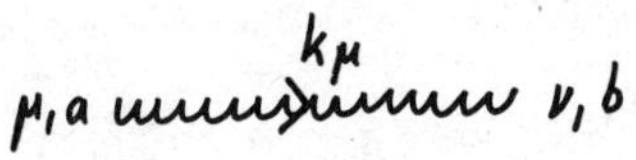

$$\frac{-i\,\delta_{ab}}{k^2+i\epsilon}\left[g_{\mu\nu} - (1-a)\frac{k_\mu k_\nu}{k^2}\right] \tag{3.46}$$

ghost

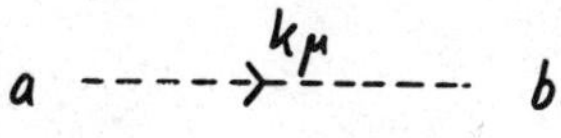

$$-i\,\delta_{ab}\,\frac{1}{k^2+i\epsilon} \tag{3.47}$$

Factors:

i) $\int \frac{d^4 l}{(2\pi)^4}$ for each loop integration

ii) (-1) for closed fermion (ghost) loops

iii) Statistical factors, like: $\frac{1}{2}$  ; $\frac{1}{6}$ 

FOOTNOTES OF SECTION 3

1) The reader interested in the details of these questions in QED can consult my lectures in ref. [R. 14], where earlier references can be found.

2) This is precisely the way that the problem was tackled by Feynman for the first time, see ref. (3.1). The formal general solution was later obtained by Fadeev-Popov, ref. (3.2).

3) Other references where B-R-S transformations are discussed, are Marciano and Pagels in [R. 7] and Brandt, ref. (3.4).

REFERENCES OF SECTION 3

1) R.P. Feynman, Acta Phys. Polon. 26(1963)697.

2) L.D. Fadeev and Y.N. Popov, Phys. Letters, 25B(1967)29.

3) C. Becchi, A. Rouet and R. Stora, Phys. Letters 52B(1974)344. Comm. Math.Phys. 42(1975)127.

4) R.A. Brandt, Nuclear Phys. B116(1976)413.

5) A.A. Slavnov, Sov. J. Particles Nucl. 5, nº3(1975)303.

6) J.C. Taylor, Nucl. Phys. B33(1971)436.

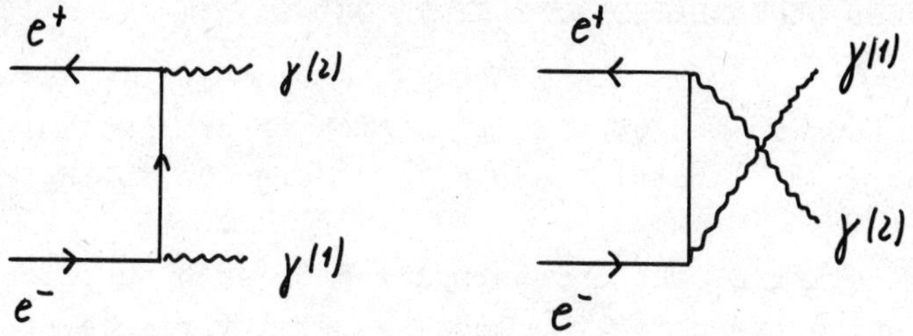

Fig. 3.1 Lowest order Feynman diagrams contributing to the process $e^+e^- \rightarrow \gamma\gamma$

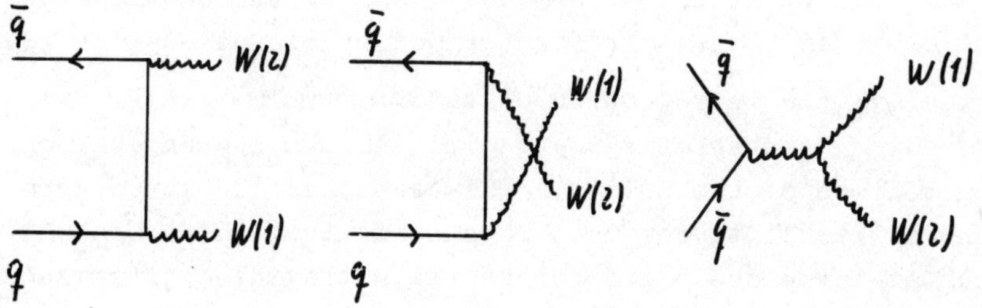

Fig. 3.2 Lowest order Feynman diagrams contributing to the process $q\bar{q} \rightarrow WW$

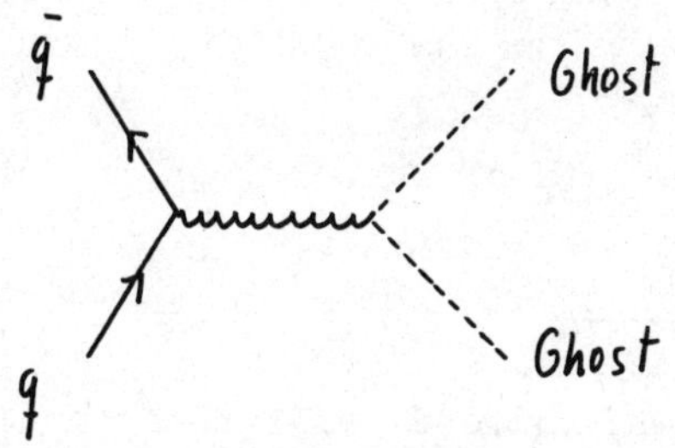

Fig. 3.3 Lowest order Feynman diagram contributing to the process $q\bar{q} \rightarrow$ Ghost Ghost.

4. RENORMALIZED PERTURBATION THEORY IN QCD

The steps in the construction of a renormlized perturbation theory of QCD at the practitioner's level are the following:

1. Regularization of the unrenormalized Green's functions by a procedure which preserves the local gauge invariance of the theory as embodied in the Slavnov-Taylor identities discussed in section 3.
2. Identification of the primitive divergences. This is done by inspection of all one-loop Feynman diagrams.
3. Introduction of one-loop renormalization subractions which render finite the primitive divergences. Again, the subtractions are not arbitrary, they are constrained by Slavnov-Taylor identities.
4. Formally, the introduction of the subtractions in 3. can be achieved adding counterterms to the initial QCD Lagrangean corresponding to each of the primitive divergences. The actual proof of renormalizability consists in showing that there are all the counter-terms needed to remove the ultraviolet divergences of the theory at an arbitrary order.
5. Physical observables must be independent of the particular substraction method used to calculate them. This invariance is called the renormalization group invariance. As we shall see it constrain the scaling properties of the Green's functions of the theory in a way which is most easily expressed in terms of linear partial differential equations.

Let us now discuss each of these steps in further detail.

4a. Dimensional Regularization

This technique was introduced by 't Hooft and Veltman (4.1); and independently by Bollini and Giambiagi (4.2). It has been discussed in detail in various review articles[1)]; therefore, here we shall only outline the method and give a list of useful formulae to reproduce the calculations we shall do later on.

The crucial point of the method consists in giving a meaning to the divergent integrals which appear in the calculation of unrenormalized Feynman diagrams by changing the dimensions of space-time. Consider the typical 4-dimensional integral

$$I(4,m) = \int \frac{d^4k}{(2\pi)^4} \frac{1}{(k^2 - \mathbb{R}^2 + i\epsilon)^m} \tag{4.1}$$

where m is a non-negative integer. It is ultraviolet divergent for $2m < 4$. If we now change the space-time dimension $4-n$ the corresponding integral

$$I(n,m) = \int \frac{d^nk}{(2\pi)^n} \frac{1}{(k^2 - \mathbb{R}^2 + i\epsilon)^m} \tag{4.2a}$$

exists for $n < 2m$ and is given by

$$I(n,m) = \frac{i^{1-2m}}{(2\sqrt{\pi})^n} \frac{\Gamma(m-\frac{n}{2})}{\Gamma(m)} \mathbb{R}^{2(\frac{n}{2}-m)} \tag{4.2b}$$

where $\Gamma(n)$ is Euler's gamma-function

$$\Gamma(z) = \int_0^\infty dt\, t^{z-1} e^{-t} \qquad Re\, z > 0 \tag{4.3a}$$

For our purposes it is more convenient to use the Weierstrass's analytic continuation for the Euler's function

$$\Gamma(z) = \sum_{n=0}^{\infty} \frac{(-1)^n}{n!\,(n+z)} + \int_1^\infty dt\, t^{z-1} e^{-t} \tag{4.3b}$$

which shows the analyticity of $\Gamma(z)$ in the entire z-plane except for poles in the negative real axis for $z = 0, -1, -2, \ldots$. Notice that this analytic continuation of the Euler function is unique since it clearly overlaps that of eq. (4.3a)[2].

The r.h.s. of eq. (4.2b) can thus be taken as a definition of the integral in eq. (4.2a) when the integral does not exist. The divergences of the integral for

$n = 2m$ .... logarithmic

and

$$n = 2m + 2 \quad \ldots\ldots \quad \text{quadratic}$$

now appear as poles at $z \to 0$ and $z \to -1$ of the Γ -function.

The great advantage of this method is that no new couplings have been introduced in the Lagrangean and hence all the symmetry properties, in particular, local gauge invariance, are preserved. The only change appears in the dimensions of the fields and coupling constant.

In order that the action in $n=4-\epsilon$ dimensions remains dimensionless we have to attribute the following dimensions:

gluon field	W_μ	...	dimension	$1-\epsilon/2$
quark field	ψ	...	dimension	$3/2-\epsilon/2$
ghost field	φ	...	dimension	$1-\epsilon/2$
coupling constant	g	...	dimension	$\epsilon/2$
masses	m_i	...	dimension	1
gauge parameter	a	...	dimension	0

The formal manipulations like the shift of integration variables; symmetric integration; and partial integrations are now unambiguously defined, provided that the Dirac algebra is performed for an arbitrary n, i.e.,:

$$g^{\mu\nu} g_{\mu\nu} = n \quad , \quad \gamma^\mu \gamma_\mu = n\, I$$

$$\gamma_\mu \gamma^\alpha \gamma^\mu = (2-n)\gamma^\alpha \, , \quad \gamma^\mu \gamma^\alpha \gamma^\beta \gamma_\mu = 4 g^{\alpha\beta} I + (n-4)\gamma^\alpha\gamma^\beta \tag{4.4}$$

$$\gamma_\mu \gamma^\alpha \gamma^\beta \gamma^\lambda \gamma^\mu = -2\gamma^\lambda\gamma^\beta\gamma^\alpha - (n-4)\gamma^\alpha\gamma^\beta\gamma^\lambda$$

$$k^\mu k^\nu \longrightarrow \frac{1}{n} k^2 g^{\mu\nu} \tag{4.5}$$

$$Tr\ 1 = 2^{n/2} \tag{4.6}$$

$$Tr(\gamma^\rho\gamma^\lambda\gamma^\mu\gamma^\nu) = 2^{n/2}(g^{\rho\lambda}g^{\mu\nu} - g^{\rho\mu}g^{\lambda\nu} + g^{\rho\nu}g^{\lambda\mu})$$

Some useful properties of the Γ-function are:

$$n\,\Gamma(n) = \Gamma(n+1) \quad \text{, valid for all } n \tag{4.7}$$

For integer n, $\quad \Gamma(n+1) = n\,!$

$$\Gamma(1) = 1 \quad ; \quad \Gamma(1/2) = \sqrt{\pi} \tag{4.8}$$

Expansion of the Γ function around n = 4 :

$$\Gamma\left(\frac{n}{2}-1\right) = 1 - \gamma/2\,(n-4) + \frac{1}{8}\left(\gamma^2 + \frac{\pi^2}{6}\right)(n-4)^2 + \cdots \tag{4.9a}$$

where γ is Euler's constant

$$\gamma = \lim_{n\to\infty}\left(1 + \frac{1}{2} + \cdots + \frac{1}{n} - \log n\right) = 0.572 \tag{4.9b}$$

Then, using (4.9) and (4.7)

$$\Gamma(\epsilon) = \frac{1}{\epsilon}(1 - \gamma\epsilon + \cdots) \tag{4.10}$$

In the process of combining denominators of propagators with Feynman parameters it is sometimes useful to have a generalization of the formula

$$\frac{1}{ab} = \int_0^1 dx\, \frac{1}{[ax + b(1-x)]^2} \tag{4.11a}$$

to arbitrary powers

$$\frac{1}{a^\alpha b^\beta} = \frac{\Gamma(\alpha+\beta)}{\Gamma(\alpha)\Gamma(\beta)} \int_0^1 dx\, \frac{x^{\alpha-1}(1-x)^{\beta-1}}{[ax + b(1-x)]^{\alpha+\beta}} \tag{4.11b}$$

If α (or β) = ϵ , I suggest using the trick

$$\frac{1}{a^{\epsilon} b^{\beta}} = \frac{a}{a^{1+\epsilon} b^{\beta}} = a \frac{\Gamma(1+\beta+\epsilon)}{\Gamma(1+\epsilon)\Gamma(\beta)} \int_0^1 dx \frac{x^{\epsilon}(1-x)^{\beta-1}}{[ax+b(1-x)]^{1+\beta+\epsilon}} \tag{4.11c}$$

A useful generalization of eq. (4.2b) which practically covers all the one-loop calculation requirements is

$$I(r,m) = \int \frac{d^n k}{(2\pi)^n} \frac{(k^2)^r}{(k^2 - \mathbb{R}^2)^m} = i \frac{(-1)^{r-m}}{(16\pi^2)^{n/4}} \frac{\Gamma(r+\frac{n}{2})\Gamma(m-r-\frac{n}{2})}{\Gamma(\frac{n}{2})\ \Gamma(m)} \mathbb{R}^{2r-2m+n} \tag{4.12}$$

In performing the integrals over Feynman parameters it is useful to know a few parametric representations of special functions; e.g.,

the beta-function

$$B(x,y) = \int_0^1 dt\ t^{x-1} (1-t)^{y-1} \tag{4.13a}$$

where

$$B(x,y) = \frac{\Gamma(x)\Gamma(y)}{\Gamma(x+y)} = B(y,x) \tag{4.13b}$$

and the hypergeometric function

$$\begin{aligned} {}_2F_1(a,b,c;z) &= \frac{\Gamma(a)}{\Gamma(b)\Gamma(c-b)} \int_0^1 dt\ t^{b-1}(1-t)^{c-b-1}(1-tz)^{-a} \\ &\mathrm{Re}\,c > \mathrm{Re}\,b > 0 \ ; \quad |\arg(1-z)| < \pi \end{aligned} \tag{4.14a}$$

where

$$\begin{aligned} {}_2F_1(a,b,c;z) &= 1 + \frac{a \cdot b}{1 \cdot c} z + \frac{a(a+1)b(b+1)}{1 \cdot 2 \cdot c(c+1)} z^2 + \cdots \\ &= \frac{\Gamma(c)}{\Gamma(b)\Gamma(a)} \sum_{n=0}^{\infty} \frac{\Gamma(a+n)\Gamma(b+n)}{\Gamma(c+n)} \frac{z^n}{n!} \end{aligned} \tag{4.14b}$$

4b. Primitive Divergences

The one loop Feynman diagrams which lead to ultraviolet divergences in QCD can be classified in two classes:

1. Self Energy Diagrams

Quark self energy	Fig. 4.1
Gluon self energy	Fig. 4.2
Ghost self energy	Fig. 4.3

2. Vertex Interaction Diagrams

Gluon - Quark - Quark coupling	Fig. 4.4
3 - Gluon coupling	Fig. 4.5
Gluon - Ghost - Ghost coupling	Fig. 4.6
4 - Gluon coupling	Fig. 4.7

Associated to these diagrams there will be renormalization constants which we have written in brackets in the figures:

δm : quark mass renormalization

Z_{2F} : quark wave-function renormalization

Z_{3YM} : gluon wave-function renormalization

$\tilde{Z}_3$: ghost wave-function renormalization

Z_{1F} : quark vertex renormalization

Z_{1YM} : 3-gluon vertex renormalization

$\tilde{Z}_1$: Ghost vertex renormalization

Z_5 : 4 - gluon vertex renormalization

To see how the divergences appear let us do the explicit calculation of the gluon self-energy due to quark loops i.e., the diagram in Fig. 4.8. We choose this diagram for two reasons. Pedagogically it is one of the simplest to calculate (it is the analog of the photon self energy in QED). From the physical point of view it plays a crucial role when finite quark mass effects in QCD calculations are taken into account. Using the Feynman rules of section 3 we can write

$$i\,\Pi^{(2)}_{\mu\nu}(q) = \sum_j \int \frac{d^np}{(2\pi)^n}(-1)\Big\{ ig\,T^{(A)}_{ij}\,\gamma_\mu \frac{i}{\not p - m_j + i\epsilon}\, ig\,T^{(B)}_{ji}\,\gamma_\nu \frac{i}{\not p - \not q - m_j + i\epsilon}\Big\} \tag{4.15a}$$

$$= (ig)^2\,\delta_{AB}\,\frac{1}{2}\sum_j \int \frac{d^np}{(2\pi)^n}\,\frac{Tr\,(\gamma_\mu(\not p + m_j)\gamma_\nu(\not p - \not q + m_j))}{(p^2 - m_j^2 + i\epsilon)((p-q)^2 - m_j^2 + i\epsilon)}$$

where we have used eq. (2.14). Next we combine the two factors in the denominator using the identity in eq. (4.11a) with

$$a \equiv p^2 - m_j^2 + i\epsilon \qquad\qquad b \equiv (p-q)^2 - m_j^2 + i\epsilon$$

Then

$$i\,\Pi^{(2)}_{\mu\nu}(q) = (ig)^2\,\delta_{AB}\,\frac{1}{2}\sum_j \int_0^1 dx \int \frac{d^np}{(2\pi)^n}\,\frac{Tr\,\{\gamma_\mu(\not p + m_j)\gamma_\nu(\not p - \not q + m_j)\}}{[(p-\ell)^2 - R^2]^2} \tag{4.15b}$$

where

$$\ell_\mu = q_\mu(1-x)\,, \qquad R^2 = -x(1-x)q^2 + m_j^2 - i\epsilon \tag{4.15c,d}$$

Next we shift the d^np integration by a change of variables

$$p - \ell = \tilde p \tag{4.16}$$

Using the fact that upon integration

$$\tilde p_\mu \tilde p_\nu \longrightarrow \frac{1}{n}\,\tilde p^2 g_{\mu\nu} \tag{4.17}$$

and that integrals linear in p vannish, we get after some algebra[3]

$$i\,\Pi^{(2)}_{\mu\nu}(q) = (ig)^2\,\delta_{AB}\,\frac{1}{2}\sum_j \int_0^1 dx\; 2^{n/2}\,2x(1-x)\,(g_{\mu\nu}q^2 - q_\mu q_\nu)\cdot$$

$$\cdot \int \frac{d^n\tilde{p}}{(2\pi)^n} \frac{1}{(\tilde{p} - \mathbb{R}^2)^2} \qquad (4.18)$$

i.e., using eq. (4.12)

$$i\,\Pi^{(2)}_{\mu\nu}(q) = -i\,(g_{\mu\nu}q^2 - q_\mu q_\nu)\;\Pi^{(2)}(q^2,\epsilon) \qquad (4.19)$$

and

$$\Pi^{(2)}(q^2,\epsilon) = \frac{\alpha_s}{\pi}\,\delta_{AB}\,\frac{1}{2}\,(2\pi)^{\epsilon/2}\,\Gamma(\epsilon/2)\sum_j \int_0^1 dx\,2x(1-x)\left(\frac{\nu^2}{\mathbb{R}^2}\right)^{\epsilon/2} \qquad (4.20)$$

where ν is an arbitrary mass scale which has been absorbed in the definition of a dimensionless coupling constant α_s

$$\alpha_s = \frac{g^2}{4\pi}\,(\nu^2)^{-\epsilon/2} \qquad (4.21)$$

The ultraviolet divergence of the one-loop gluon self-energy diagram appears now as a singularity of the type $\frac{1}{\epsilon}$ when $\epsilon \to 0$.

The $\frac{1}{\epsilon}$ singularity in eq. (4.20) can be removed by substracting $\Pi^{(2)}(q^2,\epsilon)$ at an arbitrary value of q^2 . For example, in the μ -renormalization scheme which we shall discuss later on, the substraction point is chosen at an arbitrary euclidean point $q^2 = -\mu^2$, and the renormalized gluon self-energy $\Pi^{(2)}(q^2)$ is defined as

$$\begin{aligned}\Pi^{(2)}(q^2) &= \lim_{\epsilon\to 0}\; \Pi^{(2)}(q^2,\epsilon) - \Pi^{(2)}(-\mu^2,\epsilon)\\ &= -\frac{\alpha_s}{\pi}\,\delta_{AB}\,\frac{1}{2}\sum_j \int_0^1 dx\,2x(1-x)\log\frac{-q^2x(1-x)+m_j^2-i\epsilon}{\mu^2x(1-x)+m_j^2}\end{aligned} \qquad (4.22)$$

We shall discuss other substraction procedures later on.

Formally, the substraction procedure can be achieved by adding a counterterm of the type

$$\Delta_3\,\frac{1}{4}\,(\partial_\mu \vec{W}_\nu - \partial_\nu \vec{W}_\mu)\cdot(\partial^\mu \vec{W}^\nu - \partial^\nu \vec{W}^\mu) \qquad (4.23)$$

to the QCD Lagrangean, with Δ_3 understood as a power series in α_s. The normalization of Δ_3 is fixed by the choice of renormalization scheme. For example in the μ-scheme, and to one loop, $\pi^{(2)}(-\mu^2;\epsilon)$ will be part of the contribution to Δ_3. (There will also be other contributions coming from the subtraction constants associated to the other one loop gluon self-energy diagrams in fig. 4.2).

If we systematically carry out this procedure, which we have illustrated for the quark loop contribution to the gluon self-energy, to all the other one loop diagrams we shall get a modified QCD Lagrangean of the type

$$\begin{aligned} \mathcal{L}_{QCD} \to \mathcal{L}_{QCD} &+ \Delta_3 \frac{1}{4} (\partial_\mu \vec{W}_\nu - \partial_\nu \vec{W}_\mu)\cdot(\partial^\mu \vec{W}^\nu - \partial^\nu \vec{W}^\mu) \\ &+ \Delta_1 \frac{1}{2} (\partial_\mu \vec{W}_\nu - \partial_\nu \vec{W}_\mu)\, g\, \vec{W}^\mu \times \vec{W}^\nu \\ &+ \Delta_5 \frac{1}{4} g^2\, \vec{W}_\mu \times \vec{W}_\nu \cdot \vec{W}^\mu \times \vec{W}^\nu \\ &- \Delta_2\, i \sum_j \bar{\psi}_j \gamma^\mu \partial_\mu \psi_j + \Delta_4 \sum_j m_j \bar{\psi}_j \psi_j \\ &- \Delta_{1F}\, g\, \bar{\psi} \frac{\vec{\lambda}}{2} \gamma^\mu \psi \vec{W}_\mu + \Delta_6 \frac{1}{2a} (\partial_\mu \vec{W}^\mu)^2 \\ &+ \tilde{\Delta}_3\, \partial_\mu \vec{\bar{\varphi}}\, \partial^\mu \vec{\varphi} + \tilde{\Delta}_1\, g\, \partial_\mu \vec{\bar{\varphi}} \cdot \vec{W}^\mu \times \vec{\varphi} \\ &\equiv \mathcal{L}_{QCD} + \Delta \mathcal{L}_{QCD} \end{aligned} \tag{4.24}$$

where we have used the short-hand notation

$$\vec{W}_\mu \times \vec{W}_\nu \equiv f_{abc}\, W_\mu^{(b)}\, W_\nu^{(c)} \qquad a = 1,2 \cdots 8 \tag{4.25}$$

Notice that all the added counterterms are of the same form as the different terms in the initial $\mathcal{L}_{QCD}$.

The proof of renormalizability consists in showing that with the Δ's in eq. (4.24) understood as power series in α_s, the corresponding counterterms are all the extra terms needed to remove the ultraviolet divergences of the theory at an arbitrary order of the perturbation expansion.

We shall now show that it is possible to rescale the fields and parameters of the theory and define new fields

$$\vec{W}_0^{\mu} \quad , \quad \vec{\varphi}_0 \quad , \quad \psi_0 \tag{4.26a}$$

and new parameters

$$g_0 \; , \quad m_{j0} \quad , \quad a_0 \tag{4.26b}$$

in terms of which

$$\begin{aligned} \mathcal{L}_{QCD} + \Delta \mathcal{L}_{QCD} = & - \frac{1}{4} \vec{F}_{\mu\nu}^{0} \vec{F}_0^{\mu\nu} + \frac{i}{2} \sum_j \bar{\psi}_{j0} \gamma^{\mu} D_{\mu}^{0} \psi_{j0} \\ & - \sum_j m_{j0} \bar{\psi}_{j0} \psi_{j0} - \frac{1}{2a_0} (\partial_{\mu} \vec{W}_0^{\mu})^2 \\ & - \partial_{\mu} \vec{\bar{\varphi}}_0 \, D_0^{\mu} \vec{\varphi}^{0} \end{aligned} \tag{4.27}$$

with

$$\vec{F}_{\mu\nu}^{0} = \partial_{\mu} \vec{W}_{\nu}^{0} - \partial_{\nu} \vec{W}_{\mu}^{0} + g_0 \vec{W}_{\mu}^{0} \times \vec{W}_{\nu}^{0} \tag{4.28a}$$

and

$$D_{\mu}^{0} = \partial_{\mu} - i g_0 \vec{T} \cdot \vec{W}_{\mu}^{0} \tag{4.28b}$$

This can be seen as follows. The Δ's in eq. (4.24) determine a set of renormalization constants

$$Z_{2F} = 1 - \Delta_2 \qquad Z_{1F} = 1 - \Delta_{1F} \tag{4.29a,b}$$

$$Z_{3YM} = 1 - \Delta_3 \qquad Z_{1YM} = 1 - \Delta_1 \tag{4.30a,b}$$

$$\tilde{Z}_3 = 1 - \tilde{\Delta}_3 \qquad \tilde{Z}_1 = 1 - \tilde{\Delta}_1 \tag{4.31a,b}$$

$$Z_i = 1 - \Delta_i \quad , \quad i = 4,5,6 \tag{4.32a,b,c}$$

Let us now rescale the fields as follows

$$Z_{3YM}^{1/2} \vec{W}_\mu \equiv \vec{W}_\mu^{\,0} \tag{4.33}$$

$$\tilde{Z}_3^{1/2} \vec{\varphi} \equiv \vec{\varphi}^{\,0} \tag{4.34}$$

$$Z_{2F}^{1/2} \psi \equiv \psi^0 \tag{4.35}$$

For each interaction term we rescale the coupling constant with the aid of the corresponding Z's as read from the Lagrangean in eq. (4.24) i.e.,

$$Z_{1YM} Z_{3YM}^{-3/2} g = g_{0YM} \tag{4.36a}$$

$$\tilde{Z}_1^{-1} \tilde{Z}_3^{-1} Z_{3YM}^{-1/2} g = \tilde{g}_0 \tag{4.36b}$$

$$Z_{1F} Z_{3YM}^{-1/2} Z_{2F}^{-1} g = g_{0F} \tag{4.36c}$$

$$Z_5 Z_{3YM}^{-2} g^2 = g_{05}^2 \tag{4.36d}$$

For each mass term we have

$$Z_4 Z_{2F}^{-1} m_j = m_{0j} \tag{4.37}$$

and for the gauge term

$$Z_6 Z_{3YM}^{-1} \frac{1}{a} = \frac{1}{a_0} \tag{4.38}$$

In terms of the new fields $\vec{W}_\mu^0$, $\vec{\varphi}^0$ and ψ^0 and the new couplings in eq. (4.36); masses eq. (4.37) and gauge coupling eq. (4.38) we have

$$\begin{aligned}\mathcal{L}_{QCD} + \Delta\mathcal{L}_{QCD} &= -\frac{1}{4}\vec{F}^0_{\mu\nu}\vec{F}^{\mu\nu}_0 - \frac{1}{2}(\partial_\mu \vec{W}^0_\nu - \partial_\nu \vec{W}^0_\mu)\, g_{0YM}\, \vec{W}^\mu_0 \times \vec{W}^\nu_0 \\ &+ \frac{1}{4} g^2_{0S}\, \vec{W}^0_\mu \times \vec{W}^0_\nu \cdot \vec{W}^\mu_0 \times \vec{W}^\nu_0 \\ &+ i\sum_j \bar{\psi}^0_j \gamma^\mu \partial_\mu \psi^0_j - \sum_j m^0_j \bar{\psi}_{j0}\psi_{j0} + g_{0F}\bar{\psi}_0 \frac{\vec{\lambda}}{2}\gamma^\mu \psi_0 \vec{W}^0_\mu \\ &- \frac{1}{2a_0}(\partial_\mu \vec{W}^\mu_0)^2 - \partial_\mu \bar{\vec{\varphi}}^0 \partial^\mu \vec{\varphi}^0 - \tilde{g}_0\, \partial_\mu \bar{\vec{\varphi}}_0\, \vec{W}^\mu_0 \times \vec{\varphi}_0\end{aligned} \qquad (4.39)$$

This new form of the Lagrangean is not quite of the Yang Mills type. It does not have the BRS invariance discussed in section 3 unless

$$g_{0YM} = \tilde{g}_0 = g_{0S} = g_{0F} \qquad (4.40)$$

i.e., unless the scaled coupling constants in eqs. (4.36) are all equal. This means that in order that the renormalization procedure preserves the local gauge invariance of the theory, the normalization of the counterterms i.e., the Δ's (and hence the Z's) cannot be all arbitrary. We can normalize Z_{3YM} ; Z_3 and $Z_{2F}^{1/2}$ as we wish. Then, once we fix one of the vertex renormalization constants all the others are constrained by the identities

$$\frac{Z_{3YM}}{Z_{1YM}} = \frac{\tilde{Z}_3}{\tilde{Z}_1} \; ; \quad \frac{Z_{3YM}}{Z_{1YM}} = \frac{Z_{2F}}{Z_{1F}} \qquad (4.41a,b)$$

and

$$Z_5 = Z_{1YM}\,\frac{Z_{1YM}}{Z_{3YM}} \qquad (4.41c)$$

which follow from eqs. (4.36) when restricted to eq. (4.40). These crucial identities were first derived by Taylor (4.3) and Slavnov (4.4). It is clear that $\mathcal{L}_{QCD} + \Delta\mathcal{L}_{QCD}$ in eq. (4.39) when restricted to eq. (4.40) can be written in the form announced in eq. (4.27). We shall refer to this form as the <u>bare QCD Lagrangean</u>.

Let us denote by

$$\Gamma_0 (p_1 \cdots p_N ; g_0 , a_0 , m_j^0 , \epsilon)$$

an arbitrary amputated and connected Green's function with N external legs calculated with the bare QCD Lagrangean in eq. (4.27). All the UV divergences are regulated via the ϵ-regularization. The renormalizability of the theory is equivalent to the statement that once some of singularities in $\frac{1}{\epsilon}$ have been absorbed in the redefinition of the paramenters of the theory

$$g , a , m_j$$

all the remaining singularities can be absorbed in three-independent renormalization constants

$$Z_{3YM} , Z_{2F} , \tilde{Z}_3$$

From the identity

$$\mathcal{L}_{QCD} + \Delta \mathcal{L}_{QCD} = \mathcal{L}_0$$

we can find the relation between the renormalized Green's functions

$$\Gamma (p_1 \cdots p_N ; g , a , m_j ; \mu)$$

and the bare Green's functions

$$\Gamma (p_1 \cdots p_N ; g , a , m_j ; \mu)$$
$$= \lim_{\epsilon \to 0} \left(Z_{3YM}^{-1/2} \right)^G \left(\tilde{Z}_3^{-1/2} \right)^{FP} \left(Z_{2F}^{-1/2} \right)^Q \Gamma_0 (p_1 \cdots p_N ; g_0 , a_0 , m_j^0 , \epsilon) \tag{4.42}$$

where G, FP and Q denote respectively the number of external gluons, external ghosts and external quarks.

4c. Review of Renormalization Schemes

In this section we shall discuss various renormalization schemes which are often used in practical calculations in the literature. We shall illustrate these methods with an explicit calculation of the one-loop contribution to the gluon self-energy.

We shall consider four renormalization schemes:

i. μ-renormalization; Green's functions are subtracted at euclidean values of their invariants;

ii. Weinberg's mass-independent renormalization (4.5) (W-renormalization): In this scheme, the renormalization constants are obtained from unrenormalized (but regularized) Green's functions evaluated at some euclidean value of their invariants with all mass terms set to zero.

iii. 't Hooft's ϵ-renormalization (4.6) ('t H-renormalization): Renormalized Green's functions are obtained by subtracting the poles in $\frac{1}{\epsilon}$ of the corresponding unrenormalized ϵ-regularized Green's functions.

iv. On-shell renormalization (λ-renormalization): Green's functions are subtracted at values of their invariants on the mass-shell, or nearby (hence the λ-dependence). This is the method which is commonly used in QED. Here, the mass scale λ plays the role of an infrared regulator.

Of these methods, the 't H-renormalization scheme is the most original one and I feel it requires more of an explanation than just a few calculational examples. The rest of this subsection is a general discussion of the ϵ-renormalization scheme.

Let us consider the bare QCD Lagrangean in eq. (4.27) in $n=4-\epsilon$ dimensions. As already mentioned (in the subsection on ϵ-regularization) for the action to remain dimensionless we have attributed the following dimensions:

gluon field	W_μ^0	....	dimension	$1-\epsilon/2$
quark field	ψ^0	....	dimension	$3/2-\epsilon/2$
ghost field	φ^0	....	dimension	$1-\epsilon/2$
bare coupling constant	g_0	....	dimension	$\epsilon/2$
bare masses	m_j^0	....	dimension	1
bare gauge parameter	a_0	....	dimension	0

Next we introduce a unit of mass ν (with dimension 1) so as to express the bare coupling constant as

$$g_0 = \nu^{\epsilon/2} \times \text{dimensionless quantity}$$

So far all this is at the regularization level and therefore common to all the four schemes mentioned above.

Now, in the ϵ-renormalization, given g_o ; m_j^o ; a_o and ν , the corresponding renormalized quantities

$$g \;,\; m_j \;,\; a$$

and the renormalized fields W, Ψ and φ are fixed by requiring that all the renormalization constants Z are of the form

$$Z = 1 + \sum_{n=1}^{\infty} \frac{\hat{Z}_n(g,m,\nu)}{\epsilon^n} \tag{4.43}$$

in particular one requires

$$g_0 = \nu^{\epsilon/2} g \Big/ 1 + \sum_{n=1}^{\infty} \frac{\hat{Z}_{g,n}(g,m,\nu)}{\epsilon^n} \Big) \equiv \nu^{\epsilon/2} g\, Z_g \tag{4.44}$$

and (for each mass flavor i,j = 1,2,.... n_j)

$$m_i^o = \sum_{j=1}^{n_j} m_j \Big(\delta_{ij} + \sum_{n=1}^{\infty} \frac{\hat{Z}_{m,n}(g,m,\nu)_{ij}}{\epsilon^n} \Big) \equiv \sum_{j=1}^{n_j} (Z_m)_{ij}\, m_j \tag{4.45}$$

In fact in the case of color -SU(3), without further couplings, the matrix $(Z_m)ij$ is obviously diagonal. The poles in $\frac{1}{\epsilon}$ which appear in eqs. (4.43), (4.44) and (4.45) are defined to be precisely those needed to subract the poles in the corresponding unrenormalized Feynman diagrams. With this definition of Z the following theorem can be proved[4)].

The coefficients $Z_{g,n}$; $(Z_{m,n})_{ij}$ and Z_n when expressed in powers of the renormalized coupling constant g do not depend on ν nor on m_i . We shall not reproduce here the formal proof of this theorem[4)]. The mechanism which is at the basis of these proofs is the following: ν only appears in powers of $\nu^{\epsilon/2}$ and therefore it is reabsorbed into the definition of the dimensionless renormalized coupling constant. On the other hand, in the residues of the poles, m_R only appears in polynomials i.e., in positive powers independent of ϵ . Since the Z_n's are all dimensionless, the only way they can appear is the way stated in the theorem.

As an example, the renormalized gluon self-energy, via quark loops, in the ϵ -renormalization scheme, can be easily obtained from eq. (4.20). After subtraction of the pole term, we find

$$\Pi^{(2)}\left(\frac{q^2}{\nu^2}, \frac{m^2}{\nu^2}\right) = -\frac{\alpha_s}{\pi}\,\delta_{AB}\,\frac{1}{2}\sum_j \int_0^1 dx\, 2x(1-x)\left[\gamma - \log 2\pi + \log\frac{-q^2x(1-x)+m_j^2-i\epsilon}{\nu^2}\right] \tag{4.46}$$

where γ is Euler's constant defined in eq. (4.9b). At the approximation where $-q^2 \gg m_j^2$, we find

$$\begin{aligned}\Pi^{(2)}\left(\frac{q^2}{\nu^2}, \frac{m^2}{\nu^2}\right) = -\frac{\alpha_s}{\pi}\,\delta_{AB}\,\frac{1}{2}\sum_j \Big\{ & \frac{2}{3}\,\frac{1}{2}\log\frac{-q^2}{\nu^2} + \frac{1}{3}(\gamma - \log 2\pi) \\ & - \frac{5}{9} + 2\,\frac{m^2}{-q^2} + O\left[\left(\frac{m^2}{q^2}\right)^2 \log\frac{-q^2}{m^2}\right]\Big\}\end{aligned} \tag{4.47}$$

and at the approximation where $-q^2 \ll m_j^2$

$$\begin{aligned}\Pi^{(2)}\left(\frac{q^2}{\nu^2}, \frac{m^2}{\nu^2}\right) = -\frac{\alpha_s}{\pi}\,\delta_{AB}\,\frac{1}{2}\sum_j \Big\{ & \frac{2}{3}\,\frac{1}{2}\log\frac{m_j^2}{\nu^2} + \frac{1}{3}(\gamma - \log 2\pi) \\ & + \frac{1}{15}\,\frac{-q^2}{m_j^2} + O\left[\left(\frac{-q^2}{m_j^2}\right)^2\right]\Big\}\end{aligned} \tag{4.48}$$

The expression of the renormalized gluon self energy, induced by quark loops, in Weinberg's scheme can be easily obtained from eq. (4.20) and the definition

$$\begin{aligned}\Pi^{(2)}\left(\frac{q^2}{\mu^2}, \frac{m^2}{\mu^2}\right) &= \lim_{\epsilon\to 0}\Big\{\Pi^{(2)}(q^2, m^2; \epsilon) - \Pi^{(2)}(-\mu^2, 0, \epsilon)\Big\} \\ &= -\frac{\alpha_s}{\pi}\,\delta_{AB}\,\frac{1}{2}\sum_j \int_0^1 dx\, 2x(1-x)\log\frac{-q^2x(1-x)+m_j^2-i\epsilon}{\mu^2 x(1-x)}\end{aligned} \tag{4.49}$$

At the approximation where $-q^2 \gg m_j^2$,

$$\begin{aligned}\Pi^{(2)}\left(\frac{q^2}{\mu^2}, \frac{m^2}{\mu^2}\right) = -\frac{\alpha_s}{\pi}\,\delta_{AB}\,\frac{1}{2}\sum_j \Big\{ & \frac{2}{3}\,\frac{1}{2}\log\frac{-q^2}{\mu^2} \\ & + 2\,\frac{m^2}{-q^2} + O\left[\left(\frac{m^2}{-q^2}\right)^2 \log\frac{-q^2}{m^2}\right]\Big\}\end{aligned} \tag{4.50}$$

and at the approximation where $-q^2 \ll m_j^2$;

$$\tilde{\Pi}^{(2)}\left(q^2/\mu^2,\ m^2/\mu^2\right) = -\frac{\alpha_s}{\pi}\,\delta_{AB}\,\frac{1}{2}\sum_j \Big\{ \frac{2}{3}\,\frac{1}{2}\log\frac{m_j^2}{\mu^2} + \frac{5}{9} + \frac{1}{15}\,\frac{-q^2}{m_j^2} + O\Big[\Big(\frac{-q^2}{m_j^2}\Big)^2\Big]\Big\} \tag{4.51}$$

For completeness' sake we shall give the full expression of the gluon propagator summed at the one-loop approximation i.e.,

$$i\,D^{ab}_{\mu\nu}(k) =$$

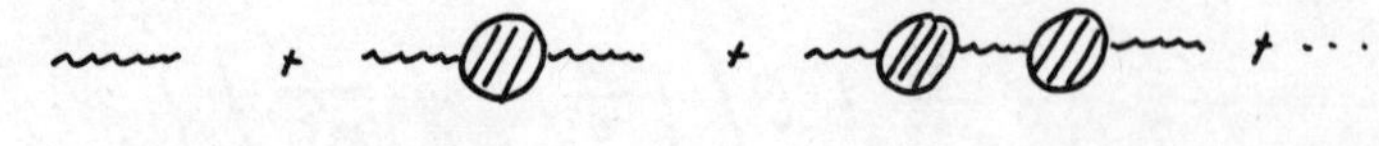

$$= -i\,\delta_{ab}\Big\{ \frac{g_{\mu\nu} - k_\nu k_\mu/k^2}{k^2+i\epsilon}\;\frac{1}{1+\tilde{\Pi}^{(2)}(k^2, m_j^2, \alpha_s, a;\mu)} + a\,\frac{k_\mu k_\nu}{k^2}\,\frac{1}{k^2+i\epsilon}\Big\} \tag{4.52}$$

where $\tilde{\Pi}^{(2)}(k^2, m_j^2, \alpha_s, a;\mu)$ denotes the proper gluon self-energy [see Fig. (4.2)]

$$i\,\Pi_{\mu\nu}(k,\alpha_s) = -i\,(g_{\mu\nu}k^2 - k_\mu k_\nu)\,\Pi(k^2, m_j^2, \alpha_s, a;\mu)$$

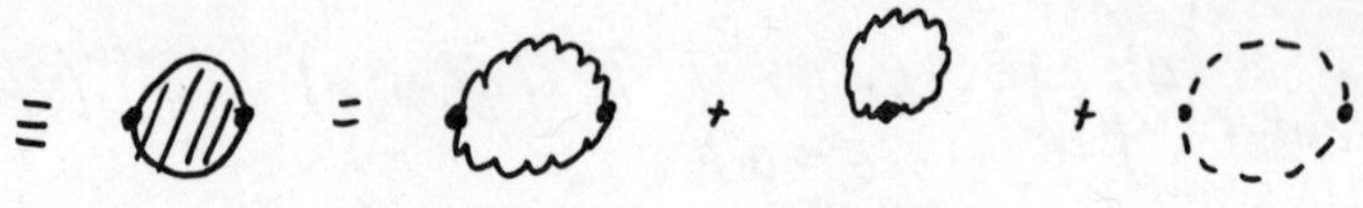

$$+ \sum_{i=1}^{n}$$

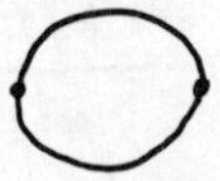

$$+\ O(\alpha_s^2) \tag{4.53}$$

Notice that the gauge term in the free gluon propagator does not get renormalized, except for the a parameter itself, by the interaction. This is a consequence of the Slavnov-Taylor identity

$$k^\mu k^\nu\, i\,D^{ab}_{\mu\nu}(k) = -ia\,\delta_{ab} \tag{4.54}$$

which we have proved in section (3.c). From the one-loop diagrams in eq. (4.53) one gets, in the μ -renormalization scheme,

$$\Pi^{(2)}(k^2, m_j^2, \alpha_s, a; \mu) = \frac{\alpha_s}{\pi} \left\{ \frac{N}{2} \left(\frac{13}{12} - \frac{a}{4} \right) \log \frac{-k^2}{\mu^2} + \left(-\frac{1}{2} \right) \sum_{j=1}^{n} \int_0^1 dx\, 2x(1-x) \log \frac{-k^2 x(1-x) + m_j^2}{\mu^2 x(1-x) + m_j^2} \right\} \qquad (4.55)$$

FOOTNOTES OF SECTION 4

1) See e.g. the review article by Leibrandt in ref. [R. 20].

2) It may be useful to some readers to give a short bibliography of textbooks in analysis where they will find useful mathematical information in connection with the ϵ-regularization. The standard textbook in the Anglo Saxon literature is Whittaker and Watson, ref. [R. 21]. In French, the best book for this purpose is, in my opinion, Dieudonné's textbook; ref. [R. 22]. A very interesting book on the subject is Carlson's, ref. [R. 23] (I wish to thank Dr. Jacques Calmet for bringing this last reference to my attention).

3) The reader should exercise himself with this calculation. Let me give some of the steps:

$$\mathrm{Tr}\{\gamma_\mu(\not p+m)\,\gamma_\nu(\not p+\not q+m)\} =$$

$$2^{2-\epsilon/2}\Big\{\frac{1}{4-\epsilon}\,\tilde{p}^2(-2g_{\mu\nu}+\epsilon g_{\mu\nu}) - x(1-x)(2q_\mu q_\nu - g_{\mu\nu}q^2) + m^2 g_{\mu\nu}\Big\}$$

where we have used (4.16), (4.17) and (4.15a). Next from (4.12) we have [using also (4.7)]

$$I(1,2) = -i\,\frac{1}{16\pi^2}\,(4\pi)^{\epsilon/2}(2-\epsilon/2)\,\frac{\Gamma(\epsilon/2)}{-1+\epsilon/2}\;\frac{-q^2x(1-x)+m^2}{(\mathbb{R}^2)^{\epsilon/2}}$$

and

$$I(0,2) = -i\,\frac{1}{16\pi^2}\,(4\pi)^{\epsilon/2}\,\Gamma(\epsilon/2)\,\frac{1}{(\mathbb{R}^2)^{\epsilon/2}}$$

4) In fact we do not know of a simple proof os this theorem. We do not consider the proof given in ref. (4.7) as complete; on the other hand the proofs which, it is claimed, follow from the work of the authors in refs. (4.8) and (4.9) are extremely difficult to follow.

REFERENCES OF SECTION 4

1) G. 't Hooft and M. Veltman, Nucl. Phys. B44(1972)189.
2) C.G. Bellini and J.J. Giambiagi, Nuovo Cimento 12B(1972)20.
3) J.C. Taylor, Nucl. Phys. B33(1971)436.
4) A.A. Slavnov, Sov. J. Particles Nucl. 5, no. 3(1975)303.
5) S. Weinberg, Phys. Rev. D8(1973)3497.
6) G.'Hooft, Nucl. Phys. B61(1973)455.
7) J.C. Collins, and A.J. Macfarlane, Phys. Rev. D10(1974)120.
8) J.C. Collins, Nucl. Phys. B92(1975)477.
9) P. Breitenlohner and D. Maison, Commun. Math. Phys. 52(1977)39.

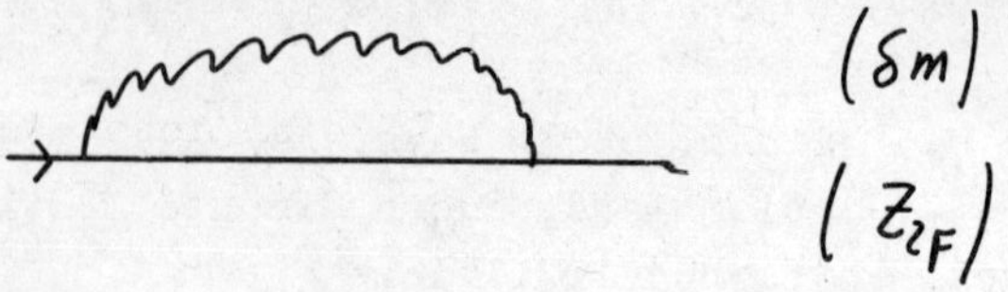

Fig. 4.1 Quark self-energy diagram at the one loop approximation.

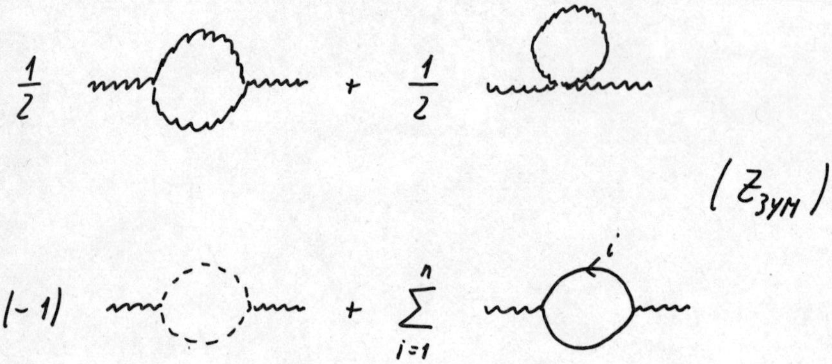

Fig. 4.2 Gluon self-energy diagrams at the one loop approximation.

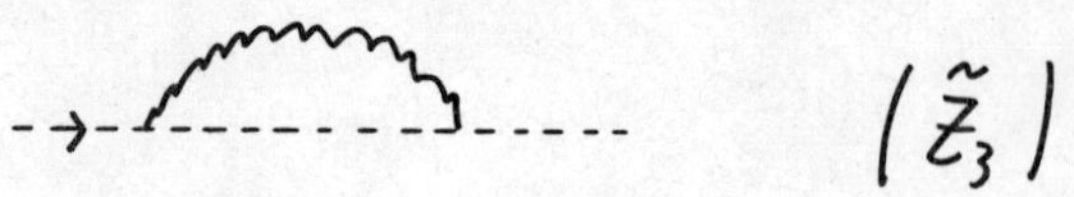

Fig. 4.3 Ghost self-energy diagram at the one loop approximation.

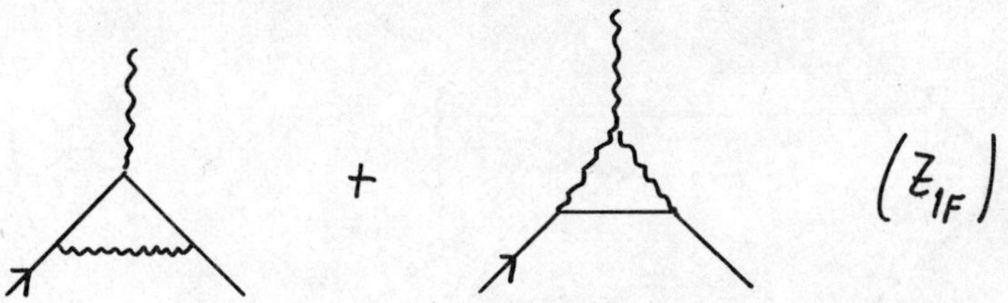

Fig. 4.4 Gluon-Quark-Quark vertex diagrams at the one loop approximation.

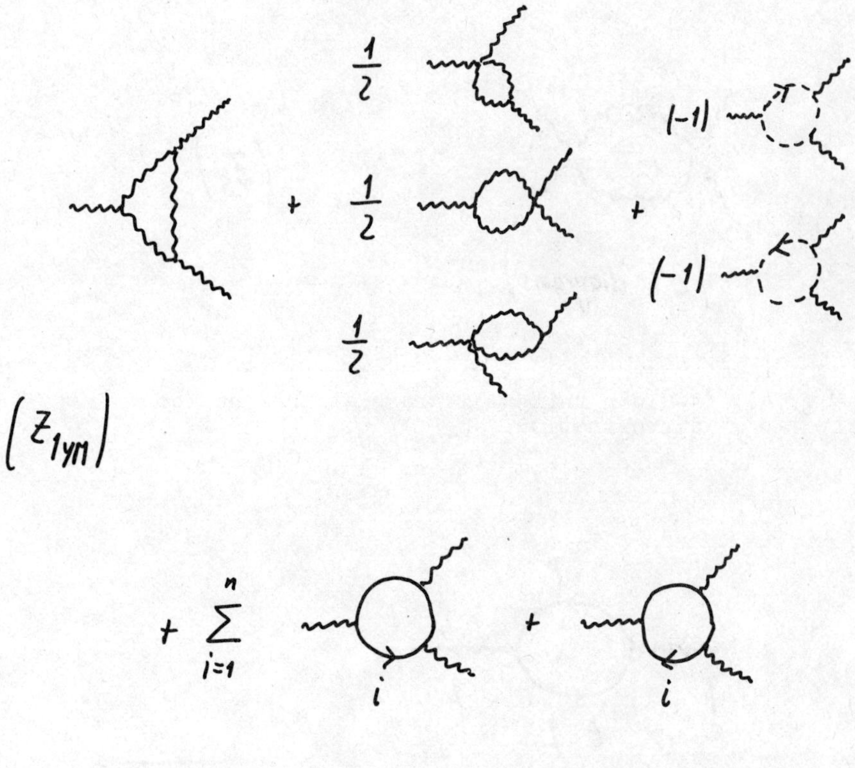

Fig. 4.5 3-Gluon vertex diagrams at the one loop approximation.

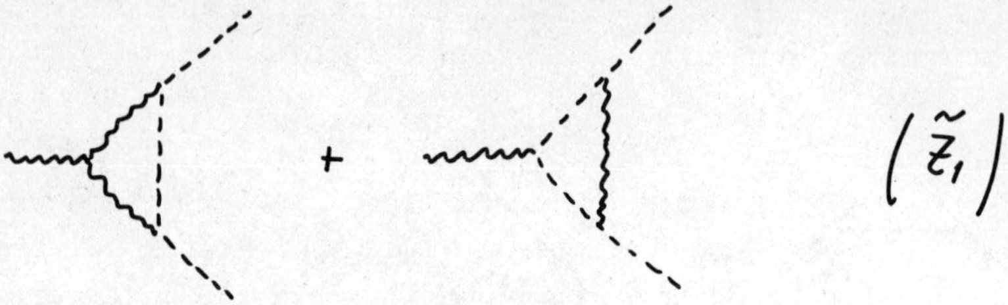

Fig. 4.5 Gluon-Ghost-Ghost vertex diagram at the one loop approximation.

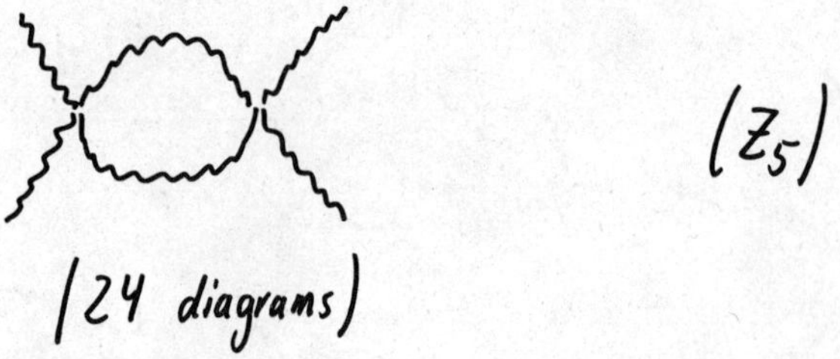

Fig. 4.7 4-Gluon vertex diagrams at the one loop approximation.

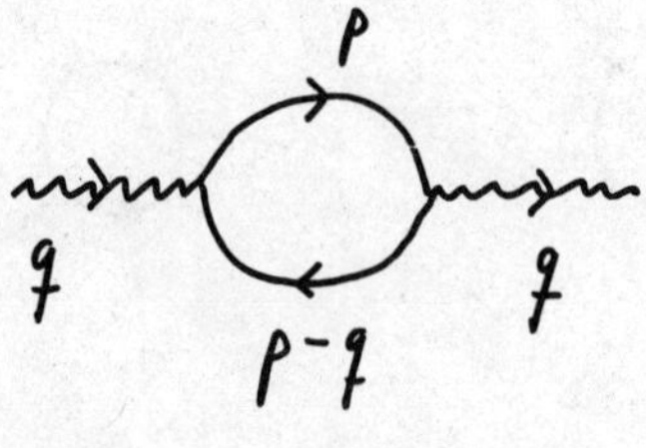

Fig. 4.8 Routing of momenta corresponding to the amplitude in eq. (4.15a).

5. RENORMALIZATION INVARIANCE. GENERAL THEORY AND QCD EFFECTIVE COUPLING CONSTANT.

Renormalization invariance is the statement that physical observables must be independent of the renormalization scheme which one chooses for their theoretical calculation. For historical reasons[1)] renormalization invariance comes under the heading of Renormalization Group Invariance. In fact, as we shall see, the group structure is only defined within classes of renormalization. The general renormalization invariance corresponds to the algebraic structure of a grupoid. Contrary to the general belief, we think that this distinction is of physical relevance, in connection with the problem of finite quark mass corrections to the asymptotic behaviour of Green's functions in perturbative QCD.

5a. The Algebraic Structure of Renormalization Invariance

Let us assume we have chosen a particular renormalization scheme, say R , and call $\Gamma_R(\ldots)$ the associated R-renormalized Green's functions. Their relation to the bare Green's functions $\Gamma_0(\ldots)$ will be

$$\Gamma_R(\cdots) = Z(R)\,\Gamma_0(\cdots) \tag{5.1a}$$

where Z(R) denotes the appropriate product of renormalization constants defined within the R-scheme [see e.g. eq. (4.24)] . If we choose another renormalization scheme, the R′-scheme, we shall have

$$\Gamma_{R'}(\cdots) = Z(R')\,\Gamma_0(\cdots) \tag{5.1b}$$

Obviously, between the Green's functions defined by the R- scheme and those defined by the R′-scheme there exists a relation

$$\Gamma_{R'}(\cdots) = Z(R',R)\,\Gamma_R(\cdots) \tag{5.2}$$

where

$$Z(R', R) = Z(R')/Z(R) \tag{5.3}$$

Let us now consider the set of all possible $Z(R', R)$ for arbitrary R and R'. Among the elements of this set there is the composition law

$$Z(R'', R) = Z(R'', R')\, Z(R', R) \tag{5.4a}$$

To each element $Z(R', R)$ there corresponds an inverse

$$Z^{-1}(R', R) = Z(R, R') \tag{5.4b}$$

and we can define the unit element as

$$Z(R, R) = 1 \tag{5.4c}$$

The composition law in eq. (5.4a) is not defined however for arbitrary pairs of Z's : the product

$$Z(R_i, R_j)\, Z(R_k, R_l)$$

is not in general an element of the set $Z(R', R)$ unless $R_j = R_k$. A set of transformations obeying eqs. (5.4) defines a grupoid structure[2].

Let us illustrate the rather abstract notion introduced above with a specific example: the lowest order contribution to Z_{3YM} from quark loops. Suppose we work in the μ-scheme and we want to relate two renormalizations performed at subtraction points μ_1 and μ_2. The relevant quantity will be $Z(\mu_1, \mu_2)$, which can be easily extracted from eqs. (4.20) and (4.22):

$$Z(\mu_2, \mu_1) = 1 + \frac{\alpha_s}{\pi}\, \delta_{AB}\, \frac{1}{2} \sum_{j=1}^{n} \int_0^1 dx\; 2x(1-x) \log \frac{\mu_2^2 x(1-x) + m_j^2}{\mu_1^2 x(1-x) + m_j^2} \tag{5.5}$$

Clearly, this representation of Z obeys the composition law of eq. (5.4a); i.e., the grupoid law. It does not obey the composition law of a group, unless $m_j=0$. Indeed, when $m_j = 0$

$$Z(\mu_2,\mu_1)\Big|_{m_j=0} = Z(\mu_2/\mu_1) = 1 + \frac{\alpha_s}{\pi}\delta_{AB}\frac{n}{2}\frac{2}{3}\log\frac{\mu_2}{\mu_1} \tag{5.6}$$

and the product

$$Z(\mu_i/\mu_j)\, Z(\mu_\ell/\mu_k) \tag{5.7a}$$

for arbitrary pairs (i,j) , (ℓ,k) is still an element of the set of Z's

$$Z(\mu_i/\mu_j)\, Z(\mu_\ell/\mu_k) = 1 + \frac{\alpha_s}{\pi}\delta_{AB}\frac{n}{2}\frac{2}{3}\log\frac{\mu_i\mu_\ell}{\mu_j\mu_k} = Z(\mu_i\mu_\ell/\mu_j\mu_k) \tag{5.7b}$$

It is interesting that the ϵ-renormalized Green's functions are related to each other by finite renormalizations, the $Z(R_i,R_j)$, which obey the group law. This is because of the property that the coefficients $\hat{Z}_{g,n}$, $\hat{Z}_{m,n}$ and $\hat{Z}_n$ in eqs. (4.45), (4.44) and (4.43) when expressed in powers of the renormalized coupling constant g do not depend on any mass scale. As a result of this it will be possible to trade the renormalized coupling constant $g(\nu)$; the renormalized masses $m_i(\nu)$ and the arbitrary mass scale ν in terms of renormalization group invariant quantities. In general, however, these quantities will not be the same invariants, as those defined by other renormalization schemes! We shall come back to this point later on[3].

5b. Renormalization Group Equations

The differential approach to renormalization invariance was pioneered by Stueckelberg and Peterman (5.1) and by Gell-Mann and Low (5.2). Later on, the study of scaling behaviour in field theory, largely motivated by the experimental observation of Bjorken scaling (5.7) in deep inelastic electron-proton scattering gave rise to the Callan

Symanzik equation, (5.8) which is a very powerfull technique to express the renormalization invariance constraints on the short distance behaviour of Green's functions. New general methods which apply to arbitrary Green's functions have been formulated by 't Hooft (5.9) and Weinberg (5.10). We shall describe them in this section.

The central idea of 't Hooft's and Weinberg's methods is to treat the coupling constant renormalization, mass renormalization and gauge renormalization in the same footing i.e., g ; m_i and a are treated as coupling constants of various interaction terms in the Lagrangean.

The starting point is eq. (4.42), which we rewrite in the form ($\alpha_s = g^2/4\pi$)

$$\Gamma(p_1\cdots p_N;\alpha_s,a,m_i;\mu) = Z_\Gamma(\mu,\epsilon)\,\Gamma_0(p_1\cdots p_N;\alpha_s^0,a_0,m_i^0;\epsilon) \tag{5.8}$$

where Z_Γ denotes the product of wave function renormalizations associated to the specific choice of Green's function Γ . The renormalized Γ and Z_Γ depend on μ . Z_Γ and Γ_0 depend on ϵ ; but Γ_0 is independent of μ i.e.,

$$\mu\frac{d}{d\mu}\,\Gamma_0(p_1\cdots p_N;g_0,a_0,m_i^0;\epsilon) = 0 \tag{5.9}$$

From (5.8) and (5.9) and after chain differentiation, we get

$$\left(\mu\frac{\partial}{\partial\mu} + \mu\frac{d\alpha_s}{d\mu}\frac{\partial}{\partial\alpha_s} + \sum_i \frac{\mu}{m_i}\frac{dm_i}{d\mu}m_i\frac{\partial}{\partial m_i} + \mu\frac{da}{d\mu}\frac{\partial}{\partial a}\right)\Gamma(p_1\cdots p_N;\alpha_s,a,m_i;\mu) \tag{5.10}$$
$$= \frac{1}{Z_\Gamma}\,\mu\,\frac{dZ_\Gamma}{d\mu}\,\Gamma(p_1\cdots p_N;\alpha_s,a,m_i;\mu)$$

By this procedure, we have introduced a set of universal functions (in principle functions of α_s , $x_i = m_i/\mu$, a , ϵ):

$$\mu\frac{d\alpha_s}{d\mu} = \alpha_s\,\beta(\alpha_s,x_i,a;\epsilon) \tag{5.11a}$$

$$\frac{\mu}{m_i}\frac{dm_i}{d\mu} = -\gamma_i(\alpha_s,x_j,a;\epsilon) \tag{5.11b}$$

$$\mu \frac{da}{d\mu} = \beta_a(\alpha_s; x_i, a; \epsilon) \tag{5.11c}$$

and

$$\frac{1}{Z_{3YM}} \mu \frac{dZ_{3YM}}{d\mu} = \gamma_{YM}(\alpha_s, x_i, a; \epsilon) \tag{5.12a}$$

$$\frac{1}{Z_{2F}} \mu \frac{dZ_{2F}}{d\mu} = \gamma_F(\alpha_s, x_i, a; \epsilon) \tag{5.12b}$$

$$\frac{1}{\tilde{Z}_3} \mu \frac{d\tilde{Z}_3}{d\mu} = \tilde{\gamma}(\alpha_s, x_i, a; \epsilon) \tag{5.12c}$$

These functions are universal in the sense that eq. (5.10) is valid for an arbitrary Green's function $\Gamma(p_1, \dots p_N; \alpha_s, a, m_i; \mu)$ with the same coefficient functions in front of the partial differentiations.

From the renormalizability of the theory it follows that the universal functions

$$\alpha_s\beta, \ \beta_a, \ \gamma_i; \ \gamma_{YM}, \ \gamma_F, \ \tilde{\gamma}$$

are non-singular when $\epsilon \to 0$. This can be seen by writing eq. (5.10) for the set of renormalized Green's functions associated to the diagrams with primitive divergences and then solving the system of algebraic equations.

The explicit form of the universal functions in the variables α_s, x_i and a depends on the choice of renormalization scheme:

i. In the μ-scheme the universal functions have a non trivial dependence in all the variables α_s, x_i and a.

ii. In the t'H-renormalization scheme and in the W-renormalization scheme they are independent of the masses x_i and in fact the β-function is gauge invariant. (In the μ-scheme there is a gauge dependence in the β-function via the dependence on x_i).

There is another constraint on the Green's functions which comes from dimensional analysis and which we next discuss. Assume we scale all the momenta $p_1 \dots p_N$ in $\Gamma(p_1 \dots p_N; \alpha_s, a, m_i; \mu)$ with a common dimensionless factor λ; and call D the dimension of Γ in units of mass. Clearly

$$\Gamma(\lambda p_1 \cdots \lambda p_N ; \alpha_s, a, m_i ; \mu) \equiv \mu^D F\left(\lambda^2 \frac{p_k \cdot p_\ell}{\mu^2} ; \alpha_s, a ; \frac{m_i}{\mu}\right) \tag{5.13}$$

where F is now a dimensionless function of dimensionless variables ($k, \ell = 1,2,\ldots N$). From Euler's theorem on homogeneous functions we have

$$\left\{ \lambda \frac{\partial}{\partial \lambda} + \sum_i m_i \frac{\partial}{\partial m_i} + \mu \frac{\partial}{\partial \mu} - D \right\} \Gamma(\lambda p_1 \cdots \lambda p_N ; \alpha_s, a, m_i ; \mu) = 0 \tag{5.14}$$

In fact, in the absence of a dynamical theory for the Green's functions this would be all the information on the scaling behaviour of the Green's functions. This naive behaviour is clearly modified by the dynamical constraint embodied in eq. (5.10). We can combine both equations as follows:

with $$t = \log \lambda \tag{5.15}$$

and eliminating $\mu \frac{\partial}{\partial \mu}$ between eqs. (5.10) and (5.14) we get, using the definitions in eqs. (5.11), (5.12) and (5.15):

$$\left\{ -\frac{\partial}{\partial t} + \beta(\alpha_s)\alpha_s \frac{\partial}{\partial \alpha_s} + \beta_a(\alpha_s)\frac{\partial}{\partial a} - \sum_i \left[1 + \gamma_i(\alpha_s)\right] x_i \frac{\partial}{\partial x_i} + D - \gamma_\Gamma(\alpha_s) \right\} \Gamma(e^t p_1 \cdots e^t p_N ; \alpha_s, a, x_i ; \mu) = 0 \tag{5.16}$$

where

$$-2\gamma_\Gamma(\alpha_s) = n_{YM}\,\gamma_{YM}(\alpha_s) + n_F\,\gamma_F(\alpha_s) + \tilde{n}\,\tilde{\gamma}(\alpha_s) \tag{5.17}$$

with

n_{YM} = number of external gluon lines in Γ

n_F = number of external quark lines in Γ

$\tilde{n}$ = number of external ghost lines in Γ

Equation (5.16) governs the behaviour of Green's functions when all the momenta are scaled up with a common factor, at fixed μ.

The general solution of eq. (5.16) can be obtained via the method of characteristics to solve linear partial differential equations[4]. First one solves the system of ordinary differential equations:

$$\frac{d\bar{\alpha}_s(t,\alpha_s)}{dt} = \bar{\alpha}_s \beta(\bar{\alpha}_s) \quad , \quad \bar{\alpha}_s(0,\alpha_s) = \alpha_s \tag{5.18a}$$

$$\frac{d\bar{x}_i(t,\alpha_s)}{dt} = -[1+\gamma_i(\bar{\alpha}_s)]\,\bar{x}_i \quad , \quad \bar{x}_i(0,\alpha_s) = x_i \tag{5.18b}$$

$$\frac{d\bar{a}(t,\alpha_s)}{dt} = \beta_a(\alpha_s) \quad , \quad \bar{a}(0,\alpha_s) = a \tag{5.18c}$$

Then, the general solution of eq. (5.16) can be expressed in terms of those of the ordinary differential equations

$\bar{\alpha}(t,\alpha_s)$, $\bar{x}_i(t,\alpha_s)$ and $\bar{a}(t,\alpha_s)$ as follows (recall $\lambda = e^t$)

$$\Gamma(e^t p_1 \cdots e^t p_N ; \alpha_s, a, x_i ; \mu) = \lambda^D \, \Gamma(p_1 \cdots p_N ; \bar{\alpha}_s, \bar{a}, \bar{x}_i ; \mu) \cdot \exp\left\{ -\int_0^t dt' \, \gamma_\Gamma[\bar{\alpha}_s(t',\alpha_s)] \right\} \tag{5.19}$$

What we learn from this general solution is that the behaviour of Green's functions when all the momenta are scaled up, at fixed μ, is governed by the flow of the effective parameters of the theory (coupling constant, gauge parameter and masses) as functions of the scaling variable; as well as by a change in the overall scaling factor:

$$\lambda^D \to \exp\left\{ tD - \int_0^t dt' \, \gamma_\Gamma[\bar{\alpha}_s(t',\alpha_s)] \right\} \tag{5.20}$$

which varies from one Green's function to another [see eq. (5.17)]. This is why the term γ_Γ is often called the anomalous dimension of Γ.

The solutions of the ordinary differential equations (5.18) are called

effective coupling constant	$\bar{\alpha}(t,\alpha_s)$
effective masses (in units of μ)	$\bar{x}_i(t,\alpha_s)$
effective gauge parameter	$\bar{a}(t,\alpha_s)$

It is clear from eq. (5.19) that in order to get further insight on the scaling properties of the QCD Green's functions Γ, first we have to work out the explicit t-dependence of these effective parameters. This we do in the next subsection; and in section 6.

5c. Effective Coupling Constant, Asymptotic Freedom; and the Renormalization Group Invariant Scale Λ.

We are concerned with the solution of the ordinary differential equation (5.18a). First we shall discuss some general qualitative features of this equation and their physical implications. Finite quark mass effects will be neglected throughout this section. We shall come back to quark masses in section 6.

In general, the solution of an equation of the type

$$\frac{dz}{dt} = z\,\beta(z) \tag{5.21}$$

is governed by the zeros of $\beta(z)$ (also called the fixed points of $\beta(z)$). In perturbation theory, with our definition of $\beta(z)$ (see eq. (5.11a)) it is clear that

$$\beta(z) = \frac{z}{\pi}\beta_1 + \left(\frac{z}{\pi}\right)^2\beta_2 + \cdots \tag{5.22}$$

where β_1, and β_2 are numerical coefficients which we can in principle calculate, and which we shall describe how to calculate later on. Clearly z=o is a fixed point, since $\beta(z=o)=0$. We have two possibilities, illustrated in Fig. 5.1, corresponding to whether the slope at the origin i.e., the coefficient β_1, in eq. (5.22), is positive or negative. The behaviour of the solution to eq. (5.21) near the origin can be seen by solving the equation

$$\int_{\alpha_s}^{\bar{\alpha}_s(t,\alpha_s)} \frac{dz}{z\beta_1 \frac{z}{\pi}} = t \tag{5.23}$$

$$\bar{\alpha}(t,\alpha_s) = \frac{\alpha_s}{1-\beta_1 \frac{\alpha_s}{\pi} t} \tag{5.24}$$

The behaviour of the effective coupling constant as the scaling parameter t increases depends crucially on the sign of β_1 (see fig. 5.2). If $\beta_1 > 0$, $\bar{\alpha}_s(t,\alpha_s)$ increases as t increases. If $\beta_1 < 0$, $\bar{\alpha}_s(t,\alpha_s)$ decreases as t increases. Theories with $\beta_1 < 0$ are called "asymptotically free", and as we shall see QCD is asymptotically free. The physical implications of the asymptotic freedom property can be best seen from eq. (5.19): the behaviour of the Green's functions when all the momenta are scaled up with a common factor is governed by a theory where $\bar{\alpha}_s \to 0$ i.e., the free field theory. This is precisely the type of dynamic behaviour that theorists have been looking for in order to interpret the success of the parton model in describing scaling phenomena at high energies. In fact, it has been proved (see ref. (5.11))that the only field theories with the asymptotic freedom property are non-abelian gauge theories of the Yang-Mills type (without spontaneous symmetry breakdown). The value of β_1 for a SU(N) gauge theory with n-quarks was first calculated by the authors of ref. (5.12). The result is

$$\beta_1 = -\left(\frac{11}{3}\frac{N}{2} - \frac{2}{3}\frac{n}{2} \right) \tag{5.25}$$

For colour SU(3) (N = 3)

$$\beta_1 = -\frac{11}{2} + \frac{n}{3} \tag{5.26}$$

i.e., the theory will be asymptotically free provided

$$n \leq 16 \tag{5.27}$$

By contrats to QCD, quantum electrodynamics (QED) is a theory with $\beta_1 > 0$ ($\beta_1 = \frac{2}{3}$) ; therefore it does not have the asympt otic freedom property. At shorter and shorter distances the effective coupling constant of QED becomes larger and larger[5)] and eventually it may enter a non perturbative regime. On the other hand the theory is stable at long distances in the sense that at low energies there is a smooth transition of the quantized perturbation theory to the classical theory. This is guaranteed by Thirring's low energy theorem (see ref. (5.15)) which states that order by order in renormalized perturbation theory, (with on-shell renormalization), the Compton scattering ($\gamma e \to \gamma e$) cross-section for photons with incident energies smaller than the electron mass coincides with the classical Thomson cross-section

$$\lim_{E_\gamma \ll m}$$

$$\to \sigma_{Thomson} = \frac{8}{3} \pi \frac{e^2}{4\pi \hbar c} \left(\frac{\hbar}{mc}\right)^2 \qquad (\simeq 6.6\ 10^{-25}\ cm^2) \qquad (5.28)$$

In fact it is this limit which in QED provides an interpretation of the on-shell renormalized coupling constant and the on-shell renormalized mass are respectively identified with the classical charge and mass of the electron.

In QCD we do not know of classical observables (equivalent to e and m in QED) in terms of which we can express the renormaliz ed parameters of the theory. How do we fix the renormalized parameters? To convince you that this is not an academic question let us consider the following situation: suppose that in order to compare the results of an experiment, say experiment I, with theory we have fixed the renormalized parameters by subtracting at an euclidean point μ_1 i.e.,

$$\text{EXP I} \xrightarrow[\text{parametrized by}]{} \alpha(\mu_1),\ m(\mu_1)$$

Somebody else who wants to compare the results of another experiment, say experiment II, with theory may find it more convenient to use another set of renormalized parameters

EXP II $\xrightarrow[\text{parametrized by}]{}$ $\alpha(\mu_2),\ m(\mu_2)$

It is clear that in order to make a prediction of

EXP II from the results of EXP I

we must know how to get the set

$$\{\alpha(\mu_2),\ m(\mu_2)\} \qquad \text{from} \qquad \{\alpha(\mu_1),\ m(\mu_1)\}$$

This is precisely what the renormalization group does for us. The crucial equation, as regards the dependence on the coupling constant, is eq. (5.18a) which we rewrite as follows

$$t = \int \frac{dz}{z\,\beta(z)} \equiv \Psi(z) + \text{const.} \tag{5.29}$$

Neither t, nor z are renormalization group invariants but the combination

$$t - \Psi(z) = t_0 - \Psi(z_0) = \text{const.} \tag{5.30}$$

must be invariant since it is precisely the arbitrary constant of the general solution of the differential equation. At the one loop approximation, which amounts to keeping the first term in the perturbation series expansion of the $\beta(z)$ function, see eqs. (5.22) and (5.23),

$$\Psi(z) = -\frac{\pi}{\beta_1}\,\frac{1}{z} \tag{5.31}$$

and we have, according to (5.30),

$$\frac{1}{2}\log\mu^2 + \frac{\pi}{\beta_1\,\alpha_s(\mu)} = \text{const.} \equiv \frac{1}{2}\log\Lambda^2 \tag{5.32}$$

In the plane α_s , $\frac{1}{2}\log\mu^2$ this is the equation of an hyperbola (recall that in QCD $\beta_1 < 0$)

$$\frac{\left(\alpha_s(\mu)+\frac{1}{2}\log\frac{\mu^2}{\Lambda^2}\right)^2}{4\pi/-\beta_1} - \frac{\left(-\alpha_s(\mu)+\frac{1}{2}\log\frac{\mu^2}{\Lambda^2}\right)^2}{4\pi/-\beta_1} = 1 \qquad (5.33)$$

as shown in Fig. (5.4). The physical branch of the hyperbola ($\alpha_s(\mu) > 0$) corresponds to $\mu^2 > \Lambda^2$. The two parameters which fix the shape of the hyperbola are β , and Λ . Instead of parametrizing the physical observables in terms of $\alpha_s(\mu)$, which obviously depends on the arbitrary choice of the renormalization point μ , we can use Λ which is a constant of motion with respect to the scale μ . From eqs. (5.24) and (5.32) we can easily express , at the one loop approximation, the effective coupling constant in terms of Λ . Since

$$\frac{1}{2}\log Q^2 + \frac{\pi}{\beta_1\bar{\alpha}(Q^2)} = \frac{1}{2}\log\mu^2 + \frac{\pi}{\beta_1\alpha_s(\mu)} = \frac{1}{2}\log\Lambda^2 \qquad (5.34)$$

we have

$$\bar{\alpha}_s\left(\frac{1}{2}\log\frac{Q^2}{\mu^2}, \alpha_s(\mu)\right) = \frac{\pi}{-\beta_1\frac{1}{2}\log\frac{Q^2}{\Lambda^2}} \qquad (5.35)$$

As we shall see later, the value of Λ which one obtains from the comparison between theory and experiment is

$$\Lambda \simeq 500 \text{ MeV} \qquad (5.36)$$

So far we have only discussed eq. (5.29) at the one-loop level. What happens when we take into account higher order terms in the expansion of $\beta(z)$ in eq. (5.22)? In connection with this question let me first show that only the first two coefficients β_1 and β_2 in the perturbation series expansion of $\beta(z)$ are, in fact, renormalization group invariant[6]. For that purpose, let us call

$$\beta_I(\alpha_I) \quad , \quad \beta_{II}(\alpha_{II})$$

the two β-functions calculated with two different renormalization prescriptions, say I and II; with corresponding renormalized coupling constants α_{I} and α_{II}. To find the functional relation between $\beta_{I}(\alpha_{I})$ and $\beta_{II}(\alpha_{II})$, consider again eq. (5.29) with the integral taken between limits α_{I} and

$$\bar{\alpha}\left(\frac{1}{2}\log\frac{Q^2}{\mu_{II}^2}, \alpha_{II}\right) \qquad \text{i.e.,}$$

$$\frac{1}{2}\log\frac{Q^2}{\mu_{II}^2} = \int_{\alpha_I}^{\bar{\alpha}\left(\frac{1}{2}\log\frac{Q^2}{\mu_{II}^2}, \alpha_{II}\right)} \frac{dz}{z\,\beta_I(z)} \tag{5.37}$$

Next, let us apply the operator $\mu_{II}\frac{d}{d\mu_{II}}$ to both sides of this equation. We get

$$-1 = -\frac{1}{\alpha_I\,\beta_I(\alpha_I)}\,\beta_{II}(\alpha_{II})\,\alpha_{II}\,\frac{\partial\alpha_I}{\partial\alpha_{II}} \tag{5.38}$$

where we have used the fact that the effective coupling constant $\bar{\alpha}\left(\frac{1}{2}\log\frac{Q^2}{\mu_{II}^2}, \alpha_{II}\right)$ obeys the equation

$$\left(\mu_{II}\frac{\partial}{\partial\mu_{II}} + \beta_{II}(\alpha_{II})\,\alpha_{II}\frac{\partial}{\partial\alpha_{II}}\right)\bar{\alpha}\left(\frac{1}{2}\log\frac{Q^2}{\mu_{II}^2}, \alpha_{II}\right) \tag{5.39}$$

The sought functional relation is

$$\beta_I(\alpha_I) = \beta_{II}(\alpha_{II})\,\frac{\alpha_{II}}{\alpha_I}\,\frac{\partial\alpha_I}{\partial\alpha_{II}} \tag{5.40}$$

Consider now the formal series expansions

$$\beta_I(\alpha_I) = \beta_1^{(I)}\frac{\alpha_I}{\pi} + \beta_2^{(I)}\left(\frac{\alpha_I}{\pi}\right)^2 + \cdots \tag{5.41a}$$

$$\beta_{II}(\alpha_{II}) = \beta_1^{(II)}\frac{\alpha_{II}}{\pi} + \beta_2^{(II)}\left(\frac{\alpha_{II}}{\pi}\right)^2 + \cdots \tag{5.41b}$$

and

$$\alpha_I = \alpha_{II} + H_2 \, \alpha_{II}^2 \tag{5.42}$$

It is easy to check that the constraints on these coefficients, which follow from the functional relation in eq. (5.40), are

$$\beta_1^{(I)} = \beta_1^{(II)} \quad , \quad \beta_2^{(I)} = \beta_2^{(II)} \tag{5.43}$$

while higher order coefficients involve a dependence on the H_2 coefficients. This finishes the proof that only the first two coefficients in the perturbation theory expansion of the β-function are renormalization invariants. It has been pointed out by t'Hooft, in ref. (5.17), that, provided $\beta(z)$ does not have a fixed point between $z = 0$ and $z = \infty$, it is always possible to choose a renormalization prescription for which all the coefficients β_i, $i \geqslant 3$ can be put to zero. This is an intriguing point which doesn't seem to have been fully analyzed.

From the phenomenological point of view it turns out that the precision of the deep inelastic scattering experiments already warrants a two-loop evaluation of the theoretical predictions. In particular, the effective coupling constant is required at the two-loop level. Let us then discuss this next.

The integration of eq. (5.29) with the first two terms in the expansion of $\beta(z)$ gives

$$t = \Psi(z) + \text{const.} \tag{5.44a}$$

with

$$\Psi(z) = \int \frac{dz}{z^2 \frac{1}{\pi} \beta_1 \left(1 + \frac{\beta_2}{\beta_1} \frac{z}{\pi}\right)} = \frac{\pi}{\beta_1} \left(-\frac{1}{z} + \frac{\beta_2}{\beta_1} \frac{1}{\pi} \log \frac{1 + \frac{\beta_2}{\beta_1} \frac{z}{\pi}}{z} \right) \tag{5.44b}$$

The Cte. in eq. (5.44a) is a renormalization group invariant which for dimensional reasons we write as $\frac{1}{2}\log\Lambda^2$; exactly as we did in the one-loop case, eq. (5.32),

$$\frac{1}{2}\log\mu^2 - \Psi[\alpha_s(\mu)] = \frac{1}{2}\log\Lambda^2 \tag{5.45}$$

In the plane $\alpha_s(\mu)$, $\frac{1}{2}\log\mu^2$, we no longer have an hyperbola, as in the one-loop approximation. The equation is now

$$\frac{\left(\alpha_s(\mu)+\frac{1}{2}\log\frac{\mu^2}{\Lambda^2}\right)^2}{-\frac{4\pi}{\beta_1}\left[1-\frac{\beta_2}{\beta_1}\frac{\alpha_s(\mu)}{\pi}\log\frac{1+\frac{\beta_2}{\beta_1}\frac{\alpha_s(\mu)}{\pi}}{\alpha_s(\mu)}\right]} - \frac{\left(-\alpha_s(\mu)+\frac{1}{2}\log\frac{\mu^2}{\Lambda^2}\right)^2}{\frac{4\pi}{-\beta_1}\left[1-\frac{\beta_2}{\beta_1}\frac{\alpha_s(\mu)}{\pi}\log\frac{1+\frac{\beta_2}{\beta_1}\frac{\alpha_s(\mu)}{\pi}}{\alpha_s(\mu)}\right]}=1 \tag{5.46}$$

which can be viewed as an hyperbola modulated by an $\alpha_s(\mu)$ dependent eccentricity. Notice that the ratio β_2/β_1 , plays a crucial role in the modulation of the eccentricity as a function of $\alpha_s(\mu)$. The actual value of β_2 for a SU(N) gauge theory has been computed independently by Jones (5.18) and Caswell (5.19),

$$\beta_2 = -\left[\frac{17}{3}\left(\frac{N}{2}\right)^2 - \frac{1}{2}\frac{N^2-1}{2N}\frac{n}{2} - \frac{5}{3}\frac{N}{2}\frac{n}{2}\right] \tag{5.47}$$

For colour -SU(3) (N=3) ;

$$\beta_2 = -\frac{51}{4} + \frac{19}{12}n \tag{5.48}$$

In the applications one wants the effective coupling constant i.e., the inverse of eq. (5.44). This was a trivial problem at the one loop approximation (see eqs. (5.31) and (5.35)). Now, however, we are faced with the inversion of a transcendental equations, which we shall do by successive approximations.

First we expand $\Psi(z)$ in eq. (5.44b) to order O(z) . We shall not keep terms of O(z) , because this is the order at which β_3 contributes and we only want to retain two-loop effects; i.e.,

$$\frac{1}{2}\log\frac{Q^2}{\mu^2} = \frac{\pi}{\beta_1}\left(\frac{1}{\alpha_s(\mu)} - \frac{1}{\bar{\alpha}_s(Q^2)}\right) - \frac{\beta_2}{\beta_1^2}\log\frac{\bar{\alpha}_s(Q^2)}{\alpha_s(\mu)} \tag{5.49}$$

Next, we solve at the one-loop level i.e.,

$$\bar{\alpha}_s^{(2)}(Q^2) = \frac{\alpha_s(\mu)}{1 - \frac{\alpha_s(\mu)}{\pi}\beta_1 \frac{1}{2}\log\frac{Q^2}{\mu^2}} \tag{5.24}$$

and insert this in the argument of the log in eq. (5.49). We get

$$\bar{\alpha}_s(Q^2) = \bar{\alpha}_s^{(2)}(Q^2) - \left(\bar{\alpha}_s^{(2)}(Q^2)\right)^2 \frac{\beta_2}{\beta_1}\frac{1}{\pi}\log\left(1 - \frac{\alpha_s(\mu)}{\pi}\beta_1\frac{1}{2}\log\frac{Q^2}{\mu^2}\right) \tag{5.50}$$

This expression is the exact sum of leading and next-to-leading terms of the perturbation series. As pointed out before (see eq. (5.45)) the combination of log μ and $\alpha_s(\mu)$ which remains renormalization group invariant, at the two loop level, is

$$\frac{1}{2}\log\mu^2 + \frac{\pi}{\beta_1}\frac{1}{\alpha_s(\mu)} - \frac{\beta_2}{\beta_1^2}\log\frac{1 + \pi\frac{\beta_2}{\beta_1}\alpha_s(\mu)}{\alpha_s(\mu)} = \frac{1}{2}\log\Lambda^2 \tag{5.51}$$

i.e., to order $O(\alpha_s(\mu))$,

$$\frac{1}{2}\log\mu^2 + \frac{\pi}{\beta_1}\frac{1}{\alpha_s(\mu)} - \frac{\beta_2}{\beta_1^2}\log\alpha_s(\mu) + O(\alpha_s(\mu)) = \frac{1}{2}\log\Lambda^2 \tag{5.52}$$

In their QCD analysis of deep inelastic electron scattering experiments, Buras, Floratos, Ross and Sachrajda (5.20) have found convenient to fix the Cte. in eq. (5.45) not to $\frac{1}{2}\log\Lambda^2$ (as everybody does at the oneloop level), but to the combination

$$\begin{aligned}\frac{1}{2}\log\mu^2 + \frac{\pi}{\beta_1}\frac{1}{\alpha_s(\mu)} - \frac{\beta_2}{\beta_1^2}\log\alpha_s(\mu) + O(\alpha_s(\mu))\\ = -\frac{\beta_2}{\beta_1^2}\log\left(\frac{-\beta_1}{\pi}\right) + \frac{1}{2}\log\Lambda^2_{BFRS}\end{aligned} \tag{5.53}$$

In terms of Λ_{BFRS}, the effective coupling constant at the two-loop level reads

$$\bar{\alpha}\left(Q^2/\Lambda^2_{BFRS}\right) = \bar{\alpha}^{(2)}\left(Q^2/\Lambda^2_{BFRS}\right) - \left(\bar{\alpha}^{(2)}\left(Q^2/\Lambda^2_{BFRS}\right)\right)^2\frac{\beta_2}{\beta_1}\log\log Q^2/\Lambda^2_{BFRS} \tag{5.54a}$$

with

$$\bar{\alpha}^{(2)}\left(Q^2/\Lambda^2_{BFRS} \right) = \frac{2\pi}{-\beta_1 \log Q^2/\Lambda^2_{BFRS}} \qquad (5.54b)$$

Since β_1 and β_2 are renormalization group invariant quantities, the parametrization in terms of Λ_{BFRS} is also renormalization group invariant. However, one must be careful when comparing the value of obtained from a one-loop analysis using eq. (5.35) and the value of Λ_{BFRS} obtained from a two-loop analysis using eq. (5.54) since they correspond to two different parametrizations of the <u>same</u> constant. The correct comparison should be between

Λ (one-loop) defined via eqs. (5.32) and (5.35)

and

$$\Lambda \text{ (two-loops)} = \left(-\frac{\beta_1}{\pi} \right)^{\frac{\beta_2}{\beta_1^2}} \Lambda_{BFRS} \qquad (5.35)$$

with Λ_{BFRS} defined via eqs. (5.35) and (5.54) .

FOOTNOTES OF SECTION 5

1) The classical papers on this subject are Stuekelberg and Peterman, ref. (5.1); and Gell-Mann and Low ref. (5.2). See also Bogoliubov and Chirkov's textbook ref. [R.25] .

2) The composition law eq. (5.4a) for renormalization transformations was first pointed out by Astaud and Jouvet, ref. (5.3). For recent discussions related to this point see refs. (5.4).

3) A nice discussion of this point in the case of an abelian gauge theory, QED, can be found in a recent preprint by Coquereaux, ref. (5.5). A detailed discussion of the same problem, within QCD, is in preparation; ref. (5.6).

4) See e.g. Valiron textbook ref. [R. 27] .

5) We wish to emphasize that this increase of the effective coupling constant in QED at shorter and shorter distances has, in fact, been experimentally checked. The precision measurements of the anomalous magnetic moment of the muon (the CERN g-2 experiments) prove the short distance behaviour of the photon propagator via diagrams of the type shown in Fig. 5.3. One can view the muon g-2 experiment as a measurement of $\bar{\alpha}_{QED}(t = \log m_\mu / m_e , \alpha_{QED})$ and indeed one finds consistency with the predicted theoretical value. For details see ref. (5.13). For a recent comparison between theory and experiments see ref. (5.14).

6) To our knowledge the first place in the literature where this remark was made, within the context of QED, is in de Rafael and Rosner's paper, ref. (5.16). The relevance of this property in connection with perturbative QCD has been emphasized by 't Hooft, ref. (5.17).

REFERENCES OF SECTION 5

1) E.C.G. Stueckelberg and A. Peterman, Helv. Phys. Acta 26(1953)499.

2) M. Gell-Mann and F.E. Low, Phys. Rev. 95(1954)1300.

3) M. Astaud and B. Jouvet, C.R. Acad. Sci. 264(1967)1433; Nuovo Cimento 53A(1968)841 ibid 63A(1969)5.

4a) Masuo Suzuki, Commun. Math. Phys. 57(1977)193.

4b) M.C. Bergere and C. Bervillier, CEN Saclay, Preprint DPh-T/78/112.

5) R. Coquereaux, Preprint CPT 2, Marseille (1979).

6) R. Coquereaux and E. de Rafael (to be published).

7) J.D. Bjorken, Phys. Rev. 179(1969)1547.

8a) C. Callan Jr., Phys. Rev. D2(1970)1541.

8b) K. Symanzik, Commun. Math. Phys. 18(1970)227.

9) G. 't Hooft, Nucl. Phys. B61(1973)455.

10) S. Weinberg, Phys. Rev. D8(1973)3497.

11) S. Coleman and D.J. Gross, Phys. Rev. Letters 31(1973)851.

12a) H.D. Politzer, Phys. Rev. Lett. 30(1973)1346.

12b) D.J. Gross and F. Wilczek, Phys. Rev. Lett. 30(1973)1346.

12c) G.'t Hooft, Unpublished remark at the Marseille Conference on Yang Mills Theories, 1972.

13) B. Lautrup and E. de Rafael, Nucl. Phys. B70(1974)317.

14) J. Calmet, S. Narison, M. Perrottet and E. de Rafael, Rev. Mod. Phys. 49(1977)21.

15) W. ~~Thirring~~, Phil. Mag. 41(1950)113.

16) E. de Rafael and J.L. Rosner, Ann. of Phys. 82(1974)369.

17) G. 't Hooft, Some Observations on Quantum Chromodynamics, Notes based on lectures given at the Orbis Scientiae 1977. University of Miami (1977).

18) D.R.T. Jones, Nucl. Phys. B75(1974)531.

19) W. Caswell, Phys. Rev. Letters 33(1974)244.

20) A.J. Buras, E.G. Floratos, D.A. Ross and C.T. Sachrajda, Nucl. Phys. B131(1977)308.

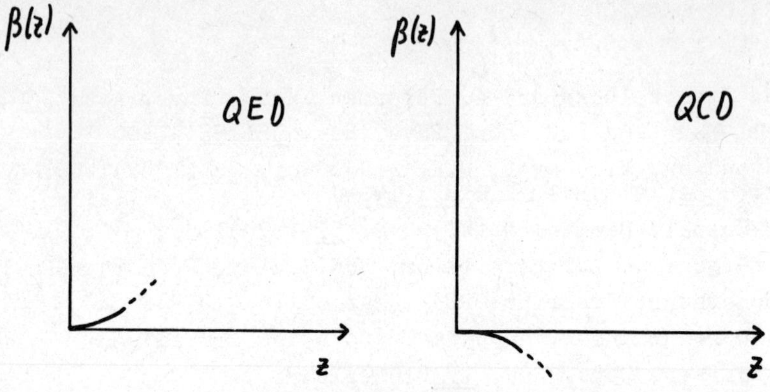

Fig. 5.1 The fixed point of the β -function at the origin. Fig. 5.1a corresponds to the case of QED; Fig. 5.1b to the case of QCD.

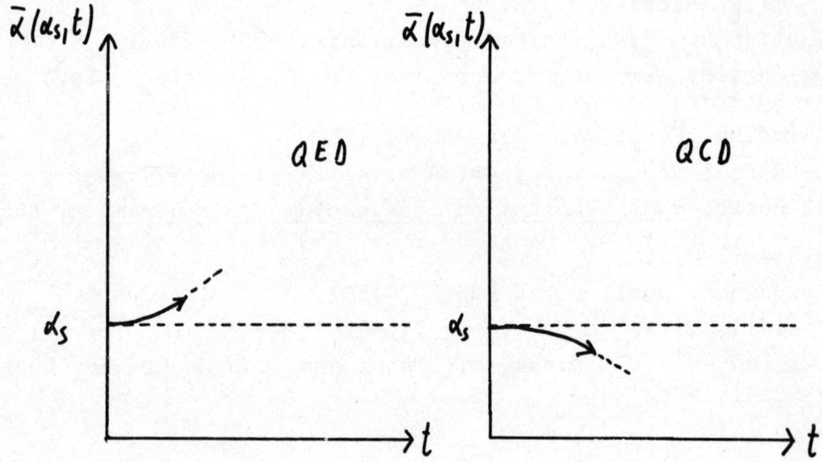

Fig. 5.2 Behaviour of the effective coupling constant as the scaling parameter increases. Fig. 5.2a corresponds to the case where $\beta_1 > 0$ [QED] . Fig. 5.2b to the case where $\beta_1 < 0$ [QCD].

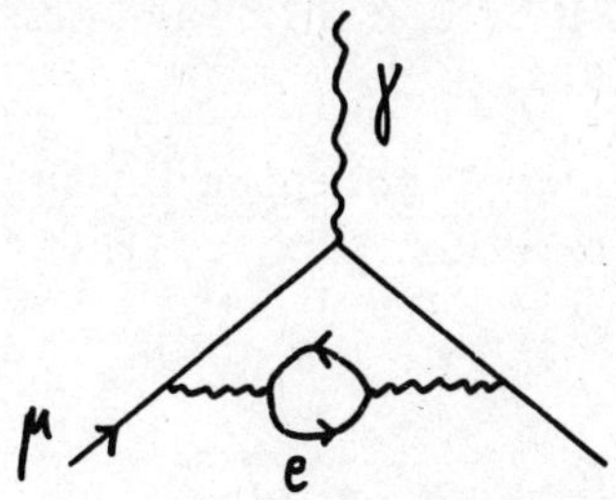

Fig. 5.3 Typical diagram which contributes to the muon g-2 and is sensitive to the short distance behaviour of QED.

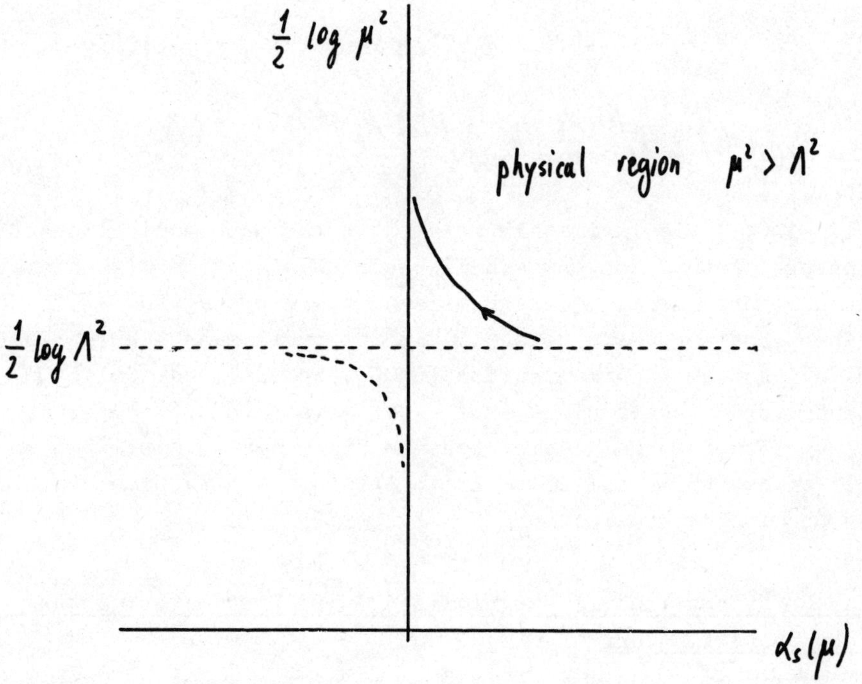

Fig. 5.4 Graphycal representation of the hyperbola eq. (5.33) in the plane $\alpha_s(\mu)$, $\frac{1}{2}\log\mu^2$.

6. RENORMALIZATION INVARIANCE. EFFECTIVE QUARK MASSES

Other than $\bar{\alpha}$, the general solution of the renormalization group equation, eq. (5.19), depends on $\bar{x}_i$, itself solution of the ordinary differential equation

$$\frac{d\bar{x}_i(t,\alpha_s)}{dt} = -[1+\gamma_i(\bar{\alpha}_s)]\,\bar{x}_i \tag{6.1}$$

with the boundary condition

$$\bar{x}_i(0,\alpha_s) = x_i \quad , \qquad x_i = \frac{m_i}{\mu} \tag{6.2a,b}$$

In perturbation theory $\gamma_i(\alpha_s)$, defined in eq. (5.11b), takes the form

$$\gamma_i(\alpha_s) = \gamma_i^{(1)}\frac{\alpha_s}{\pi} + \gamma_i^{(2)}\left(\frac{\alpha_s}{\pi}\right)^2 + \dots \tag{6.3}$$

Clearly $\gamma_i(\alpha_s)$ is intimately related to the renormalization of the fermion propagator. Hence we shall begin this section with a discussion of the unrenormalized, ϵ-regularized, quark propagator at the one loop approximation. Next we shall discuss quark mass renormalization and wave function renormalization in QCD. We shall calcualte $\gamma_i^{(1)}$, the lowest order coefficient in the α_s expansion of γ_i, eq. (6.3), in the various renormalization schemes discussed so far. Then we shall go back to eq. (5.1) and solve it to obtain the effective quark masses at the one loop approximation.

6a. The Unrenormalized, ϵ-Regularized Quark Propagator at the one loop approximation.

The general expression for the unrenormalized, ϵ-regularized, fermion propagator is

$$\frac{i}{\not p - m_0 - \Sigma_{UR}(\not p,\epsilon) + i\epsilon} \tag{6.4}$$

with $-i\,\Sigma_{UR}(\not p,\epsilon)$ the one-particle irreducible fermion self energy (also called the proper self energy) which, on invariance grounds, has the general structure

$$\Sigma_{UR}(\not p,\epsilon) = \Sigma_{1UR}(p^2;\epsilon) + (\not p - m_0)\,\Sigma_{2UR}(p^2;\epsilon) \qquad (6.5)$$

To lowest order in the coupling constant g_o, $-i\,\Sigma_{UR}$ is given by the Feynman diagram in Fig. 6.1.

$$-i\,\Sigma_{UR}(\not p,\epsilon) = (ig_0)^2 \int \frac{d^n k}{(2\pi)^n}\, \gamma^\mu \frac{i}{\not p + \not k - m_0 + i\epsilon}\, \gamma^\nu\, \frac{N^2-1}{2N} \cdot \frac{-i\left(g_{\mu\nu} - (1-a)\frac{k_\mu k_\nu}{k^2}\right)}{k^2} \qquad (6.6)$$

where we have used the gluon propagator in an arbitrary gauge. Combining the denominators with a Feynman parameter, we get

$$-i\,\Sigma_{UR}(\not p,\epsilon) = \frac{N^2-1}{2N}\, g_0^2 \int_0^1 dx \int \frac{d^n k}{(2\pi)^n}\, \frac{-1}{[(k-\ell)^2 - \mathbb{R}^2]^2} \Big\{ \gamma^\mu(\not p + \not k + m_0)\gamma_\mu + (1-a)\Big[(\not p - \not k - m_0) - \frac{2x}{(k-\ell)^2 - \mathbb{R}^2}\, 2p\cdot k\, \not k\Big]\Big\} \qquad (6.7)$$

where

$$\ell_\mu \equiv -p_\mu(1-x) \quad , \quad \mathbb{R}^2 \equiv (1-x)(m_0^2 - p^2 x) \qquad (6.8a,b)$$

Next we shift the integration variable

$$k \to k + \ell \qquad (6.9)$$

and we do the integration over k using the formulae given in section 4a. Remembering that $n = 4-\epsilon$, we get

$$\Sigma_{UR}(\not p,\epsilon) = g_0^2\, \frac{N^2-1}{2N}\, \frac{1}{(16\pi^2)^{2-\epsilon/4}} \int_0^1 dx \Big\{ \Gamma(\epsilon/2)\,(\mathbb{R}^2)^{-\epsilon/2} \cdot$$

$$\cdot\Big[(-2+\epsilon)x\not p+(4-\epsilon)m_0+(1-a)\big((2-x)\not p-m_0\big)\Big]$$
$$-(1-a)2x(\mathbb{R}^2)^{-\epsilon/2}\frac{\Gamma(3-\epsilon/2)\Gamma(\epsilon/2)}{\Gamma(2-\epsilon/2)2}2\not p\frac{1}{4-\epsilon}$$
$$+(1-a)2x(\mathbb{R}^2)^{-1-\epsilon/2}\frac{1}{2}\Gamma(1+\epsilon/2)2p^2\not p(1-x)^2\Big\} \qquad (6.10)$$

We are only intersted in terms which are singular or non vanishing when $\epsilon\to 0$; hence, after some algebra, we can rewrite $\Sigma_{UR}(\not p,\epsilon)$ in a way from which one can directly read the Σ_1 and Σ_2 contributions defined in eq. (6.5):

$$\Sigma_{1UR}(p^2,\epsilon)=(g_0\nu^{-\epsilon/2})^2\frac{N^2-1}{2N}\frac{1}{(16\pi^2)^{1-\epsilon/4}}\int_0^1 dx\Big\{\Gamma(\epsilon/2)\Big(\frac{\nu^2}{\mathbb{R}^2}\Big)^{\epsilon/2}$$
$$\cdot\big[2m_0(2-x)-m_0\epsilon(1-x)+(1-a)m_0(1-2x)\big] \qquad (6.11)$$
$$+(1-a)2x(1-x)m_0\frac{p^2}{m_0^2-p^2x}\Big\}$$

and

$$\Sigma_{2UR}(p^2,\epsilon)=(g_0\nu^{-\epsilon/2})^2\frac{N^2-1}{2N}\frac{1}{(16\pi^2)^{1-\epsilon/4}}\int_0^1 dx\Big\{\Gamma(\epsilon/2)\Big(\frac{\nu^2}{\mathbb{R}^2}\Big)^{\epsilon/2} \qquad (6.12)$$
$$\cdot\big[-2x+\epsilon x+(1-a)2(1-x)\big]+(1-a)2x(1-x)\frac{p^2}{m_0^2-p^2x}\Big\}$$

6b. <u>Quark Mass Renormalization and Wave Function Renormalization</u>

We can now proceed to an explicit discussion of mass renormalization, at the one loop approximation, in the various renormalization schemes which we have introduced in section 4C. As we shall see, at that approximation, only $\Sigma_{1UR}(p^2,\epsilon)$ is relevant for mass renormalization. For convenience we start our discussion with

a) <u>'t Hooft's ϵ-renormalization</u>

Here we are instructed to subtract the poles in $\frac{1}{\epsilon}$. Σ_{1UR} has a pole in $\frac{1}{\epsilon}$ which can be easily read from the expresssion in eq. (6.11)

$$\tilde{\Sigma}_1 \equiv \text{singular part of } \Sigma_{1\,UR}$$

$$= \frac{1}{\epsilon}\frac{\alpha_s}{\pi} m_0 \frac{3}{2}\frac{N^2-1}{2N} \tag{6.13}$$

where we have set $g_0 \nu^{-\epsilon/2} = g$ and $g^2/4\pi = \alpha_s$, to express $\tilde{\Sigma}_1$, at the required order in terms of the renormalized coupling constant g. Let us now rewrite the unrenormalized fermion propagator in eq. (6.4), using eq. (6.5), in the following form

$$\frac{1}{1-\Sigma_{2UR}(p^2,\epsilon)} \cdot \frac{i}{\not p - \left[m_0 + \dfrac{\Sigma_{1UR}(p^2,\epsilon)}{1-\Sigma_{2UR}(p^2,\epsilon)}\right]} \tag{6.14}$$

Mass renormalization in the 't H scheme is defined by requiring that the pole of the fermion propagator has no singularities in $1/\epsilon$. At the one loop approximation, and using the result in eq. (6.13), this is achieved by defining

$$m_0 = m\left[1+\frac{1}{\epsilon}\frac{\alpha_s}{\pi}\frac{N^2-1}{2N}\left(-\frac{3}{2}\right)+O\left(\left(\frac{\alpha_s}{\pi}\right)^2\right)\right] \tag{6.15}$$

From this result we can explicitely calculate the coefficient $\hat{Z}_{m,1}$ in the general expression of the mass-renormalization constant given in eq. (4.45),

$$\hat{Z}_{m,1} = \frac{\alpha_s}{\pi}\frac{N^2-1}{2N}\left(-\frac{3}{2}\right) \tag{6.16}$$

Notice that to lowest order in α_s the renormalization constant Z_m in the 't H scheme is gauge invariant.

For completeness' sake let us also work out at the one loop approximation the fermion wave function renormalization in the 't H scheme. Again this is done by requiring that the residue at the pole has no $1/\epsilon$ singularities. This requirement fixes the wave function renormalization constant to

$$Z_2 = \frac{1}{1-\hat{\Sigma}_2} \tag{6.17}$$

where

$$\hat{\Sigma}_2 \equiv \text{singular part of } \Sigma_{2\,UR}(p^2,\epsilon)$$

$$= \frac{1}{\epsilon}\frac{\alpha_s}{\pi}\frac{N^2-1}{2N}\left[-\frac{1}{2}+\frac{1}{2}(1-a)\right] \tag{6.18}$$

i.e.,

$$Z_2 = 1+\frac{1}{\epsilon}\frac{\alpha_s}{\pi}\frac{N^2-1}{2N}\left[-\frac{1}{2}+\frac{1}{2}(1-a)\right]+O\left(\left(\frac{\alpha_s}{\pi}\right)^2\right) \tag{6.19}$$

which is the explicit expression, to order α_s/π, of the general form given in eq. (4.43). We have reproduced the well known result that, at the one loop approximation, and in the Landau gauge $a=o$, the fermion wave function renormalization constant has no ultraviolet singularities.

The full ϵ-renormalized fermion propagator, in an arbitrary gauge, is then

$$\frac{i}{\not p - m - \Sigma(\not p,\nu)+i\epsilon} = Z_2^{-1}\frac{i}{\not p - m_o - \Sigma_{UR}(\not p,\epsilon)+i\epsilon}$$

$$= \frac{i}{\not p\,[1-\Sigma_2(p^2,\nu^2)] - m\,[1-\Sigma_2(p^2,\nu^2)] - \Sigma_1(p^2,\nu^2)} \tag{6.20}$$

where, to order α_s/π

$$\Sigma_1(p^2,\nu^2) = \frac{\alpha_s}{\pi}\frac{N^2-1}{2N}\frac{1}{2}m\Bigg[\frac{3}{2}(\log 4\pi-\gamma)+\int_0^1 dx\Bigg\{-(1-x)+(2-x)\log\frac{\nu^2}{(1-x)(m^2-p^2x)}$$

$$+(1-a)\left[x(1-x)\frac{p^2}{m^2-p^2x}+(1-2x)\frac{1}{2}\log\frac{\nu^2}{(1-x)(m^2-p^2x)}\right]\Bigg\}\Bigg] \tag{6.21}$$

and

$$\Sigma_2(p^2/m^2,\nu^2/m^2) = \frac{\alpha_s}{\pi}\frac{N^2-1}{2N}\frac{1}{2}\Big[\frac{1}{2}[-1+(1-a)](\log 4\pi-\gamma)+$$

$$+\int_0^1 dx \left\{ x - x\log\frac{\nu^2}{(1-x)(m^2-p^2x)} + (1-a)\left[(1-x)\log\frac{\nu^2}{(1-x)(m^2-p^2x)} + \frac{x(1-x)p^2}{m^2-p^2x}\right]\right\}\Bigg] \quad (6.22)$$

where γ is Euler's constant, defined in eq. (4.9b).

For completeness' sake we give the integrated expressions of $\Sigma_1(p^2/m^2, \nu^2/m^2)$ and $\Sigma_2(p^2/m^2, \nu^2/m^2)$ which are useful for discussions of finite renormalizations. We find[1])

$$\Sigma_1(p^2/m^2, \nu^2/m^2) = \frac{\alpha_s}{\pi}\frac{N^2-1}{2N}\frac{1}{2}m\left\{\frac{3}{2}(\log 4\pi - \gamma) + \frac{5}{2} + \frac{3}{2}\log\frac{\nu^2}{m^2-p^2}\right.$$

$$+\frac{1}{2}\frac{m^2}{-p^2} - \frac{1}{2}\frac{m^2}{-p^2}\left(4+\frac{m^2}{-p^2}\right)\log\left(1-\frac{p^2}{m^2}\right)$$

$$\left. + (1-a)\left[-\frac{1}{2} - \frac{1}{2}\frac{m^2}{-p^2} + \frac{1}{2}\frac{m^2}{-p^2}\left(1+\frac{m^2}{-p^2}\right)\log\left(1-\frac{p^2}{m^2}\right)\right]\right\} \quad (6.23)$$

and

$$\Sigma_2(p^2/m^2, \nu^2/m^2) = \frac{\alpha_s}{\pi}\frac{N^2-1}{2N}\frac{1}{2}\left\{\frac{1}{2}[-1+(1-a)](\log 4\pi - \gamma)\right.$$

$$-\frac{1}{2} + \frac{1}{2}\log\frac{m^2-p^2}{\nu^2} - \frac{1}{2}\left(\frac{m^2}{-p^2}\right)^2\log\left(1+\frac{-p^2}{m^2}\right) + \frac{1}{2}\frac{m^2}{-p^2}$$

$$\left. + (1-a)\left[\frac{1}{2} - \frac{1}{2}\log\frac{m^2-p^2}{\nu^2} + \frac{1}{2}\left(\frac{m^2}{-p^2}\right)^2\log\left(1+\frac{-p^2}{m^2}\right) - \frac{1}{2}\frac{m^2}{-p^2}\right]\right\} \quad (6.24)$$

Notice that Σ_2 vanishes in the Landau gauge (a=o). Interesting particular values of these expresssions are:

i) the value at $p^2 = -\nu^2$, when m=o ,

$$\Sigma_1(-\nu^2/m^2, \nu^2/m^2; m \to 0) = \frac{\alpha_s}{\pi}\frac{N^2-1}{2N}\frac{m}{2}\left\{\frac{3}{2}(\log 4\pi - \gamma) + \frac{5}{2} + (1-a)\left(-\frac{1}{2}\right)\right\} \quad (6.25)$$

$$\Sigma_2\left(-\nu^2/m^2, \nu^2/m^2; m\to 0\right) = \frac{\alpha_s}{\pi}\frac{N^2-1}{2N}\frac{1}{2}\left\{\frac{1}{2}\left[-1+(1-a)\right](\log 4\pi-\gamma) - \frac{1}{2} + (1-a)\frac{1}{2}\right\} \tag{6.26}$$

ii) the limit $-p^2 \gg m^2$ at arbitrary ν ,

$$\lim_{-p^2\gg m^2} \Sigma_1\left(p^2/m^2, \nu^2/m^2\right) = \frac{\alpha_s}{\pi}\frac{N^2-1}{2N}\frac{m}{2}\left\{\frac{3}{2}(\log 4\pi-\gamma) + \frac{5}{2} + \frac{3}{2}\log\frac{\nu^2}{-p^2} + (1-a)\left(-\frac{1}{2}\right) + O\left[\frac{m^2}{-p^2}\log\frac{-p^2}{m^2}\right]\right\} \tag{6.27}$$

$$\lim_{-p^2\gg m^2} \Sigma_2\left(p^2/m^2, \nu^2/m^2\right) = \frac{\alpha_s}{\pi}\frac{N^2-1}{2N}\frac{1}{2}\left\{\frac{1}{2}\left[-1+(1-a)\right](\log 4\pi-\gamma) - \frac{1}{2} + \frac{1}{2}\log\frac{-p^2}{\nu^2} + (1-a)\left[\frac{1}{2} - \frac{1}{2}\log\frac{-p^2}{\nu^2}\right] + O\left[\left(\frac{m^2}{-p^2}\right)^2\log\frac{-p^2}{m^2}\right]\right\} \tag{6.28}$$

iii) the limit $-p^2 \ll m^2$

$$\lim_{-p^2\ll m^2} \Sigma_1\left(p^2/m^2, \nu^2/m^2\right) = \frac{\alpha_s}{\pi}\frac{N^2-1}{2N}\frac{m}{2}\left\{\frac{3}{2}(\log 4\pi-\gamma) + \frac{3}{2}\log\frac{\nu^2}{m^2} + \frac{3}{4} + \frac{5}{6}\frac{-p^2}{m^2} + (1-a)\left[-\frac{1}{4} + \frac{-1}{12}\frac{-p^2}{m^2}\right] + O\left[\left(\frac{-p^2}{m^2}\right)^2\right]\right\} \tag{6.29}$$

$$\lim_{-p^2\ll m^2} \Sigma_2\left(p^2/m^2, \nu^2/m^2\right) = \frac{\alpha_s}{\pi}\frac{N^2-1}{2N}\frac{1}{2}\left\{\frac{1}{2}\left[-1+(1-a)\right](\log 4\pi-\gamma) - \frac{1}{2}\log\frac{\nu^2}{m^2} - \frac{1}{4} + \frac{1}{3}\frac{-p^2}{m^2} + (1-a)\left[\frac{1}{2}\log\frac{\nu^2}{m^2} + \frac{1}{4} - \frac{1}{3}\frac{-p^2}{m^2}\right] + O\left(\left(-\frac{p^2}{m^2}\right)^2\right)\right\} \tag{6.30}$$

We can now proceed to the calculation of $\gamma_i^{(1)}$ in eq. (6.3). Recall the definition (see eq. (5.11b))

$$-\gamma(\alpha_s) = \frac{\mu}{m}\,\frac{dm}{d\mu} \qquad (6.31)$$

For convenience we drop the index i which to lowest order is irrelevant. Also, in the ϵ-renormalization we can identify μ and ν. From this definition and the relation

$$m_0 = Z_m\, m \qquad (6.32)$$

we have

$$-\gamma(\alpha_s) = \frac{\mu}{m}\,\frac{dZ_m^{-1}}{d\mu}\, m^0 = Z_m\,\mu\,\frac{dZ_m^{-1}}{d\mu} \qquad (6.33)$$

To lowest order, the only dependence of Z_m^{-1} in μ is via the coupling constant

$$\begin{aligned}\mu\,\frac{dZ_m^{-1}}{d\mu} &\equiv \left(\mu\frac{\partial}{\partial\mu} + \beta(\alpha_s,\epsilon)\,\alpha_s\frac{\partial}{\partial\alpha_s} + \cdots\right) Z_m^{-1} \\ &= \left(-\alpha_s\epsilon\frac{\partial}{\partial\alpha_s} + \beta(\alpha_s)\,\alpha_s\frac{\partial}{\partial\alpha_s}\right) Z_m^{-1}\end{aligned} \qquad (6.34)$$

where we have used the fact that in the ϵ-renormalization the β function has a non trivial linear dependence in ϵ[2]. Combining eqs. (6.31), (6.32) we have

$$-\gamma(\alpha_s)\,Z_m^{-1} + \epsilon\,\alpha_s\frac{\partial}{\partial\alpha_s}Z_m^{-1} - \beta(\alpha_s)\,\alpha_s\frac{\partial}{\partial\alpha_s}Z_m^{-1} = 0 \qquad (6.35)$$

If we now insert the result of our calculation [eq. (6.16)]

$$Z_m = 1 + \frac{1}{\epsilon}\,\frac{\alpha_s}{\pi}\,\frac{N^2-1}{2N}\left(-\frac{3}{2}\right) + \cdots \qquad (6.36)$$

we get, to the desired order of accuracy

$$-\gamma^{(1)}\frac{\alpha_s}{\pi}+\alpha_s\,\frac{N^2-1}{2N}\,\frac{3}{2}=0 \tag{6.37}$$

i.e.,

$$\gamma^{(1)}=\frac{\alpha_s}{\pi}\,\frac{N^2-1}{2N}\,\frac{3}{2} \tag{6.38}$$

Notice that the value of $\gamma^{(1)}$, in the ϵ-renormalization scheme, is mass-independent and gauge invariant.

b) The μ-renormalization scheme

Mass renormalization in the μ-scheme is defined by requiring that the renormalized fermion propagator has a pole at $p^2 = -\mu^2$. This is ensured by defining the renormalized fermion self-energy as follows

$$\Sigma_R(\not p,\mu)=\Sigma_{1R}(p^2,\mu^2)+(\not p-m(\mu))\,\Sigma_{2R}(p^2,\mu^2) \tag{6.39}$$

with

$$\Sigma_{1R}(p^2,\mu^2)=\Sigma_{1UR}(p^2,\epsilon)-\Sigma_{1UR}(-\mu^2,\epsilon) \tag{6.40a}$$

and

$$\Sigma_{2R}(p^2,\mu^2)=\Sigma_{2UR}(p^2,\epsilon)-\Sigma_{2UR}(-\mu^2,\epsilon) \tag{6.40b}$$

In the language of renormalization constants of section 4, see eq. (4.37), we have

$$Z_4\,Z_{2F}^{-1}\,m(\mu)=m_0 \tag{6.41}$$

with

$$Z_4 Z_{2F}^{-1} = 1 - \frac{1}{m(\mu)} \Sigma_1 (p^2 = -\mu^2, \epsilon) \tag{6.42}$$

and

$$Z_{2F} = \frac{1}{1 - \Sigma_2 (p^2 = -\mu^2, \epsilon)} \tag{6.43}$$

Notice that in the μ-scheme, the expresssion for $\Sigma_1(-\mu^2, \epsilon)$ coincides with that for $\Sigma_{1UR}(p^2 = -\mu^2, \epsilon)$ in eq. (6.11) except for the replacement $m_o \to m(\mu)$. This is because $\Sigma_{1UR}(p^2, \epsilon)$ was calculated, using the bare Lagrangean $\mathcal{L}_o$ while $\Sigma_1(-\mu^2, \epsilon)$ comes from the counterterm part $\Delta\mathcal{L}$ in eq. (4.24). The quantity of interest is [see eq. (5.11b)]

$$\gamma\left(\alpha(\mu), \frac{m(\mu)}{\mu}, a\right) = -\frac{\mu}{m(\mu)} \frac{dm(\mu)}{d\mu}$$

$$= \frac{-1}{Z_4^{-1} Z_{2F}} \mu \frac{d\, Z_4^{-1} Z_{2F}}{d\mu} =$$

(to lowest order in $\alpha(\mu)$)

$$= -\mu \frac{d}{d\mu} \frac{1}{m(\mu)} \Sigma_1 (p^2 = -\mu^2, \epsilon)$$

$$= \frac{\alpha(\mu)}{\pi} \frac{N^2-1}{2N} \left\{ \frac{3}{2} - \int_0^1 dx\, (2-x) \frac{m^2}{m^2 + \mu^2 x} \right.$$

$$\left. + (1-a) \int_0^1 dx \left[x(1-x) \frac{\mu^2 m^2}{(m^2+\mu^2 x)^2} - \frac{1}{2} (1-2x) \frac{m^2}{m^2+\mu^2 x} \right] \right\}$$

$$= \frac{\alpha(\mu)}{\pi} \frac{N^2-1}{2N} \left\{ \frac{3}{2} + \frac{m^2}{\mu^2} - \frac{m^2}{\mu^2}\left(2 + \frac{m^2}{\mu^2}\right) \log\left(1 + \frac{\mu^2}{m^2}\right) \right.$$

$$\left. + (1-a)\left[-\frac{m^2}{\mu^2} + \frac{1}{2} \frac{m^2}{\mu^2} \left(1 + 2\frac{m^2}{\mu^2}\right) \log\left(1 + \frac{\mu^2}{m^2}\right)\right] \right\} \tag{6.44}$$

This expression coincides with the one previously obtained by Nachtmann and Wetzel in ref. (6.1). We see that in the μ-scheme, the lowest order value for the γ-function associated to the mass term in the QCD Lagrangean is both mass dependent and gauge dependent. Two asymptotic expressions are useful to have explicitly:

i)
$$\lim_{m(\mu) \ll \mu} \gamma^{(1)}\left(\alpha(\mu), \frac{m(\mu)}{\mu}, a\right) = \tag{6.45}$$
$$= \frac{\alpha(\mu)}{\pi} \frac{N^2-1}{2N} \left\{ \frac{3}{2} + 2\frac{m^2}{\mu^2} \log\frac{m^2}{\mu^2} + (1-a)\left(-\frac{1}{2}\right)\frac{m^2}{\mu^2} \log\frac{m^2}{\mu^2} + O\left[\left(\frac{m^2}{\mu^2}\right)^2 \log\frac{m^2}{\mu^2}\right] \right\}$$

i.e., at the limit $m \to o$ we find again the ϵ-renormalization result.

ii)
$$\lim_{m(\mu) \gg \mu} \gamma^{(1)}\left(\alpha(\mu), \frac{m(\mu)}{\mu}, a\right) =$$
$$= \frac{\alpha(\mu)}{\pi} \frac{N^2-1}{2N} \left\{ \frac{2}{3} \frac{\mu^2}{m^2} + (1-a)\frac{1}{12}\frac{\mu^2}{m^2} + O\left(\left(\frac{\mu^2}{m^2}\right)^2\right) \right\} \tag{6.46}$$

We see that in the μ-scheme, large quark mass values decouple from renormalization effects by terms proportional to μ^2/m^2.

c) Weinberg's μ-renormalization

In this scheme, mass renormalization is defined by requiring that the renormalized fermion propagator has a pole at $p^2 = -\mu^2$, when all mass terms are set to zero. The renormalized fermion self-energy is defined as follows

$$\Sigma_R(\not{p}, \mu) = \Sigma_{1R}(p^2, \mu^2) + (\not{p} - m(\mu))\, \Sigma_{2R}(p^2, \mu^2) \tag{6.47}$$

with

$$\Sigma_{1R} = \Sigma_{1UR}(p^2, \epsilon) - \lim_{m^2 \to 0} \Sigma_{1UR}(-\mu^2, \epsilon) \tag{6.48a}$$

and

$$\Sigma_{2R} = \Sigma_{2UR}(p^2, \epsilon) - \lim_{m^2\to 0} \Sigma_{2UR}(-\mu^2, \epsilon) \qquad (6.48b)$$

Clearly, in this scheme

$$\gamma^{(1)} = \frac{\alpha(\mu)}{\pi} \frac{N^2-1}{2N} \frac{3}{2} \qquad (6.49)$$

which is the same result that we obtained in the ϵ-renormalization scheme.

6c. Effective Quark Masses and Renormalization Group Invariant Masses

Let us next integrate eq. (6.1), at the one loop approxiamtion, using either the ϵ-renormalization scheme or Weinberg's μ-renormalization; i.e.,

$$\bar{x}_i = x_i \exp\left\{ -\left[t + \int_0^{t'} dt' \gamma_i[\bar{\alpha}(t', \alpha_s)] \right] \right\} \qquad (6.50)$$

with $\bar{\alpha}$ given by eq. (5.35) and

$$\gamma_i(\bar{\alpha}) = \frac{\bar{\alpha}}{\pi} \frac{N^2-1}{2N} \frac{3}{2} + O\left(\frac{\bar{\alpha}^2}{\pi^2}\right) \qquad (6.51)$$

We find

$$\bar{x}_i = x_i \left(\frac{\mu^2}{-q^2}\right)^{1/2} \exp\left\{ - \frac{N^2-1}{2N} \frac{3}{2} \frac{1}{-\beta_1} \log\left(\frac{\frac{1}{2} \log {-q^2}/{\Lambda^2}}{\pi/-\alpha_s\beta_1} \right) \right\} \qquad (6.52)$$

We can now introduce a renormalization group invariant mass $\hat{m}_i$, much the same as the scale Λ in eq. (5.32) associated to the coupling constant,[3)]

$$\hat{m}_i = m_i(\mu) \left(\frac{\pi}{-\beta_1 \alpha_s(\mu)} \right)^{\frac{N^2-1}{2N}\frac{3}{2}/-\beta_1} \qquad (6.53)$$

in terms of which

$$\bar{x}_i = \frac{\hat{m}_i}{\sqrt{-q^2}} \frac{1}{\left(\frac{1}{2}\log {-q^2}/{\Lambda^2}\right)^{\frac{N^2-1}{2N}\frac{3}{2}/-\beta_1}} \tag{6.54}$$

That $\hat{m}_i$ in eq. (6.53) is indeed a renormalization group invariant, within the class of ϵ -renormalizations or Weinberg's μ -renormalizations, can be seen as follows. Froms eqs. (5.11b) and (6.3) we have

$$\int \frac{dm_i}{m_i} = -\frac{N^2-1}{2N}\frac{3}{2}\int \frac{d\mu}{\mu}\frac{\alpha_s(\mu)}{\pi} + const \tag{6.55a}$$

i.e.,

$$\log m_i(\mu) = \frac{N^2-1}{2N}\frac{3}{2}\frac{1}{\beta_1}\int \frac{d\,\mu/\Lambda}{\mu/\Lambda}\frac{1}{\log \mu/\Lambda} + \text{const} \tag{6.55b}$$

where we have used eq. (5.32). Clearly, the Cte in these equations is a renormalization group invariant which we write as

$$\text{const} \equiv \log \hat{m}_i \tag{6.55c}$$

Then

$$m_i(\mu) = \hat{m}_i \left(\log \mu/\Lambda\right)^{\frac{N^2-1}{2N}\frac{3}{2}\frac{1}{\beta_1}} \tag{6.56}$$

which is the same result as eq. (6.53).

These renormalization group invariant masses $\hat{m}_i$ provide a well defined way of parametrizing all finite quark mass effects in QCD. In terms of the $\hat{m}_i$, the effective quark masses at a given Q^2 - value are (for color- SU(3))

$$\bar{m}_i(Q^2) = \frac{\hat{m}_i}{\left(\frac{1}{2}\log Q^2/\Lambda^2\right)^{2/-\beta_1}} \tag{6.57}$$

As $Q^2 \to \infty$, $\bar{m}_i(Q^2) \to 0$, but the ratio of effective masses goes to a constant

$$\lim_{Q^2 \to \infty} \frac{\bar{m}_i(Q^2)}{\bar{m}_j(Q^2)} = \frac{\hat{m}_i}{\hat{m}_j} \tag{6.58}$$

Notice that all these results i.e., eqs. (6.53); (6.57) and (6.58) are also valid in the μ -renormalization scheme provided that[4)]

$$m_i^2 \ll \mu^2 < Q^2 \tag{6.59}$$

The crucial result of this section is contained in eq. (6.54). When inserted into the general solution of the renormalization group equation, eq. (5.19), it tells us that the asymptotic behaviour of The Green's functions when all the momenta are scaled is governed by the massless theory ($\bar{x}_i \to 0$ when $\lambda = e^t \to \infty$). This means that provided $\Gamma(\ldots \bar{x}_i \to 0 \ldots)$ is not singular at the limit $\bar{x}_i \to 0$, the massless theory (no $\sum m_i \bar{\Psi}_i \Psi_i$ terms in the Lagrangean) will correctly give the leading asymptotic behaviour of the Green's functions. The existence of domains of momenta $p_1, \ldots, p_N$ for which the limit $\bar{x}_i \to 0$ is not singular (the so called non-exceptional momenta regions) has been discussed by Symanzik. They are the regions where all p_i are space-like and no partial sum of momenta vanishes. However, in other regions the Green's functions need not actually be singular. In fact, the existence of processes with time-like momenta, and yet non-singular observables when $\bar{x}_i \to 0$, is precisely the physics of Jets which is the topic of this school[6)].

We propose to identify the masses

$$\hat{m}_i \qquad i = \text{up, down, strange, charm, bottom,} \ldots$$

with the so called "current algebra quark masses"; i.e., the $\hat{m}_i$ are a set of parameters which parametrize the breaking of chiral SU(n) x SU(n) . From phenomenological applications of current algebra and PCAC it is estimated that[7)]

$$\frac{\hat{m}_s}{\hat{m}_d} \simeq 20 \quad ; \quad \frac{\hat{m}_s}{\hat{m}_u} \simeq 36 \tag{6.60}$$

$$\frac{\hat{m}_s}{\frac{\hat{m}_u + \hat{m}_d}{2}} \simeq 26 \quad ; \quad \frac{\hat{m}_d - \hat{m}_u}{\frac{\hat{m}_d + \hat{m}_u}{2}} \simeq 0.57$$

A more difficult question is to fix the absolute scale of quark masses $\hat{m}_i$. With some reasonable assumptions on the relation between strange particle masses and their quark content, Weinberg in ref. (6.6) gets

$$\hat{m}_s \simeq 0.15 \quad GeV \tag{6.61}$$

and therefore, using the ratios in (6.60),

$$\hat{m}_d \simeq 7.5 \; MeV \quad , \quad \hat{m}_u \simeq 4 \; MeV \tag{6.62a,b}$$

One can obtain a value for $\hat{m}_c$ from the rather successful non relativistic potential models of charmonium. One gets

$$\hat{m}_c \simeq 1.2 \; GeV \tag{6.63}$$

Another way to fix the mass $\hat{m}_c$ has been from the suppression of strangeness changing neutral currents [see ref. (6.7d)] from which

$$\hat{m}_c \simeq 1.5 \; \text{to} \; 2 \; GeV \tag{6.64}$$

One question we would like to answer is what is the relation between the current algebra masses $\hat{m}_i$, and the constituent quark masses which, from phenomenological arguments, are expected to be of the order of hadron masses. The only tool we have at present to answer this question is via the expression, eq. (6.57), for the effective quark mass. As Q^2 decreases, $\overline{m}_i(Q^2)$ increases. In particular, we can find the Q_o^2 for which

$$2 \, \overline{m}_i \, (Q_o^2) = Q_o \tag{6.65}$$

When we apply QCD to the annihilation of e^+e^- into hadrons there will be a threshold value Q_{th} for the production of hadrons with a given flavor. In QCD language this means the production of hadrons with a given flavor. In QCD language this means the production of a pair $q_i\bar{q}_i$ with masses $\bar{m}_i(Q_{th})$. A possible definition for the associated quark constituent masses emerges from this picture; mainly, to define

$$M_i\,(\text{constituent}) \equiv \bar{m}_i\,(Q_{th} = 2\,\bar{m}_i(Q_{th})) \tag{6.66}$$

The threshold for charm production is

$$Q_{th}\,(\text{charm}) \simeq 3\ \text{GeV} \tag{6.67}$$

and we get

$$\hat{m}_c = 1.98\ \text{GeV} \quad ; \quad M_c \simeq 1.5\ \text{GeV} \tag{6.68a,b}$$

At 3 GeV, the value of the effective coupling constant $\bar{\alpha}_s$ for 4 flavors [see eq. (5.35)] is

$$\bar{\alpha}_s\,(3\ \text{GeV}) \simeq 0.42 \tag{6.69}$$

where we have used $\Lambda = 500$ MeV. This is a reasonalble value at which to expect perturbative arguments still to work.

The threshold for the "new" flavor bottom[8] is

$$Q_{th}\,(\text{bottom}) \simeq 10\ \text{GeV} \tag{6.70}$$

Here, (with 5-flavors and $\Lambda = 500$ MeV)

$$\bar{\alpha}_s\,(10\ \text{GeV}) \simeq 0.087 \tag{6.71}$$

and

$$\hat{m}_b \simeq 8.9\ GeV \quad ; \quad M_b \simeq 5\ GeV \tag{6.72a,b}$$

It is more questionable to apply these arguments to strangeness and up and down quarks. The threshold for strange particles is

$$Q_{th}\ (\text{strange}) \simeq 1\ GeV \tag{6.73}$$

for which (3 flavors and Λ = 500 MeV)

$$\bar{\alpha}_s\ (1\ GeV) \simeq 1 \tag{6.74}$$

Here we get

$$\hat{m}_s \simeq 425\ MeV \quad , \quad M_s \simeq 500\ MeV \tag{6.75a,b}$$

6d. Effective Gauge Dependence

Other than $\bar{\alpha}_s$ and $\bar{x}_i$, the general solution of the renormalization group equation, eq. (5.19), depends on $\bar{a}$, itself solution of the ordinary differential equation

$$\frac{d\,\bar{a}\,(t,\alpha_s)}{dt} = \beta_a\,(\alpha_s) \tag{6.76}$$

with the boundary condition

$$\bar{a}\,(0,\alpha_s) = a \tag{6.77}$$

Remember that, in particular,

a = 1 .... Feynman gauge

a = 0 .... Landau gauge

The explicit form of $\beta_a(\alpha_s)$ can be obtained from the application of the renormalization group equation, eq. (5.16), to the gluon-gluon vertex one particle irreducible Green's function [see eq. (4.52)]

$$\Gamma_{2,0}(k^2, m_i^2, \alpha_s, a; \mu^2) = 1 + \Pi(k^2, m_i^2, \alpha_s, a; \mu^2) \tag{6.78}$$

i.e.,

$$\Big\{ -\frac{\partial}{\partial t} + \beta(\alpha_s)\alpha_s \frac{\partial}{\partial \alpha_s} + \beta_a(\alpha_s)\frac{\partial}{\partial a} - \sum_{i=1}^{n_f} [1 + \gamma_i(\alpha_s)] x_i \frac{\partial}{\partial x_i} + \gamma_{YM}(\alpha_s) \Big\} \Gamma_{2,0}\Big(t = \frac{1}{2}\log\frac{-k^2}{\mu^2}, x_i = \frac{m_i}{\mu}, \alpha_s, a\Big) = 0 \tag{6.79}$$

From this and the one-loop expression for $\Gamma_{2,0}$, eq. (4.55), we get the gluon anomalous dimension $\gamma_{YM}(\alpha_s)$ at lowest order,

$$\gamma_{YM}(\alpha_s) = \frac{\alpha_s}{\pi}\gamma_{YM}^{(1)} + O\Big(\Big(\frac{\alpha_s}{\pi}\Big)^2\Big) \tag{6.80}$$

with

$$\gamma_{YM}^{(1)} = -\frac{N}{2}\Big(-\frac{13}{6} + \frac{1}{2}a\Big) - \frac{1}{3}n_f + \frac{1}{2}\sum_{i=1}^{n_f}\int_0^1 dy\, 4y(1-y)\frac{x_i^2}{x_i^2 + y(1-y)} \tag{6.81}$$

Once $\gamma_{YM}(\alpha_s)$ is known, $\beta_a(\alpha_s)$ is also known. Again, this follows from the Slavnov-Taylor identity eq. (4.54) which implies

$$a = Z_{3YM}^{-1}\, a_0 \tag{6.82}$$

hence

$$Z_6 = 1 \tag{6.83}$$

in eq. (4.38), to all orders of perturbation theory. From eqs. (6.79), (4.54) and (6.83) it clearly follows that

$$\beta_a(\alpha_s) = -a\,\gamma_{YM}(\alpha_s) \tag{6.84}$$

therefore, at the one-loop approximation

$$\beta_a(\alpha_s) = -a\frac{\alpha_s}{\pi}\gamma_{YM}^{(1)} + a\,O\left(\left(\frac{\alpha_s}{\pi}\right)^2\right) \tag{6.85}$$

with $\gamma_{YM}^{(1)}$ given by eq. (6.81).

From these results it follows that the Landau gauge (a=o) is the simplest choice as far as gauge dependences are concerned, since then $\beta_a(\alpha_s) = 0$ and the a-partial derivative in the renormalization group equation altogether drops out. If we insist on working in an arbitrary gauge, then we have to integrate eq. (6.76). At the one loop approximation, and expressing the result in terms of renormalization group invariants, we find in the μ-scheme, with neglect of fermion masses $x_i^2 \to 0$:

$$\bar{a}\left(\frac{Q^2}{\Lambda^2}\right) = \frac{\hat{a}}{\left(\frac{1}{2}\log\frac{Q^2}{\Lambda^2}\right)^{\frac{1}{-\beta_1}\left(\frac{13}{12}N-\frac{1}{3}n_f\right)}}\left[1+\frac{3N}{13N-4n_f}\,\frac{\hat{a}}{\left(\frac{1}{2}\log\frac{Q^2}{\Lambda^2}\right)^{\frac{1}{-\beta_1}\left(\frac{13}{12}N-\frac{1}{3}n_f\right)}}\right]^{-1}$$

where

$$\hat{a} = \frac{a}{1-\frac{3N}{13N-4n_f}a}\left(\frac{\pi}{-\beta_1\alpha_s(\mu)}\right)^{-\frac{1}{\beta_1}\left(\frac{13}{12}N-\frac{1}{3}n_f\right)}$$

i.e., the effective gauge parameter $\bar{a}\,(Q^2/\Lambda^2)$ decreases as $Q^2/\Lambda^2 \to \infty$ provided the number of flavors

$$n_f \leq 9 \quad (N=3) \tag{6.87}$$

and the initial choice of gauge is $a \neq o$ and/or $a \neq \frac{13}{3} - \frac{4}{9} n_f$. When $n_f > 9$, $\bar{a}\,(Q^2/\Lambda^2)$ increases as Q^2/Λ^2 increases. There are two exceptional choices of the initial gauge for which

$$\bar{a}\,(Q^2/\Lambda^2) = a \tag{6.88}$$

the Landau gauge itself $a=o$; and the "peculiar" gauge

$$a = \frac{13}{3} - \frac{4}{9} n_f \tag{6.89}$$

There is a difference however between these two gauges. In the case of the Landau gauge $\bar{a} = a$ to all orders of perturbation theory; which in the case of the peculiar gauge $a = \frac{13}{3} - \frac{4}{9} n_f$ only at the one loop approximation. Note that the "peculiar" gauge corresponds to $\gamma^{(1)}_{YM} = 0$ i.e., the anomalous dimension of the gluon is of order $(\alpha_s)^2$.

FOOTNOTES OF SECTION 6

1) I wish to thank R. Coquereaux for checking these results.

2) Let us prove that, in the ϵ -renormalization,

$$\alpha_s \beta(\alpha_s,\epsilon) = -\epsilon \alpha_s + \alpha_s \beta(\alpha_s)$$

The starting point is eq. (4.44). Also recall that

$$\alpha_0 = \frac{g_0^2}{4\pi} = \nu^{\epsilon} \alpha_s Z_g^2$$

By definition

$$\alpha_s \beta(\alpha_s,\epsilon) = \nu \frac{d\alpha_s}{d\nu} = \nu \frac{d}{d\nu} \left(\alpha_0 \nu^{-\epsilon} Z_g^{-2} \right)$$

$$= -\epsilon \alpha_s - 2\alpha_s \frac{1}{Z_g} \nu \frac{dZ_g}{d\nu}$$

but the dependence of Z_g on ν is only via α_s

$$\left\{ \alpha_s \beta(\alpha_s,\epsilon) + \epsilon \alpha_s + 2\alpha_s^2 \beta(\alpha_s,\epsilon) \frac{\partial}{\partial \alpha_s} \right\} Z_g = 0$$

If we now insert the ansatz [see eq. (4.43)]

$$Z_g = 1 + \sum_{n=1}^{\infty} \frac{1}{\epsilon^n} Z^{(n)}$$

in the previous differential equation we find

$$\alpha_s \beta(\alpha_s,\epsilon) = -\epsilon \alpha_s + \text{Finite quantity} \equiv \alpha_s \beta(\alpha_s)$$

and

$$\beta(\alpha_s) = 2\alpha_s \frac{\partial Z^{(1)}}{\partial \alpha_s}$$

3) This definition of renormalization group invariant masses has been first introduced in a paper by Floratos, Narison and de Rafael, ref. (6.2).

4) Quark masses in the μ -renormalization schemes were first discussed by Georgi and Politzer, ref. (6.3).

5) See K. Symanzik, ref. (6.4). Other references relevant to the question of mass-singularities are listed in refs. (6.5).

6) For references see the other lectures at this school.

7) These are the ratios estimated by Weinberg in ref. (6.6). For recent discussions on the problem of quark masses see also refs. (6.7).

8) For a recent review see refs. (6.8) and (6.9).

REFERENCES OF SECTION 6

1) O. Nachtmann and Wetzel, Heidelberg Preprint, HD-THEP-78-3(1978).

2) E.G. Floratos, S. Narison and E. de Rafael, Spectral Function Sum Rules in Quantum Chromodynamics. I Charged Currents Sector. Marseille Preprint 78/P.1021 (to be published in the Nuclear Phys. B).

3) H. Georgi and H.D. Politzer, Phys. Rev. Lett. 36(1976)1281, Erratum Phys. Rev. Lett. 37(1976)68; Phys. Rev. D14(1976)1829.

4) K. Symanzik, Commun. Math. Phys.23(1971)49.

5a) D. Ruelle, Nuovo Cimento 19(1961)356.

5b) S. Weinberg, Phys. Rev. 118(1960)838.

5c) T. Kinoshita, J. Math. Phys. 3(1962)650.

6) S. Weinberg, in Festschrift for I.I. Rabi, ed. by Lloyd Motz (N.Y. Academy of Sciences, N.Y. 1977).

7a) A. Zepeda, Phys. Rev. Letters 41(1978)139.

7b) P. Langacker and H. Pagels, DESY preprint 78/33.

7c) F.J. Yndurain, FTUAM Preprint 78-5, (1978).

7d) M.K. Gaillard and B.W. Lee, Phys. Rev. D10(1974)897.

8) L. Lederman, Rapporteur Talk, XIX International Conference on High Energy Physics, Tokyo (1978).

9) G. Flügge, Rapporteur Talk, XIX International Conference on High Energy Physics, Tokyo (1978).

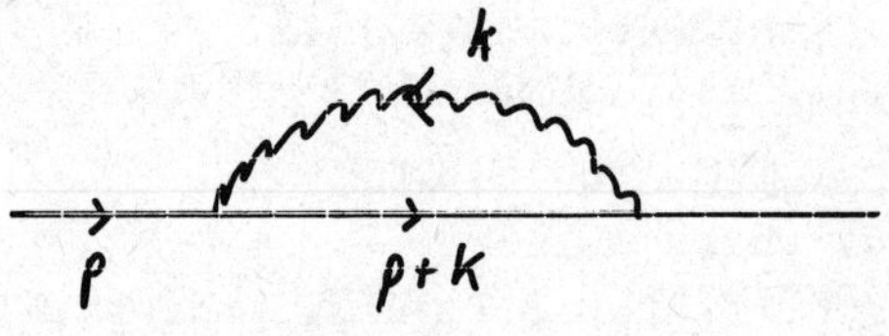

Fig. 6.1 The Quark self-energy diagram at the one-loop approximation.

7. APPLICATION TO $e^+e^- \rightarrow$ HADRONS

We shall be concerned with the QCD predictions for the ratio

$$R(t) = \frac{\sigma(e^+e^- \rightarrow hadrons)}{\sigma(e^+e^- \rightarrow \mu^+\mu^-)}$$

as a function of $t = 4E^2$ the total CM energy squared. This application of QCD is a good introduction to the recent jet ideas which are at the core of the subjects discussed in this school.

To lowest order in the electromagnetic coupling, the total e^+e^- annihilation cross-section into hadrons is governed by the absorptive part of the hadronic vacuum polarization (see figure (7.1)).

$$\sigma(e^+e^- \rightarrow hadrons) = \frac{1}{\sqrt{q^2(q^2-4m_e^2)}} \frac{e^2}{q^4} L^{\mu\nu}(k_1,k_2)\, \Pi_{\mu\nu}^{(abs)}(q) \tag{7.2}$$

where $L^{\mu\nu}$ is the leptonic tensor associated to the incoming e^+e^- beams (average over lepton polarizations assumed)

$$\begin{aligned} L^{\mu\nu}(k_1,k_2) &= \frac{1}{4}\, Tr\left[\gamma^\mu(\not{k}_1 + m_e)\gamma^\nu(\not{k}_2 - m_e)\right] \\ &= \frac{1}{2}\left[q^\mu q^\nu - q^2 g^{\mu\nu} - (k_1-k_2)^\mu(k_1-k_2)^\nu\right] \end{aligned} \tag{7.3}$$

and $\Pi_{\mu\nu}^{(abs)}$ the absorptive part of the hadronic vacuum polarization tensor

$$\Pi_{\mu\nu}^{(abs)} \equiv \frac{1}{2} \sum_{\Gamma} (2\pi)^4 \delta^{(4)}(q - P_\Gamma) \langle 0| J_\mu^{EM}(0)|\Gamma\rangle\langle\Gamma| J_\nu^{EM}(0)|0\rangle \tag{7.4a}$$

$$= \frac{1}{2}\int d^4x\, e^{iqx} \langle 0|\left[J_\mu^{EM}(x), J_\nu^{EM}(0)\right]|0\rangle \tag{7.4b}$$

$$= (q_\mu q_\nu - q^2 g_{\mu\nu})\, e^2\, Im\, \Pi(t \equiv q^2) \tag{7.4c}$$

where $\sum_{\Gamma}$ also includes summation over the full phase space of a given hadronic final state. J_{μ}^{EM} denotes the electromagnetic hadronic current. Equation (7.4b) expresses the fact that $\Pi_{\mu\nu}^{(abs)}$ is the Fourier transform of the vacuum expectation value of the commutator of the electromagnetic current with itself. The general tensor structure of $\Pi_{\mu\nu}^{(abs)}$ in eq. (7.4c) follows from the conservation of the electromagnetic current.

With neglect of the electron mass m_e with respect to q^2 we get from eqs. (7.2), (7.3) and (7.4c)

$$\sigma(e^+e^- \to hadrons) = \frac{4\pi^2\alpha}{q^2} e^2 \frac{1}{\pi} \, Im \, \Pi(t=q^2) \tag{7.5}$$

In order to eliminate the QED factors in this expression it is convenient to normalize the hadronic e^+e^- annihilation cross section to the $\mu^+\mu^-$ pair production cross section. To lowest order in the electromagnetic coupling (see figure (7.2).

$$\sigma(e^+e^- \to \mu^+\mu^-) = \frac{4\pi^2\alpha}{t} \frac{\alpha}{\pi} \frac{1}{3} \left(1 + 2\frac{m_\mu^2}{t}\right) \sqrt{1 - \frac{4m_\mu^2}{t}} \; \theta(t - 4m_\mu^2) \tag{7.6a}$$

which, for $t \gg 4m_\mu^2$, becomes

$$\sigma(e^+e^- \to \mu^+\mu^-)\Big|_{t \gg 4m_\mu^2} = \frac{4\pi\alpha^2}{3t} \tag{7.6b}$$

The electromganetic hadronic current $J_{\mu}^{EM}(x)$ associated to the QCD Lagrangean is given by the expression

$$J_\mu^{EM}(x) = -e\,\bar{\Psi}(x)\,\gamma_\mu\, Q\, \Psi(x) \tag{7.7a}$$

where $\Psi(x)$ denotes a vector in flavor space with components

$$\Psi(x) = \begin{pmatrix} u(x) \\ c(x) \\ \vdots \\ d(x) \\ s(x) \\ \vdots \end{pmatrix} \tag{7.7b}$$

and Q the diagonal matrix of quark electric charges in units of e,

$$Q = \begin{pmatrix} \frac{2}{3} & & & & \\ & \frac{2}{3} & & & \\ & & \ddots & & \\ & & & -\frac{1}{3} & \\ & & & & -\frac{1}{3} \ddots \end{pmatrix} \tag{7.7c}$$

A trace over the color indices for each flavor component is understood in eq. (7.7a).

To the approximation where quark-gluon interactions are neglected (the free field limit) we can easily evaluate $\mathrm{Im}\tilde{\Pi}(t)$ in eq. (7.4). From each flavor we shall get a contribution like the one from $\mu^+\mu^-$ pair production in eq. (7.6a) modulated by a factor 3 from the trace over the color degree of freedom times 4/9 or 1/9 depending on the quark charge. At energies far away from flavor thresholds we can neglect the effect of quark masses and we get the simple prediction

$$R = \frac{\sigma(e^+e^- \to \text{hadrons})}{\sigma(e^+e^- \to \mu^+\mu^-)} = 3 \sum_{i=1}^{n} Q_i^2 \tag{7.8}$$

values of R corresponding to various choices of n are:

$$n = 3 \quad u, d, s \; ; \quad R = 2 \tag{7.9a}$$

$$n = 4 \quad u, d, s, c \; ; \quad R = \frac{10}{3} \tag{7.9b}$$

$$n = 6 \quad u, d, s, c, b, t \; ; \quad R = 5 \tag{7.9c}$$

7a. QCD corrections to the ratio R

From the point of view of QCD, eq. (7.8) is the prediction for the value of R at the asymptotic free limit. There will be corrections to this value from the coupling of quarks to gluons and gluons to themselves as described by the QCD Lagrangean.

The first evaluation of these corrections was made by Appelquist and Georgi (7.1a) and independently by Zee (7.1b). These authors did not work with $\mathrm{Im}\,\Pi(t)$ directly. Instead they consider the full hadronic vacuum polarization tensor[1)]

$$\Pi_{\mu\nu} = i\int d^4x\, e^{iqx} \langle 0 | T(J_\mu^{EM}(x)\, J_\nu^{EM}(0)) | 0 \rangle \tag{7.10a}$$

$$= (q_\mu q_\nu - q^2 g_{\mu\nu})\, \Pi(q^2) \tag{7.10b}$$

and calculate $\Pi(q^2)$ in the deep euclidean region (for large negative values of q^2) far away from the physical region $q^2 \geqslant 4m_\pi^2$ (see figure (7.3)). There is a technical reason for that. It is only in the deep euclidean region that the Callan-Symanzik equation for $\Pi(q^2)$ as used by the authors of ref. (7.1), has a simple solution governed by asymptotic freedom. The key to the passage from $\Pi(q^2)$ in the deep euclidean region to $\mathrm{Im}\,\Pi(t)$ at large physical t-values is provided by the dispersion relation

$$\Pi(q^2) = \frac{1}{\pi}\int_{4m_\pi^2}^{\infty} dt \left(\frac{1}{t-q^2-i\epsilon} - \frac{1}{t}\right) \mathrm{Im}\,\Pi(t) \tag{7.11}$$

This is a once subtracted dispersion relation; the subraction at $q^2=0$ being related to the electric charge renormalization.

Here, we shall follow another procedure to calculate the QCD corrections to the ratio R , which is closer to the Jet philosophy initiated by Sterman and Weinberg (7.2). We shall work directly with the absorptive part of the hadronic vacuum polarization. In perturbative QCD, $\mathrm{Im}\,\Pi(t)$ is a function of t ; the renormalization scale μ ; the renormalized quark masses $m_i(\mu)$, and the renormalized coupling constant $\alpha_s(\mu)$:

$$\mathrm{Im}\,\Pi(t) = F\left(\tau = \frac{1}{2}\log\frac{t}{\mu^2},\, \alpha_s(\mu),\, x_i = \frac{m_i(\mu)}{\mu}\right) \tag{7.12}$$

This function obeys the renormalization group equation (see section 5)

$$\left\{-\frac{\partial}{\partial\tau} + \beta(\alpha_s)\alpha_s\frac{\partial}{\partial\alpha_s} - \sum_{i=1}^{n}\left[1+\gamma_i(\alpha_s)\right] x_i\frac{\partial}{\partial x_i}\right\} F(\tau,\alpha_s,x_i) = 0 \tag{7.13}$$

The solution to this equation is

$$F(z, \alpha_s, x_i) = F(0, \bar{\alpha}_s(z,\alpha_s), \bar{x}_i(z, \alpha_s, x_j)) \tag{7.14}$$

with $\bar{\alpha}_s$ and $\bar{x}_i$ the effective coupling constant and the effective normalized quark masses discussed in sects. 5c; and 6c. The lower order Feynman diagrams relevant to our calculation are shown in Fig. (7.4). Their contribution to $F(z, \alpha_s, x_i)$ can be obtained from an old QED calculation by Jost and Luttinger (7.3). The only modifications required are trivial color factors. With neglect of the quark masses $m_i \to 0$ the result is very simple

$$F(z,\alpha_s,x_i) = \frac{3}{4\pi} \sum_{i=1}^{n} Q_i^2 \left\{ \frac{1}{3} + \frac{\bar{\alpha}_s}{\pi} \frac{1}{4} \frac{4}{3} + O\left(\left(\frac{\bar{\alpha}_s}{\pi}\right)^2\right) \right\} \tag{7.15}$$

The first term reproduces the well known result for the constant value of R in eq. (7.8). The second term gives the expression for the leading correction to R

$$R = 3 \sum_{i=1}^{n} Q_i^2 \left\{ 1 + \frac{1}{-\beta_1 \frac{1}{2} \log \frac{Q^2}{\Lambda^2}} + O\left[\frac{\log \log Q^2/\Lambda^2}{\log^2 Q^2/\Lambda^2} \right] \right\} \tag{7.16}$$

where (see eq. (5.25)) $-\beta_1 = \frac{11}{2} - n/3$; n the number of flavors. The correction is positive (provided $\beta_1 < 0$) and therefore the approach to a plateau for the ratio must be from above (see Fig. 7.5). Ideally, in order to test this QCD prediction one should be in regions of energies far away from flavor thresholds. We shall discuss problems of thresholds, as well as the comparison theory/experiment, later on.

The ratio R is the best example of a "good observable" from the point of view of the Jet philosophy. There are no infrared divergences and the limit $m_i \to 0$ is non singular. Perturbation theory with QUARKS and GLUONS as physical states and the replacement $\alpha_s \to \bar{\alpha}_s(t)$ gives an accurate description of the corresponding hadronic observable i.e.,

$$\sigma(e^+e^- \to \text{hadrons})\Big|_{\text{large } t}$$

$$= \sigma \ (e^+e^- \to \text{hadrons + gluons})\Big|_{\text{large } t} \tag{7.17}$$

During the last year there has been a lot of interest on the question of defining "good hadronic observables" like the ratio R for which perturbative on shell calculations in QCD can make reliable predictions. An example, which is discussed in detail in the lectures of A. Morel at this school is the angular distribution of hadronic jets from e^+e^- annihilations. In fact it is this example which has triggered the so-called "Jet philosophy": observables which in perturbative QCD are finite when all quark masses vanish will be reliable predictions of QCD.

7b. Problem of thresholds and comparison theory versus experiment

Without neglect of quark masses the exact expression for $F(z, \alpha_s, x_i)$ in eq. (7.15) to order $\bar{\alpha}$ is

$$F(z,\alpha_s,x_i) = \frac{1}{4\pi} \sum_{i=1}^{n} Q_i^2 \, \theta(t-4\bar{m}_i^2) v_i \left(\frac{3-v_i^2}{2}\right)\left\{1+\frac{4}{3}\bar{\alpha}_s(t) f(v_i)\right\} \tag{7.18a}$$

where the dependence on $\bar{m}_i$ also appears via the variable

$$v_i = \sqrt{1-4\bar{x}_i^2}$$

with $\bar{x}_i$ defined in eq. (6.54). The function $f(v_i)$ is well approximated by the expression

$$f(v) \simeq \frac{\pi}{2v} - \left(\frac{3+v}{4}\right)\left(\frac{\pi}{2} - \frac{3}{4\pi}\right) \tag{7.18c}$$

Again, the result in eq. (7.18a) can be obtained from an old QCD calculation by Källen and Sabry (7.4). The approximate form for the function $f(v)$ in eq. (7.18c) is due to Schwinger ref [R.30]. It is exact at v=o and v=1 and accurate to ± 1% elsewhere. The step function $\theta(t-4\bar{m}_i^2)$ defines the constituent mass associated to a given flavor [see eqs. (6.65), (6.66)] . For $t \gg 4M_i$ (constituent), $\bar{x}_i \to 0$ and $v \to 1$. Then eq. (7.18a) reproduces the result of eq. (7.15) correspond ing to the massless case. However for $\bar{x}_i \sim \frac{1}{2}$, $v \to o$ and the function $f(v)$ becomes singular. The presence of this $\frac{1}{v}$ singularity is due to the Coulomb like gluon exchange diagram in Fig. (7.6a). Higher order corrections in $\bar{\alpha}_s$ also have $(\bar{\alpha}_s/v)^n$ singularities arising from the Coulomb like gluon ladder diagrams in Fig. (7.6b). In QED these singularities are the perturbation theory reflection of bound states below threshold. The location of these bound states in QED is of Coulombic nature and well described by the formula

$$M_n(\alpha) = 2m - \frac{\alpha^2 m}{4n^2} \quad (n = 1,2,\cdots) \tag{7.19}$$

In perturbative QCD, other diagrams than those of Fig. (7.6) give in general logarithmic corrections to the $\frac{1}{v}$ singularities i.e., terms like $(\alpha/v)^p(\alpha \log v)^q$; $p+q = n$. Clearly, we can only expect the quark-antiquark bound states to be of Coulombic nature if and only if $\bar{\alpha}_s \log v \ll 1$.

The moral from the paragraph above is that the perturbation theory solution eq. (7.18a) becomes useless in regions nearby physical thresholds where all the terms in the perturbation expansion become of the same order of magnitude

$$1 \sim \bar{\alpha}_s \frac{1}{v} \sim \left(\bar{\alpha}_s \frac{1}{v}\right)^2 \cdots \tag{7.20}$$

This is a precise example of a "bad observable" from the point of view of the Jet philosophy: the predictive power of the perturbation theory is destroyed by the presence of quark mass singularities.

The question now is how to confront the theory (i.e., perturbative QCD which does not like thresholds) with the experimental $\sigma(e^+e^- \to \text{hadrons})$ which is full of threshold effects? The key to the solution is the use of analiticity (and positivity) properties of the hadronic vacuum polarization to smooth the effect of physical thresholds in the experimental cross-section. There exist various propositions in the literature to deal with this problem[2)]. A very straightforward and elegant method is the one suggested by Shankar (7.5d) which we next discuss.

Let us consider $\Pi(q^2)$ in eq. (7.10b) as a complex function in the complex q^2-plane (see Fig. (7.3)) and apply Cauchy's theorem to $z^p\Pi(z)$, $z = q^2$ with the choice of integration contour shown in Fig. (7.7)

$$\frac{1}{2\pi i} \oint dz \, z^p \, \Pi(z) = 0 \tag{7.21a}$$

The integration contour can be split in two components: the line parallel to the real axis and the big circle. From the first we get an integral over the discontinuity across the cut i.e., and integral over the absorptive part of the vacuum polarization. This has to be compensated by the integral over the big circle i.e.,

$$\int_{4m_\pi^2}^{Q^2} dt \, t^p \frac{1}{\pi} \, \mathrm{Im}\, \Pi(t) = - \frac{(Q^2)^{p+1}}{2\pi} \int_{-\pi}^{\pi} d\theta \, e^{i(p+1)\theta} \, \Pi(-Q^2 e^{i\theta}) \tag{7.21b}$$

where Q^2 denotes the radius of the big circle. Shankar proposes to use this relation as follows: the l.h.s. is evaluated using the experimental input for $\mathrm{Im}\,\Pi(t)$ via eq. (7.5); the r.h.s. is calculated using the QCD prediction for $\mathrm{Im}\,\Pi(t)$ and integrating. In principle, the choice of Q^2 is arbitrary provided $Q^2 > \Lambda^2$ the perturbative QCD scale.

A detailed phenomenological analysis of the $\sigma(e^+e^- \to \text{hadrons})$ data using Shankar's proposition has been made by Moorhouse, Pennington and Ross (7.6). Their best fit to the data is obtained with

$$\Lambda \simeq 500 \text{ MeV} \tag{7.22}$$

which corresponds to

$$\bar{\alpha}_s (Q^2 = 1 \text{ GeV}) \simeq 1 \quad , \qquad \text{for 4 flavours.}$$

Their input values for the renormalized quark masses (they use the μ-scheme off shell renormalization) is

$$m_u = m_d = 10 \text{ MeV} \tag{5.24a}$$

$$m_s = 0.35 \text{ GeV} , \quad m_c = 1.85 \text{ GeV} , \quad m_b = 2.4 \text{ GeV} \tag{5.24b,c,d}$$

In their analysis they have also taken into account the presence of hadronic events due to the existence of a heavy lepton with a mass $m_L = 1.95$ GeV.

FOOTNOTES OF SECTION 7

1) T in eq. (10a) denotes time ordered product:

$$T(J_\mu(x)\, J_\nu(0)) = \theta(x)\, J_\mu(x)\, J_\nu(0) + \theta(-x)\, J_\nu(0)\, J_\mu(x)$$

2) See e.g. ref. (7.5).

REFERENCES OF SECTION 7

1a) T. Appelquist and H. Georgi, Phys. Rev. D8(1973)4000.

1b) A. Zee, Phys. Rev. D8(1973)4038.

2) G. Sterman and S. Weinberg, Phys. Rev. Letters 39(1977)1436

3) R. Jost and J.M. Luttinger, Helv. Phys. Acta, 23(1950)201

4) G. Källen and A. Sabry, K. Danske Vidensk Selsk. Mat. Fys. Medd. 29(1955)no. 17.

5a) E.C. Poggio, M.R. Quinn, S. Weinberg, Phys. Rev. D13(1976)1958.

5b) A. de Rujula and H. Georgi, Phys. Rev. D13(1976)1296.

5c) F.J. Yndurain, Phys. Letters, 63B(1976)211.

5d) R. Shankar, Phys. Rev. D15(1977)755.

6) R.G. Moorhouse, H.R. Pennington and G.C. Ross, Nucl. Phys. B124(1977)285.

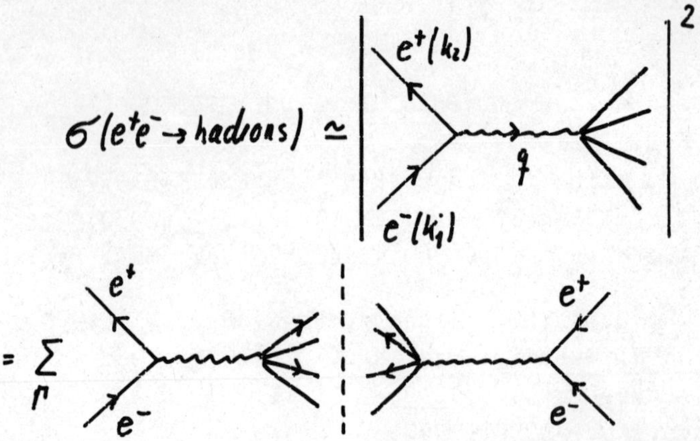

Fig. 7.1 e^+e^- annihilation to hadrons to lowest order in the electromagnetic coupling.

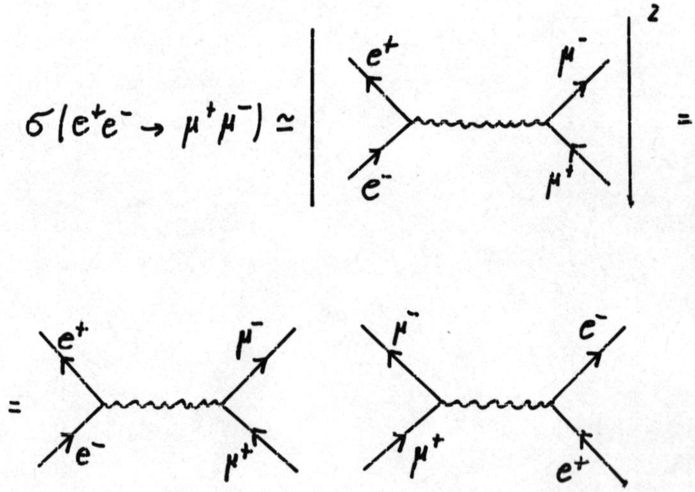

Fig. 7.2 $e^+e^- \to \mu^+\mu^-$ as a normalization to $e^+e^- \to$ HADRONS.

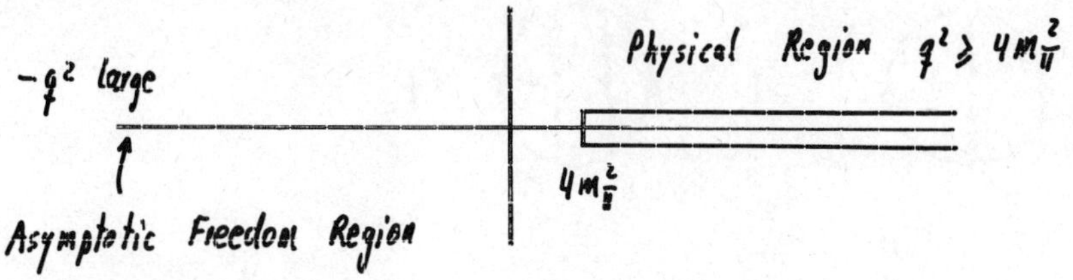

Fig. 7.3 The complex plane in the q^2 variable

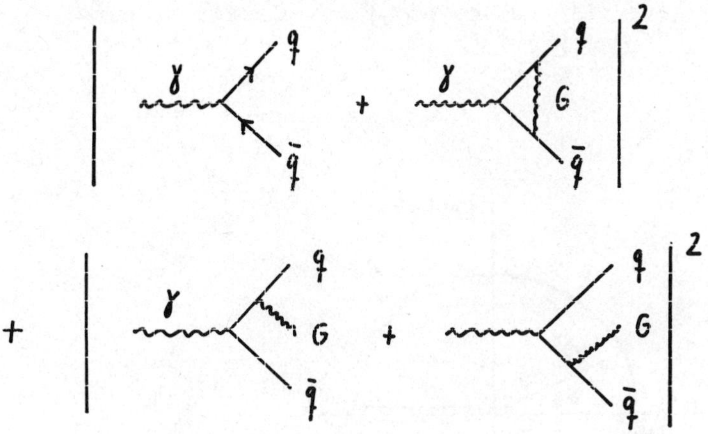

Fig. 7.4 Lower order Feynman diagrams relevant to the calculation of the function $F(\tau, \alpha_s, x_i)$ in eq. (7.15).

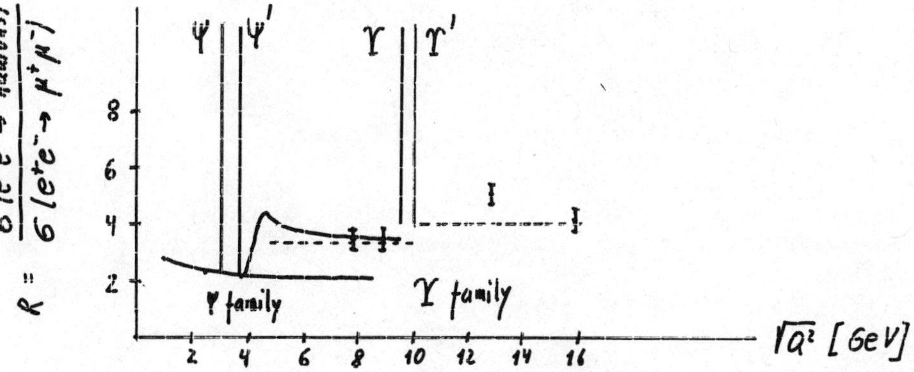

Fig. 7.5. QCD Prediction for the ratio R from the asymptotic result in eq. (7.16).

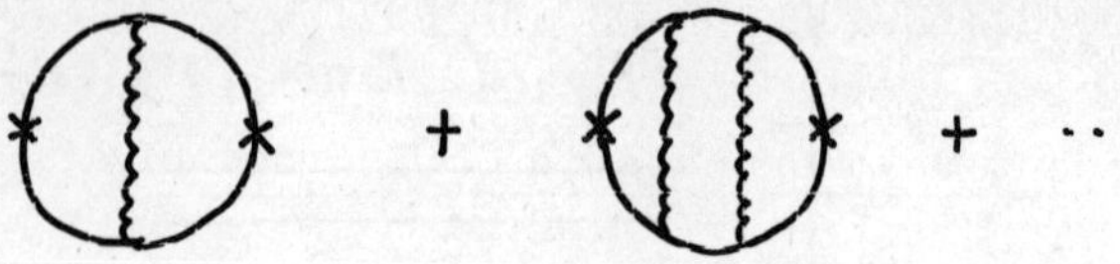

Fig. 7.6 Coulomb like gluon exchange diagrams which give rise to the $(\alpha_s/v)^n$ singularity in perturbation theory.

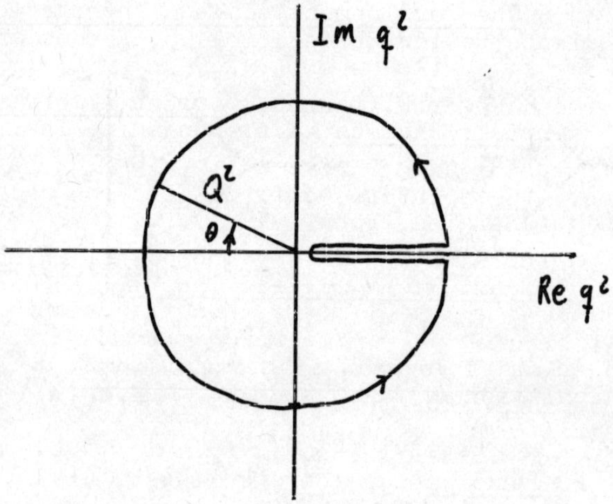

Fig. 7.7 Contour of integration in eq. (7.21a)

REFERENCES TO TEXTBOOKS, LECTURE NOTES AND REVIEW ARTICLES

1) T. Kinoshita, Recent developments in QED, Proc. of the High Energy Physics Conference, Tokyo, 1978.

2) E. Abers and B.W. Lee, Phys. Reports 9(1973)1.

3) M.A.B. Beg and A. Sirlin, Ann. Rev. Nucl. Sci. 24(1974)379.

4) S. Weinberg, Rev. Mod. Phys. 46(1974)255.

5) J.C. Taylor, Gauge Theories of Weak Interactions (Cambridge University Press, Cambridge, England) 1976.

6) H.D. Politzer, Phys. Reports 14c(1974)129.

7) W. Marciano and H. Pagels, Phys. Reports 36(1978)no.3.

8) M. Gell-Mann and Y. Neeman, The Eightfold Way: a review with a collection of reprints. N.A. Benjamin Inc., New York-Amsterdam (1964).

9) L. Montanet, (Experimental Review of Baryonium) Phenomenology of QCD Vol. 1 (Moriond 1978 - Haute Savoie France)

10) P.K. Kabir, editor, The Development of Weak Interaction Theory, International Science Review Series, vol. 5, Gordon and Breach (1963).

11) Stephen L. Adler and Roger F. Dashen, Current Algebras and Applications to particle Physics, W.A. Benjamin, Inc., New York-Amsterdam (1968).

12) M. Veltman, Proc. of 6th inter. Conf. on Electron and Photon Interactions at High Energies, Bonn, (1973).

13) Stephen L. Adler, Perturbation Theory Anomalies, in Lectures on Elementary Particles and Quantum Field Theory, 1970 Brandeis Lectures, Vol. 1, edited by J. Deser, M. Grisaru and H. Pendleton.

14) E. de Rafael, Lectures in Quantum Electrodynamics, Universidad Autónoma de Barcelona, UAB-FT-DI (1977).

15) H. Pagels, Phys. Reports 16(1975)no. 5.

16) B. W. Lee, Chiral Dynamics, Gordon and Breach, (1972).

17) Sidney Coleman, The Uses of Instantons, lectures delivered at the 1977 International School of Subnuclear Physics Ettore Majorana, HUTP- 78/004.

18) R.J. Crewther, Effects of Topological Charge in Gauge Theories, Schladming Lectures 1978; CERN preprint Th 2522.

19) J.R. Schriefer, Superconductivity, Benjamin, New York, 1964.

20) G. Leibrandt, Introduction to the technique of dimensional regularization, Rev. Mod . Phys. 47(1975)849.

21) E.T. Whittaker and G.N. Watson, A course of Modern Analysis, Cambridge University Press, 4th edition, 1927.

22) J. Dieudonne, Calcul Infinitésimal, Hermann, Paris 1968.

23) B.C. Carlson, Special Functions of Applied Mathematics, Academic Press, 1977.

24) G't Hooft and M. Veltman, Diagramar , CERN Yellow Report 73-9 (1973).

25) N.N. Bogoliubov and D.V. Shirkov, Introduction to the Theory of Quantized fields, Interscience, New York (1959).

26) R.J. Crewther, Asymptotic Behaviour in Quantum Field Theory, in Weak and Electromagnetic Interactions at High Energy, Cargèse 1975, M. Lévy, J.L. Basdevant, D. Speiser and R. Gastman, eds. (Plenum Press, New York, 1976).

27) G. Waliron, Equations Fonctionnclles Applications, Masson et cie, 1950.

28) D.J. Gross, Applications of the Renormalization Group, Les Houches Lectures 1975, published by North-Holland, eds. R. Balian and J. Zinn-Justin.

29) J. Ellis, Deep Hadronic Structure, Les Houches Lectures, 1976, published by North-Holland (1977), ed. by C.H. Llewellyn Smith.

30) J. Schwinger, Particles, Sources and Fields, Vol. II Chapter 5-4, Addison Wesley (1973).

31) E. de Rafael, Infrared structure and large - P_T behaviour of Quantum Chromodynamics, Proc. of the France-Japan Joint Seminar on the New Particles and Neutral Currents, Tokyo 1977, ed. K. Fujikawa, y. Hara; T. Mashawa, K. Terazawa.

32) C.T. Sachrajda, Parton model ideas and quantum chromodynamics, Proc. of the XIIIth rencontre de Moriond, 1978.

33) A. Peterman, Renormalization Group and the deep structure of the proton, CERN Preprint TH 2581, to appear in Physics Reports (1978).

34) V.A. Novikov, L.B. Okun, M.A. Shifman, A.I. Vainshtein, M.B. Voloshin and V.I. Zakharov, Phys. Rep. 41C(1978)no.1.

APPLICATIONS OF PERTURBATIVE QCD TO HARD SCATTERING PROCESSES

C.T. Sachrajda

CERN -- Geneva

ABSTRACT

We present a survey of the ideas and techniques necessary to apply QCD to hard scattering processes.

Lectures presented at the
GIFT School on "Quantum Chromodynamics"
Jaca, Spain, 4-8 June 1979

Geneva, 3 September 1979.

1. INTRODUCTION

QCD is emerging as the leading candidate for the theory of the strong interactions. In calculating predictions of QCD for physically measurable processes, it is hoped that, at least for hard scattering processes, i.e., processes in which there is a large momentum transfer, perturbation theory makes sense. Since the coupling constant in QCD decreases logarithmically with the value of the momentum transfer,

$$\alpha_s(Q^2) \equiv \frac{g^2(Q^2)}{4\pi} \sim \frac{1}{\log Q^2/\Lambda^2} \tag{1.1}$$

where Λ is a fixed parameter (to be determined from the data), if we can make the predictions in terms of a power series in $\alpha_s(Q^2)$, then after calculating just the first term (or the first few terms) in this series we are able to make predictions which are asymptotically meaningful and useful. There is a marked difference between QCD and QED or the standard model of weak interactions, however, since in the last cases the coupling constant is so small that for experimentally accessible values of the kinematic variables the loop integrations do not give large enough factors to spoil the validity of the perturbation series, each term in the series is still much smaller than the preceding one. In QCD however, each loop integration can give a factor of $\log Q^2$, so that we end up with a power series in $\alpha_s(Q^2) \log Q^2$ in which every term is of the same order of magnitude, and so we have to sum the series. The light-cone techniques, which use the operator product expansion and renormalization group, enable us to perform this summation in the case of deep inelastic lepton-hadron scattering. There are very many excellent reviews about this [see, for example, Refs 1)-7)], so that we will only outline very briefly the main features of this approach -- this we do in Section 2.

The corrections to the asymptotic predictions of QCD, for the behaviour of the deep inelastic structure functions with Q^2, are only suppressed by one (or more) powers of $\log Q^2$, and hence they may be relevant at present energies. Moreover, there are also new features expected in the data because of the presence of higher order corrections which would not be present if the whole answer were the lowest order result. These provide subtle tests of QCD. The "next-to-leading" order corrections have been calculated, and we review these calculations and their consequences in Section 3.

For other hard scattering processes the light cone techniques are not directly applicable (see, however, Refs 8) and 9)), so we have to find some other means to sum the series of "large logarithms", i.e., the series of the type

$$\sum_{i}^{\infty} a_i \left(\alpha_s(Q^2) \log Q^2 \right)^i \tag{1.2}$$

which emerge in perturbation theory. A careful study of the coefficients a_i shows that they are very simply related (the exact relation depends on the process) to the analogous coefficients in the calculation of the deep inelastic lepton-hadron scattering, we can use this knowledge to obtain predictions for other processes. This will be explained in detail in Section 4.

Before we knew how to apply QCD to the various hard scattering processes, the standard way to think about these processes was in terms of the parton model[10),11)]. The large momentum (or energy transfer) was assumed to take place between constituent partons of the particles involved (these partons were usually assumed to be the quarks), in what was called the "hard subprocesses". Of course, in general, not very much was known about the dynamics of quark interactions but, under the assumption that they are scale invariant, at least some "dimensional counting rules" which predicted[12),13)] the asymptotic behaviour of hard scattering cross-sections with the momentum transfer were derived. We give examples of two relevant processes, and the way they were theoretically discussed before QCD.

The first of these processes is the inclusive production of a massive lepton pair in hadronic collisions. The usual explanation of this process is due to Drell and Yan[14)], and is that a quark from one of the initial hadrons annihilates an antiquark from the other, the resulting massive photon then decays into the observed lepton pair (Fig. 1a). This is represented by the equation

$$\frac{d\sigma}{dQ^2} = \frac{4\pi\alpha^2}{3nQ^2} \sum_{\substack{\text{quark} \\ \text{flavours} \\ a}} Q_a^2 \int_0^1 dx_1\, dx_2\, \delta(x_1 x_2 - \tau)\, x_1 x_2$$

$$\cdot \left[G_{q^a/h_1}(x_1)\, G_{\bar{q}^a/h_2}(x_2) + G_{\bar{q}^a/h_1}(x_1)\, G_{q^a/h_2}(x_2) \right] \tag{1.3}$$

where n is the number of colours and $G_{q/h}(x)$ is the probability of finding a quark q in a hadron h with a fraction x of its longitudinal momentum. In QCD, however, in addition to the simple diagram of Fig. 1a, there exist diagrams such as those of Figs 1b and 1c, and naïvely it would seem that these diagrams would spoil the simple probabilistic picture of Eq. (1.3). These diagrams are of higher order in the

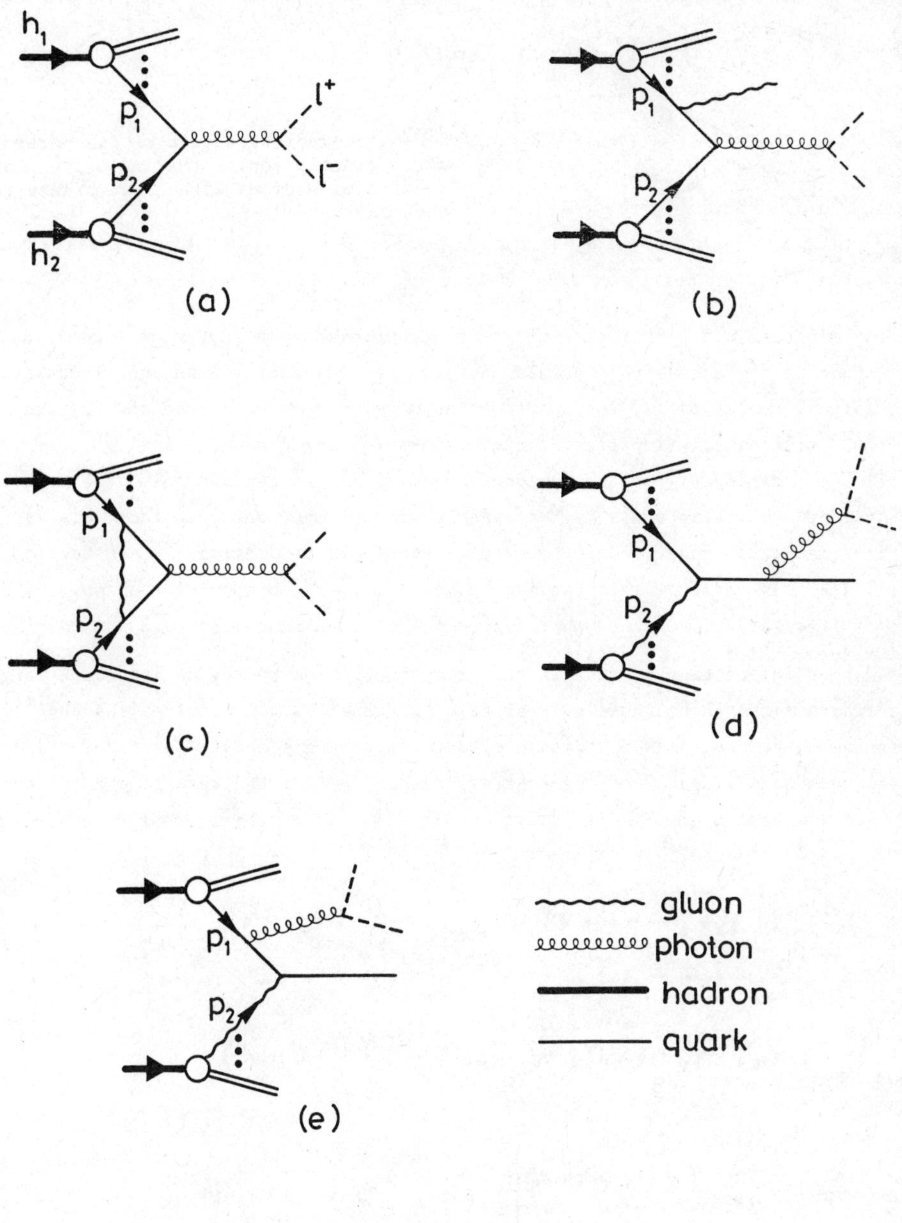

Fig. 1 Sample diagrams which contribute to massive lepton pair production in hadronic collisions

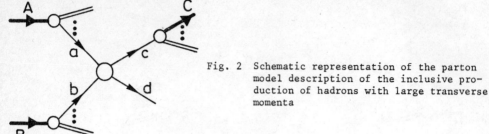

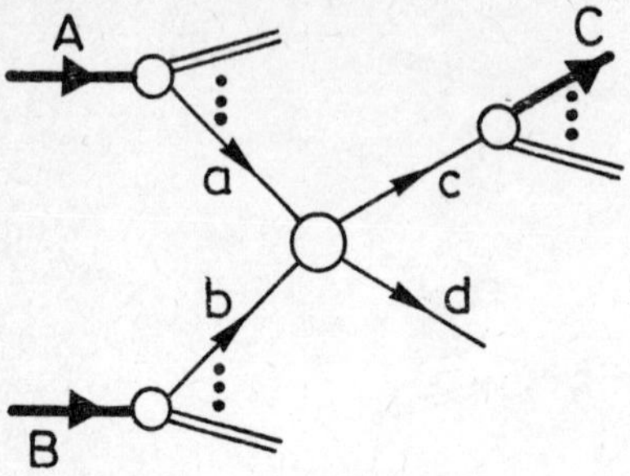

Fig. 2 Schematic representation of the parton model description of the inclusive production of hadrons with large transverse momenta

coupling constant but each factor of g is accompanied by a factor of log Q^2 and so they cannot be neglected. We can also ask what is so special about quark-antiquark annihilation. For example, the subprocesses $qg \to \ell^+\ell^- X$ *) (Figs 1d and 1e) and $qq \to \ell^+\ell^- X$ give equally important contributions ($g^2 \log Q^2$ and $g^4 \log^2 Q^2$, respectively). Moreover, for the subprocess $qq \to \ell^+\ell^- x$ we can use only valence quarks and therefore this subprocess may be expected to be important. In fact this was the first mechanism suggested for massive lepton pair production[15]. In Section 4 we will discover why the Drell-Yan model [Eq. (1.3)] is so special, and why the diagrams of Figs 1b-1e, or other subprocesses do not invalidate it.

Another interesting process is the production of particles or jets with large transverse momenta in hadronic collisions. The usual parton model philosophy[1),11)] concerning these processes is that an elastic large angle scattering takes place between two "constituents" of the initial hadrons, one of the resulting constituents then fragments, forming the trigger particle. This is represented by Fig. 2 and by

$$E_c \frac{d\sigma}{d^3p_c}(A+B\to C+X) = \sum_{a,b;c,d} \int_0^1 dx_a\, dx_b\, \frac{dx_c}{x_c^2} \cdot G_{a/A}(x_a)\, G_{b/B}(x_b)\, \tilde{G}_{C/c}(x_c) \cdot \delta(s'+t'+u') \cdot \frac{s'}{\pi} \frac{d\sigma}{dt'}(a+b\to c+d)\Big|_{\substack{s' = x_a x_b s \\ t' = x_a/x_c\, t\, ;\; u' = x_b/x_c\, u}} \tag{1.4}$$

where $\tilde{G}_{C/c}(x_c)$ is the probability of constituent c to fragment into particle C, C carrying a fraction x_c of the longitudinal momentum of c. Dimensional counting

*) Throughout these lectures $q(\bar{q})$ signifies a quark (antiquark), g signifies a gluon and ℓ a lepton.

rules have been derived from which the asymptotic behaviour of the cross-section corresponding to each subprocess (i.e., choice of a, b, c, d) can be calculated. We will see in Section 4 that within the context of QCD we can predict both the normalization and asymptotic behaviour of the cross-section.

From the studies of processes such as these two examples, we will conclude that there is a surprisingly simple ansatz with which one can adapt the predictions of the parton model to those of QCD. This ansatz is valid up to logarithmic corrections, and in Section 5 we will demonstrate how to calculate these corrections. An interesting problem which arises is whether one should take into account now the interactions which involve "spectator" partons. Naïvely, these interactions are only suppressed by one power of log Q^2 and therefore they should be included once we consider these corrections. We will demonstrate, however, that this is no so, these spectator interactions (which incidentally in any particular process are gauge dependent), are universal (the same in every process) and hence when comparing two processes, such as the Drell-Yan process and the quark distributions measured in deep inelastic scattering [as in Eq. (1.3)], they cancel out.

In the last few months, for the first time, some progress has been made in understanding the applications of QCD to exclusive processes, in particular to the form factors of hadrons at large momentum transfers, and, to a lesser extent so far, to elastic scattering at fixed angle. We review this progress in Section 6.

We should also make some comments about several interesting topics we will not have time to discuss in these lectures. One such topic is the classical physics of jets in e^+e^- annihilation experiments[16),17)]. A second topic we will not discuss concerns the interesting class of processes in which there are two large variables Q_1^2, Q_2^2 and the appropriate limit is Q_1^2, Q_2^2, and Q_1^2/Q_2^2 all $\to \infty$. An example of such a process is the production of lepton pairs at large transverse momentum, Q_1^2 is the mass squared of the lepton pair, and Q_2 its transverse momentum. For a review of this process we refer the reader to Ref. 18). We will also do very little actual phenomenology.

All of the preliminary material which we need has been covered by de Rafael[19)]. Since we will be encountering the coefficients in the perturbation series for the β function a lot, let me define our notation. We write

$$\beta(g) \equiv \frac{\partial}{\partial \ln \mu} g(\mu)$$

$$\equiv -\beta_0 \frac{g^3(\mu)}{16\pi^2} - \beta_1 \frac{g^5(\mu)}{(16\pi^2)^2} \qquad (1.5)$$

where $g(\mu)$ is the renormalized coupling constant, defined at the subtraction point μ. As $\mu^2 \to \infty$, the behaviour of the coupling constant can be deduced from Eq. (1.5) and is

$$g^2(\mu) = \frac{16\pi^2}{\beta_0} \frac{1}{\log \mu^2/\Lambda^2} \tag{1.6}$$

in agreement with (1.1).

We now proceed to discuss the applications of QCD. We advise the reader who is not familiar with concepts such as asymptotic freedom and the renormalization group to study the lectures of de Rafael[19)]. In these lectures he will also find the Feynman rules of QCD, and also a discussion of the techniques to evaluate the Feynman diagrams.

2. LIGHT CONE TECHNIQUES AND THE ASYMPTOTIC BEHAVIOUR OF DEEP INELASTIC STRUCTURE FUNCTIONS

2.1 General formalism

In this section we very briefly review the standard light cone analysis, using the operator product expansion and renormalization group, for deep inelastic lepton-hadron scattering. For details we refer the reader to Refs 1)-7). The kinematics of the process is defined in Fig. 3, and we are interested in the Bjorken limit

$$|q^2| \to \infty, \quad p.q \to \infty, \quad x \equiv \frac{-q^2}{2p.q} \text{ fixed} \tag{2.1}$$

In Fig. 3, ℓ may be any lepton. Since we believe we know the lepton-photon and lepton-weak boson couplings we can factor this vertex off, and all tests of QCD or any

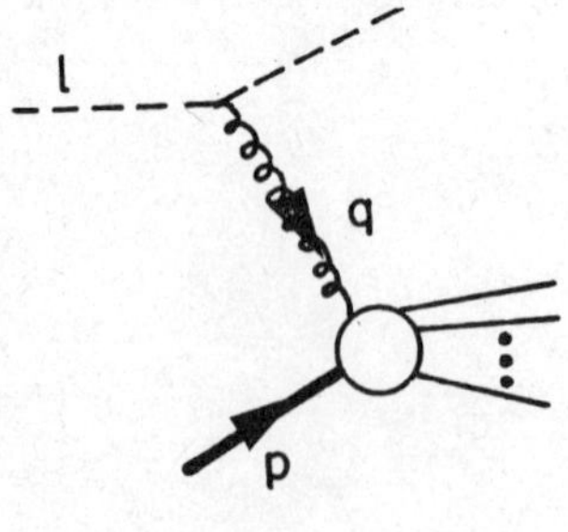

Fig. 3 Deep inelastic lepton-hadron scattering

other theory of the strong interactions can be done on the absorptive deep inelastic Compton amplitude $W_{\mu\nu}$, defined by*)

$$W_{\mu\nu}(p,q) = \frac{1}{4\pi}\int d^4x\, e^{iq\cdot x} \langle p|[J_\mu^+(x), J_\nu(0)]|p\rangle \quad (2.2)$$

We define the three structure functions W_1, W_2 and W_3 by

$$W_{\mu\nu}(p,q) = -\left(g_{\mu\nu} - \frac{q_\mu q_\nu}{q^2}\right) W_1(q^2,x) + \frac{1}{m_H^2}\left(p_\mu - q_\mu \frac{p\cdot q}{q^2}\right)\left(p_\nu - q_\nu \frac{p\cdot q}{q^2}\right) W_2(q^2,x) - i\,\frac{\varepsilon_{\mu\nu\alpha\beta}\, p^\alpha q^\beta}{2m_H^2} W_3(q^2,x) \quad (2.3)$$

where m_H is the mass of the hadronic target. By Lorentz and CP invariance and charge conservation we can only have these three structure functions, moreover by parity invariance $W_3 = 0$ for electron and muon scattering. In the parton model the assumption of Bjorken scaling was

$$\begin{aligned} W_1(x,q^2) &\xrightarrow[q^2\to-\infty]{} F_1(x) \\ \frac{p\cdot q}{m_H^2} W_2(x,q^2) &\xrightarrow[q^2\to-\infty]{} F_2(x) \\ \frac{p\cdot q}{m_H^2} W_3(x,q^2) &\xrightarrow[q^2\to-\infty]{} F_3(x) \end{aligned} \quad (2.4)$$

---

*) We give x here in two different meanings (i) Bjorken x defined in (2.1), and (ii) the symbol for the co-ordinate in configuration space (as in (2.2)). Both these are conventional uses, and we trust they will not get confused.

where the functions $F_i(x)$ are independent of q^2. These functions are just the sum over the quark distributions in the hadron, each quark being weighted by the square of its charge to the weak or electromagnetic current. Thus, measuring different structure functions (F_1 and F_3, say, in ν scattering), and also using different beams gives us considerable information about the quark distribution functions. We will also use the structure function F_L, defined by $F_L = F_2 - 2xF_1$.

We now study this process in QCD. The first relevant observation is that the Bjorken limit (2.1) corresponds to the light cone limit in configuration space, namely $x^2 \to 0$ [see, e.g., Ref. 2)]. Regions far away from the light cone give a rapidly oscillating exponential factor in the integrand in (2.2), and hence a negligible contribution to the integral.

We will now calculate the QCD prediction for the violation of Bjorken scaling. There are a number of steps in this calculation:

1) We notice that the forward elastic amplitude $T_{\mu\nu}$, ($W_{\mu\nu}$ is related to $T_{\mu\nu}$ by $W_{\mu\nu} = 1/2\pi \text{ Im } T_{\mu\nu}$) is defined by

$$T_{\mu\nu}(p,q) = i\int d^4x \; e^{iq\cdot x} \langle p | T(J_\mu^+(x) J_\nu(0)) | p\rangle \tag{2.5}$$

We now expand the product of the two currents as an infinite series of local operators, and take the Fourier transform giving us

$$\begin{aligned} & i\int d^4x \; e^{iq\cdot x} \, T(J_\mu^+(x) J_\nu(0)) = \\ & \sum_{N,i} \Big\{ -\left(g_{\mu\mu_1} g_{\nu\mu_2} q^2 - g_{\mu\mu_1} q_\nu q_{\mu_2} - g_{\nu\mu_2} q_\mu q_{\mu_1} + g_{\mu\nu} q_{\mu_1} q_{\mu_2}\right) \\ & \cdot \bar{F}_{2,i}\left(\frac{q^2}{\mu^2}, g^2\right) + \left(g_{\mu\nu} - \frac{q_\mu q_\nu}{q^2}\right) q_{\mu_1} q_{\mu_2} F_{L,i}^N\left(q^2/\mu^2, g^2\right) \\ & - i\,\varepsilon_{\mu\nu\alpha\beta}\, g_{\alpha\mu_1} q_\beta q_{\mu_2} F_{3,i}^N\left(q^2/\mu^2, g^2\right)\Big\} q_{\mu_3} \cdots q_{\mu_N} \left(\frac{2}{-q^2}\right)^N O_i^{\mu_1 \cdots \mu_N}(0) \end{aligned} \tag{2.6}$$

where i labels the different operators with N Lorentz indices. The zero argument of the operator indicates it is a local operator. Substituting (2.6) into (2.5) and working in the Bjorken limit (2.1), we readily find

$$T_{\mu\nu}(q^2,x) = \sum_{i,N} \left(\frac{1}{x}\right)^N \cdot$$

$$\Big\{ \left(g_{\mu\nu} - \frac{q_\mu q_\nu}{q^2}\right) \bar{F}^N_{L,i}\left(q^2/\mu^2, g^2\right) -$$

$$\left(g_{\mu\nu} - \frac{p_\mu q_\nu + p_\nu q_\mu}{p\cdot q} + \frac{p_\mu p_\nu}{(p\cdot q)^2} q^2\right) \bar{F}^N_{2,i}\left(q^2/\mu^2, g^2\right) - \tag{2.7}$$

$$i\,\varepsilon_{\mu\nu\alpha\beta} \frac{p_\alpha q_\beta}{p\cdot q} \bar{F}^N_{3,i}\left(q^2/\mu^2, g^2\right)\Big\} \langle p|O^N_i|p\rangle$$

where

$$\langle p|O_i^{\mu_1\ldots\mu_N}(0)|p\rangle \equiv p^{\mu_1}\ldots p^{\mu_N} \langle p|O^N|p\rangle \tag{2.8}$$

The reason why such an expansion is useful is that in the Bjorken limit for each N only three operators contribute to the right-hand side of (2.6) and (2.7). In this limit we would like the mass dimension of the $\bar{F}$'s to be as large as possible and therefore we want the $\langle p|O^N|p\rangle$ to have as small a mass dimension as possible. Going back to (2.6) this now means that we want $O^{\mu_1\ldots\mu_N}(0)$ to have as low a twist ($\equiv$ dimension - spin) as possible. Thus we look for the lowest twist operators with the correct quantum numbers and we find for each N three twist-2 operators

$$\bar{\psi}\gamma^{\mu_1}D^{\mu_2}\ldots D^{\mu_N}\psi\,\tau^k \quad - \text{ traces} \tag{2.9a}$$

$$\bar{\psi}\gamma^{\mu_1}D^{\mu_2}\ldots\ldots D^{\mu_N}\psi \quad - \text{ traces} \tag{2.9b}$$

and

$$G^{\mu_1\nu} D^{\mu_2} \ldots D^{\mu_N} G^{\mu_N\nu} \quad - \quad \text{traces} \tag{2.9c}$$

where

$$D^{\mu} = \partial^{\mu} - ig \frac{\lambda^a}{2} A_a^{\mu} \;;$$

and $\lambda^a/2$ is the generator of the colour SU(3) group and symmetrization over the indices $\mu_1 \ldots \mu_N$ is implied. $G^{\mu\nu}$ is the field strength tensor of QCD. τ^K is a flavour matrix. Hence the label i in (2.6) and (2.7) runs over these three operators. Moreover, if we choose a combination of structure functions which is a flavour non-singlet the most obvious example is the difference $F_2^{ep} - F_2^{en}$, where p(n) stands for proton (neutron); however, $F_3^{\nu A}$, where A is an isoscalar target, is also dominated only by the operator (2.9a) , then only the non-singlet (NS) operator (2.9a) contributes.

2) We now use a dispersion relation to relate $T_{\mu\nu}$ and its imaginary part $W_{\mu\nu}$. We could write down the normal dispersion relation in $s \equiv (p + q)^2$, but we prefer to change variables and write it in x. The normal threshold cuts in s and u now correspond to a cut in x from -1 to 1. The precise form of the dispersion relation depends on which structure function we look at, for scalar currents this relation is

$$\begin{aligned} T(q^2,x) &= \int_{-1}^{1} x'^{\,s-1} dx' \frac{W(q^2,x')}{x^{s-1}(x'-x)} + \text{subtractions} \\ &= \sum_{N=s}^{\infty} \left(\frac{1}{x}\right)^N \int_{-1}^{1} dx'\, x'^{\,N-1}\, W(q^2,x') + \text{subtractions} \end{aligned} \tag{2.10}$$

where s is the number of subtractions. Using the crossing properties of W under $x \to -x$ we can write

$$\int_0^1 dx'\, x'^{\,N-1}\, W(q^2,x') = \sum_i \bar{F}_i^N(q^2) \langle p|O_i^N|p\rangle \tag{2.11}$$

For our case of spin 1 currents the analogous equations are

$$\int_0^1 dx\, x^{N-2} F_\alpha(x,q^2) = \sum_i \bar{F}_{\alpha,i}^N (q^2/\mu^2, g^2) \langle p|O_i^N|p\rangle \qquad \alpha = 2, L \tag{2.12a}$$

and

$$\int_0^1 dx\, x^{N-1} F_3(x,q^2) = \sum_i \bar{F}^N_{3,i}(q^2/\mu^2, g^2) \langle p|O^N_i|p\rangle \tag{2.12b}$$

3) We now calculate the q^2 behaviour of the Wilson coefficient functions $\bar{F}$, thus calculating the violation of Bjorken scaling in the moments of the structure functions. Since the moments of νW_2 for example, which we call M_2^N, are clearly physically measurable quantities, they must be independent of μ, the renormalization point. Thus

$$\mu \frac{d}{d\mu} M_2^N(q^2) = 0 \tag{2.13}$$

As an example, let us take the non-singlet combination of structure functions $F_2^{ep} - F_2^{en}$, then there is only one term in the summation on the right-hand side of (2.12a). Thus the μ dependence of $\bar{F}$ must be compensated for by the μ dependence of $\langle p|O_i^N|p\rangle$,

$$\mu \frac{d}{d\mu} \bar{F}_2^N \cdot O^N = 0 \tag{2.14}$$

The relevant operator, which is given in (2.9a), is multiplicatively renormalizable

$$O^N \equiv Z_{O^N} \cdot O^{N,u} \tag{2.15}$$

where the label u signifies the bare (unrenormalized) operator. Equation (2.14) can now be rewritten

$$\mu \frac{d}{d\mu} \bar{F}_2^N = \bar{F}_2^N \mu \frac{d}{d\mu} (\ln Z_{O^N})$$

$$\equiv \bar{F}_2^N \gamma_{O^N} \tag{2.16}$$

where γ_{0N} is called the anomalous dimension of the operator 0^N. Hence we get

$$\left[\mu \frac{d}{d\mu} - \gamma_{0^N}\right] \bar{F}_2^N = 0 \tag{2.17}$$

from which, neglecting mass terms and working in the Landau gauge (in this gauge the gauge parameter does not get renormalized), we immediately get

$$\left[\mu \frac{\partial}{\partial\mu} + \beta(g) \frac{\partial}{\partial g} - \gamma_{0^N}\right] \bar{F}_2^N (q^2/\mu^2, g) = 0 \tag{2.18a}$$

or

$$\left[2 \frac{\partial}{\partial \ln q^2} - \beta(g) \frac{\partial}{\partial g} + \gamma_{0^N}\right] \bar{F}_2^N (q^2/\mu^2, g) = 0 \tag{2.18b}$$

since, because $F(q^2/\mu^2, g)$ is dimensionless and only depends on q^2 and μ^2, it is a function of q^2/μ^2. Thus we have a differential equation for the moments of a structure function, whose solution is

$$\bar{F}^N (q^2, g(\mu^2)) = \bar{F}^N(q^2{=}\mu^2, g(q^2)) \exp\left[-\int_{g(\mu^2)}^{g(q^2)} \frac{\gamma_{0^N}(g')}{\beta(g')} dg'\right] \tag{2.19}$$

$\gamma_{0N}(g')$ is a calculable power series in g'; we have to calculate the μ dependence of Z_{0N}, which we can do order by order in perturbation theory, so we can write

$$\gamma_{0^N}(g) = \gamma_0^N \frac{g^2}{16\pi^2} + O(g^4) \tag{2.20}$$

The asymptotic prediction is obtained from (2.19) by keeping only the first term in γ_{0N} and β, we readily obtain

$$\frac{M_2^N(q^2)}{M_2^N(q_0^2)} = \left[\frac{g^2(q^2)}{g^2(q_0^2)}\right]^{\gamma_0^N/2\beta_0} \tag{2.21}$$

which, combined with (1.6), proves that Bjorken scaling is violated logarithmically in QCD. Scale breaking of the type predicted in (2.21) seems to agree fairly well with the data[20)-22)]. Higher-order terms in γ and β give contributions to the right-hand side of (2.21) which are logarithmically suppressed relative to the lowest order term.

Equation (2.19) gives the q^2 behaviour of the moments of flavour non-singlet combinations of structure functions. Flavour singlet combinations of structure functions are dominated by the two operators (2.9b) and (2.9c) which mix under renormalization. Thus the anomalous dimension is now a matrix, and the equation analogous to (2.18b) is a matrix equation. A compact way to write the solution for moments of singlet combinations of structure functions is

$$M^N(q^2) = A^N\left(g^2(q^2)\right)^{\gamma_+^N/2\beta_0} + B^N\left(g^2(q^2)\right)^{\gamma_-^N/2\beta_0} \tag{2.22}$$

where γ_+^N, γ_-^N are the two eigenvalues of the anomalous dimension matrix. A^N and B^N are unpredicted constants.

In general a structure function such as F_2^{ep} is a combination of singlet and non-singlet terms so that its moments are given by

$$M^N(q^2) = A^N\left(g^2(q^2)\right)^{\gamma_+^N/2\beta_0} + B^N\left(g^2(q^2)\right)^{\gamma_-^N/2\beta_0} + C^N\left(g^2(q^2)\right)^{\gamma_0^N/2\beta_0} \tag{2.23}$$

and for each moment we have three constants A^N, B^N and C^N to determine from the data.

A word now about the calculation of the anomalous dimensions of the operators. These calculations are very similar to those of the β function (see Ref. 19)) and in the dimensional regularization scheme we find[23)]

$$\gamma_{0^N} = -g^2\,\frac{\partial Z_1}{\partial g^2} \tag{2.24}$$

where Z_1 is the coefficient of $1/\varepsilon$ ($\varepsilon = 4-n$, n is the number of dimensions in which the integrals are performed). The diagrams which have to be calculated in one loop are drawn in Fig. 4. We notice that because of the covariant derivatives in (2.9a) at the operator vertex for the non-singlet operator, for example, we can have two quark external legs, or two quarks and one gluon, and so on. The result of the calculation is[24)]

$$\gamma_0^N = \frac{8}{3}\left[1 - \frac{2}{N(N+1)} + 4\sum_{j=2}^{N}\frac{1}{j}\right] \tag{2.25}$$

for the non-singlet case, and is

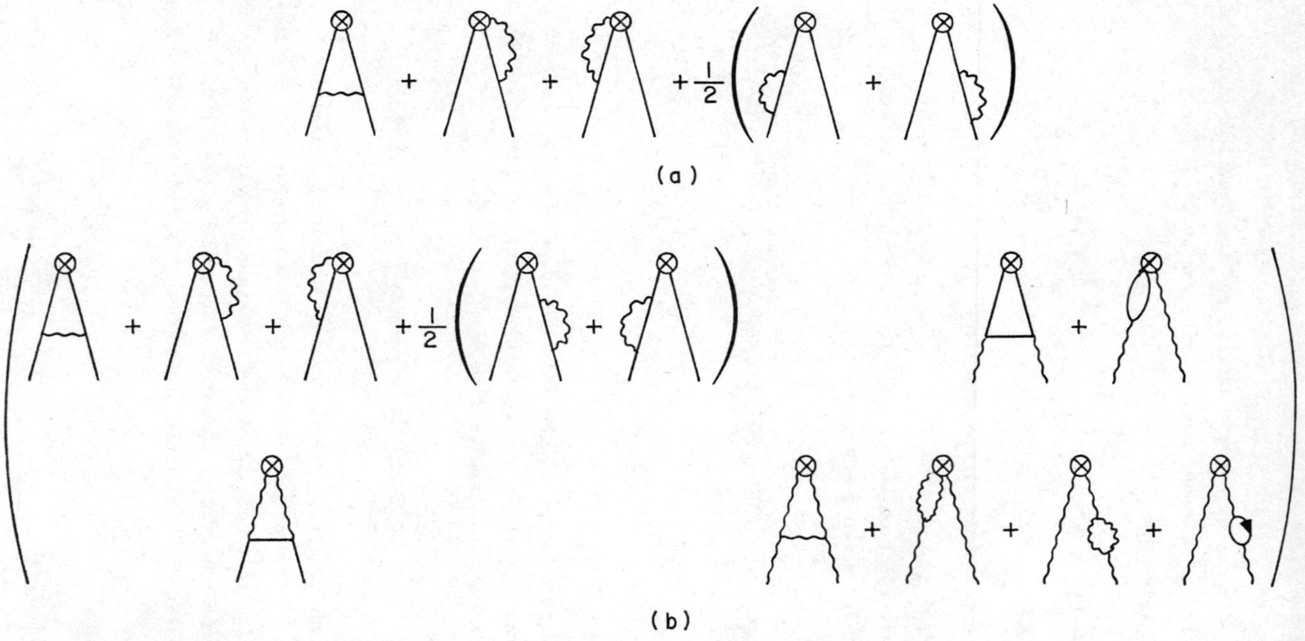

Fig. 4 One-loop diagrams contributing to the anomalous dimension of the twist-two (a) non-singlet and (b) singlet operators

$$\begin{pmatrix} \frac{8}{3}\left[1-\frac{2}{N(N+1)}+4\sum_{j=2}^{N}\frac{1}{j}\right] & -4f\,\frac{N^2+N+2}{N(N+1)(N+2)} \\ -\frac{16}{3}\,\frac{N^2+N+2}{N(N^2-1)} & 6\left[\frac{1}{3}-\frac{4}{N(N-1)}-\frac{4}{(N+1)(N+2)}+4\sum_{j=2}^{N}\frac{1}{j}\right]+\frac{4}{3}f \end{pmatrix} \tag{2.26}$$

for the singlet case. This completes our brief discussion of the calculation of the asymptotic behaviour of deep inelastic structure functions in QCD.

2.2 Probabilistic interpretation of the results

An extremely interesting and physically appealing interpretation of the above results was given by Altarelli and Parisi[25]. We start by considering non-singlet combinations of structure functions, later we will return to the singlet case. We write Eq. (2.21) in terms of the variable $t \equiv \ln q^2/q_0^2$ as

$$M^N(t) = M^N(0)\left[\frac{\alpha_s(t)}{\alpha_s(0)}\right]^{\gamma_0^N/2\beta_0} \tag{2.27}$$

which satisfies the following equation

$$\frac{dM^N(t)}{dt} = -\frac{\alpha(t)}{8\pi}\,\gamma_0^N\,M^N(t) \tag{2.28}$$

Since the moment of a convolution of two functions is the product of the moments of the functions, if we can find a function whose moments are γ_0^N, we can invert Eq. (2.28). Thus we define a function $p_{q\to q}(z)$, such that

$$\int_0^1 dz\; z^{N-1}\,P_{q\to q}(z) = -\frac{\gamma_0^N}{4} \tag{2.29}$$

and then inverting (2.28) we get

$$\frac{d\tilde{q}(x,t)}{dt} = \frac{\alpha(t)}{2\pi}\int_x^1 \frac{dy}{y}\,\tilde{q}(y,t)\,P_{q\to q}\!\left(\frac{x}{y}\right) \tag{2.30}$$

where $\tilde{q} \equiv q - \bar{q}$ which is a non-singlet combination of structure functions. Eq. (2.30) is the "evolution equation" for $\tilde{q}$. $p_{q\to q}(x)$ is the variation/unit t of the probability density of finding a quark in a quark with fraction x/y of its momentum. In its infinitesimal form Eq. (2.30) is

$$\tilde{q}(x,t) + d\tilde{q}(x,t) = \int_0^1 dy \int_0^1 dz \; \delta(zy-x)\; \tilde{q}(y,t) \cdot \left[\delta(z-1) + \frac{\alpha(t)}{2\pi} P_{q\to q}(z)\, dt\right] \qquad (2.31)$$

so that to calculate $p_{q\to q}(x)$ directly we have to calculate the coefficient of $\log q^2$ in the one-loop diagrams for deep inelastic scattering on a quark. This function must now satisfy (2.29), where the γ_0^N are given in (2.25). The second term in the square brackets is there because in QCD, quarks can radiate gluons (with non-zero energy), see Fig. 5a.

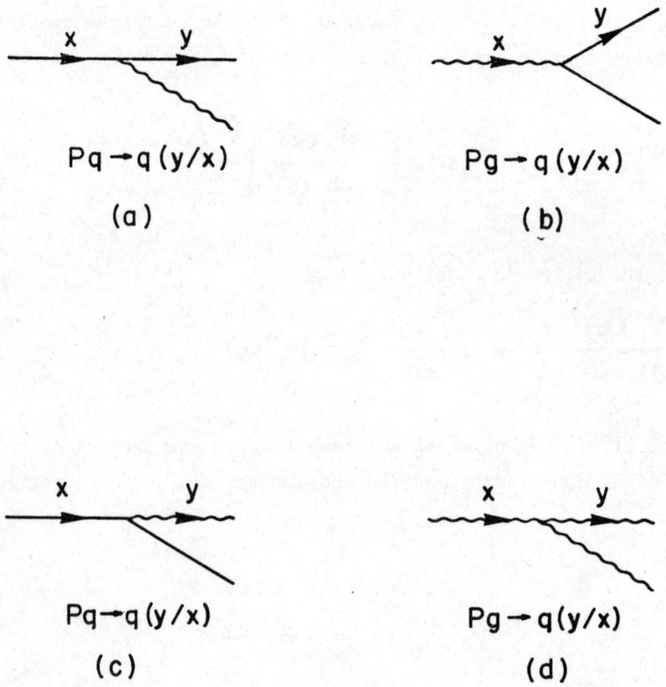

Fig. 5 Vertices which define the Altarelli-Parisi probability functions

We now study the singlet combinations of structure functions. We define Q^s to be the flavour singlet quark distribution

$$Q^s(x,q^2) = \sum_{i=1}^{f} \left[q^i(x,q^2) + \bar{q}^i(x,q^2) \right] \tag{2.32}$$

and the gluon distribution is denoted by $G(x,q^2)$. We now get a coupled set of evolution equations:

$$\frac{dQ^s(x,t)}{dt} = \frac{\alpha_s(t)}{2\pi} \int_x^1 \frac{dy}{y} \left[Q^s(y,t)\, P_{q\to q}\left(\frac{x}{y}\right) + G(y,t)\, P_{g\to q}\left(\frac{x}{y}\right) \right] \tag{2.33a}$$

$$\frac{dG(x,t)}{dt} = \frac{\alpha_s(t)}{2\pi} \int_x^1 \frac{dy}{y} \left[Q^s(y,t)\, P_{q\to g}\left(\frac{x}{y}\right) + G(y,t)\, P_{g\to g}\left(\frac{x}{y}\right) \right] \tag{2.33b}$$

The Q^s distribution varies with t now not only because of gluon bremsstrahlung as in the non-singlet case, but also because gluons can convert into $q\bar{q}$ pairs (Fig. 5b) in a flavour independent way, so that this variation of Q^s now depends also on the gluon distribution G. Similarly the variation of G with t depends on Q^s (because of bremsstrahlung Fig. 5c) and G (because gluons can also radiate gluons, Fig. 5d). The moments of $p_{q\to q}$, $p_{g\to q}$, $p_{q\to g}$ and $p_{g\to g}$ are, up to a factor of -4 as in (2.29), equal to the entries in the anomalous dimension matrix (2.26).

Thus we have a nice probabilistic interpretation of the violations of Bjorken scaling.

3. HIGHER ORDER CORRECTIONS TO DEEP INELASTIC STRUCTURE FUNCTIONS

3.1 Introduction

Once Λ is determined, using Fig. 6, one can deduce the value of the running coupling constant by using Eq. (1.6), and it turns out that

$$\alpha_s(q^2 = 10\,\text{GeV}^2) \simeq 0.2 \text{ to } 0.4$$

Thus α_s turns out to be fairly large at accessible values of Q^2, and, therefore, it becomes important to check that all the predictions of Section 2 are not invalidated by large higher order corrections. A second motivation for the calculation of these corrections is that they may be expected to provide additional and more precise tests of QCD. The next-to-leading contributions have now been calculated[23),26)] and in this section we will outline the calculation and discuss the consequences of the results.

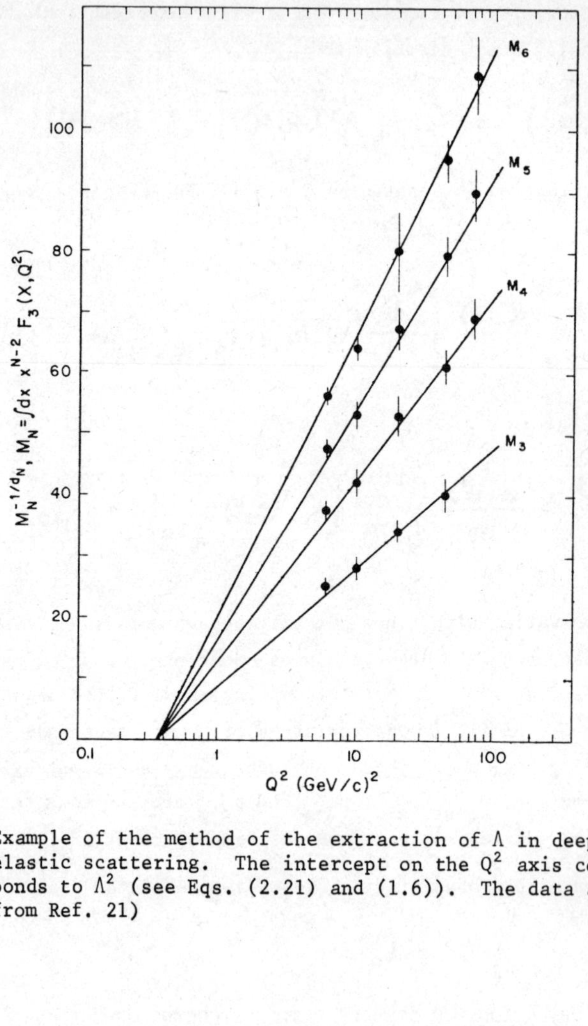

Fig. 6 Example of the method of the extraction of Λ in deep inelastic scattering. The intercept on the Q^2 axis corresponds to Λ^2 (see Eqs. (2.21) and (1.6)). The data are from Ref. 21)

In QED higher order corrections to various measurable quantities have been calculated for several decades now, so why do we consider it necessary to outline the calculations of the corrections in QCD? The answer is that whereas in QED, when we calculated the contributions from higher order terms in the perturbative expansion, we evaluate just the Feynman diagrams of the corresponding order in the coupling constant, in QCD at each stage we must include new contributions from every order of perturbation theory (this will be explained in detail in Section 4), and to sum these contributions. This summation is done most conveniently by using the operator product expansion and the renormalization group. Thus, although the techniques used in the evaluation of individual Feynman diagrams are the same as in QED, the quantities which have to be evaluated are somewhat different.

3.2 Calculation of the ingredients necessary to make predictions which include the higher order corrections

For simplicity we restrict ourselves here to the case of "non-singlet" combinations of structure functions (such as $F_2^{\mu p} - F_2^{\mu n}$ or $F_3^{\nu A}$, where A is an isoscalar target). The generalization to the singlet case is conceptually straightforward and will be discussed below. For the Nth moment of these non-singlet contributions, $M_i^N(q^2)$, we have seen in Section 2 that we can write

$$M_i^N(q^2) = \bar{F}_i^N(1, g(q^2)) \exp \int_{g(\mu^2)}^{g(q^2)} \frac{\gamma_{O^N}(g')}{\beta(g')} dg' \langle p|O^N|p\rangle \tag{3.1}$$

where γ_{O^N}, β and $\bar{F}_i^N$ are calculable power series in their arguments and $\langle p|O^N|p\rangle$ is independent of q^2. We write

$$\gamma_{O^N}(g) = \gamma_0^N \frac{g^2}{16\pi^2} + \gamma_1^N \frac{g^4}{(16\pi^2)^2} + \cdots \tag{3.2}$$

$$\beta(g) = -\beta_0 \frac{g^3}{16\pi^2} - \beta_1 \frac{g^5}{(16\pi^2)^2} + \cdots \tag{3.3}$$

and

$$\bar{F}_i^N(1, g) = 1 + \varepsilon_i^N \frac{g^2}{16\pi^2} + \cdots \tag{3.4}$$

From (3.1) to (3.4) we can see that as $q^2 \to \infty$, the behaviour with q^2 is predicted to be

$$\begin{aligned} M_i^N(q^2) &= \text{Const} \cdot [\alpha(q^2)]^{\gamma_0^N/2\beta_0} \\ &= \text{Const} \left[\frac{1}{\log q^2/\Lambda^2}\right]^{\gamma_0^N/2\beta_0} \end{aligned} \tag{3.5}$$

which is equivalent to Eq. (2.21). There are logarithmic corrections to (3.5), the prediction including the next-to-leading corrections can be written

$$M_i^N(q^2) = \text{const}\,\left[\alpha(q^2)\right]^{\gamma_0^N/2\beta_0}\left[1+\frac{\alpha(q^2)}{4\pi}\cdot a_i^N\right] \tag{3.6}$$

where

$$a^N = \frac{\gamma_1^N}{2\beta_0} - \frac{\beta_1\gamma_0^N}{2\beta_0^2} + \varepsilon^N \tag{3.7}$$

Thus, in order to make predictions which include the effects of the "next-to-leading" terms, we must know the three new quantities β_1, γ_1^N and ε^N. We now outline the calculation of these quantities.

a) β_1

β_1 was first calculated by Caswell[27)] and by Jones[28)] who found

$$\beta_1 = 102 - \frac{38}{3}f \tag{3.8}$$

where f is the number of flavours. To calculate β_1 it is convenient to regulate the ultra-violet divergences by dimensional regularization and to perform the renormalization of the coupling constant by using the "minimal subtraction" scheme of 't Hooft and Veltman[29)] (hereafter this scheme will be called the MS scheme). We recall that in this scheme[2)]

$$\beta(g) = \frac{\varepsilon}{2}g + g^3\frac{\partial a_1}{\partial g^2} \tag{3.9}$$

where

$$\varepsilon = 4-n \tag{3.10}$$

and n is the number of dimensions in which the integrals are performed. a_1 is defined by

$$Z_g = 1+\sum_{i=1}^{\infty}\frac{a_i}{\varepsilon^i} \tag{3.11}$$

where Z_g is the renormalization constant of the coupling constant. Thus the calculation of β_1 involves the calculation of the coefficient of $1/\varepsilon$ (i.e., the subleading ultra-violet divergence) in all the two loop diagrams which renormalize the coupling constant (for example, those of Fig. 7). Since the MS scheme satisfies all the Taylor-Slavnov identities, the Z_g for the three gluon coupling and for the quark-quark-gluon coupling are equal.

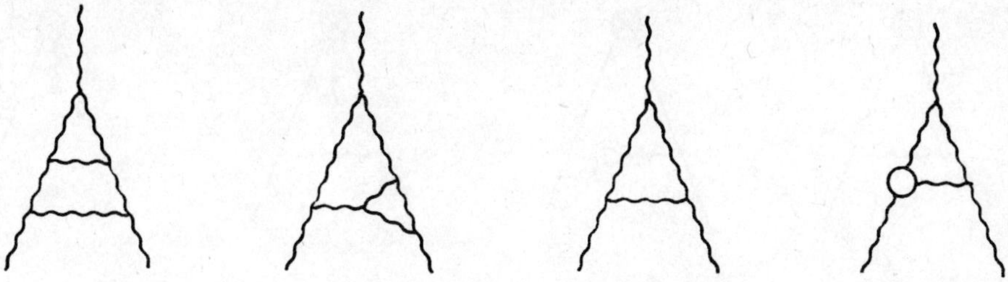

Fig. 7 Sample two-loop diagrams contributing to the renormalization constant of the coupling constant

Once we know β_1, we can find a more accurate solution of (1.5) than that given in Eq. (1.6); we now find

$$\frac{\alpha_s(q^2)}{4\pi} = \frac{1}{\beta_0 \ln q^2/\Lambda^2} - \frac{\beta_1}{\beta_0^3} \frac{\ln\ln q^2/\Lambda^2}{\ln^2 q^2/\Lambda^2} \tag{3.12}$$

where Λ has been chosen so that there are no terms of order $1/(\ln^2 q^2/\Lambda^2)$.

b) γ_1^N

The calculation of γ_1^N [23], the two loop contribution to the anomalous dimensions of the twist-two non-singlet operators (the operators are given by Eq. (2.9)), has many similarities with that of β_1. We recall that

$$\gamma^N = \mu \frac{\partial}{\partial\mu} \ln Z_{O^N} \tag{3.13}$$

where Z_{O^N} is the renormalization constant of the operator, and on the right-hand side all bare quantities are kept fixed. In the MS scheme[23]

$$\gamma^N = -g^2 \frac{\partial Z_1^N}{\partial g^2} \tag{3.14}$$

where Z_1^N is the coefficient of $1/\varepsilon$ in Z_{O^N}. Thus, to obtain γ_1^N we must calculate the coefficient of $1/\varepsilon$ (subleading ultra-violet divergence) in all the two loop diagrams which renormalize O^N (e.g., those of Fig. 8). The Feynman rules for the vertices are given in Refs. 23) and 26). The results of the calculation of γ_1^N are presented diagram by diagram (in the Feynman gauge)[23], and these results have recently been rewritten in the relatively compact form[30]

Fig. 8 Sample two-loop diagrams contributing to the anomalous dimension of the twist-two non-singlet operators

$$
\begin{aligned}
\gamma_1^N = & \left(C_F^2 - \frac{C_F C_A}{2}\right)\Big\{ 16\, S_1(N) \frac{2N+1}{N^2(N+1)^2} + 16\left[2S_1(N) - \frac{1}{N(N+1)}\right] \cdot \\
& \cdot \left[S_2(N) - S_2'\left(\frac{N}{2}\right)\right] + 64\, \tilde{S}(N) + 24\, S_2(N) - 3 - 8\, S_3'\left(\frac{N}{2}\right) \\
& - 8 \frac{3N^3+N^2-1}{N^3(N+1)^3} - 16(-1)^N \frac{2N^2+2N+1}{N^3(N+1)^3}\Big\} + \\
& C_F C_A \Big\{ S_1(N)\left[\frac{536}{9} + 8\frac{2N+1}{N^2(N+1)^2}\right] - 16\, S_1(N)\, S_2(N) - \frac{43}{6} \\
& + S_2(N)\left[-\frac{52}{3} + \frac{8}{N(N+1)}\right] - 4\,\frac{151N^4+263N^3+97N^2+3N+9}{9N^3(N+1)^3}\Big\} \\
& + C_F T_R \Big\{ -\frac{160}{9} S_1(N) + \frac{32}{3} S_2(N) + \frac{4}{3} + 16\,\frac{11N^2+5N-3}{9N^2(N+1)^2}\Big\}
\end{aligned}
\qquad (3.15)
$$

where

$$S_i(N) = \sum_{j=1}^{N} \frac{1}{j^i} \tag{3.16}$$

$$\tilde{S}(N) = \sum_{j=1}^{N} \frac{(-1)^j S_1(j)}{j^2} \tag{3.17}$$

and

$$S_i'\left(\frac{N}{2}\right) = \frac{1+(-1)^N}{2} S_i\left(\frac{N}{2}\right) + \frac{1-(-1)^N}{2} S_i\left(\frac{N-1}{2}\right) \tag{3.18}$$

C_F, C_A are the eigenvalues of the quadratic Casimir operators for the fundamental and adjoint representations (they are equal to 4/3 and 3 for QCD) and T_R is equal to half the number of flavours.

The above calculation was performed in the MS scheme, and it should be remembered that γ_1^N depends on the prescription used to renormalize the operators[23]. To see this let us define the two schemes:

Scheme (a):

$$Z_{O^N,a} = 1 - \frac{1}{2} \frac{g^2}{16\pi^2} \gamma_0^N \ln P^2/\mu^2 + O(g^4) \tag{3.19a}$$

and scheme (b):

$$Z_{O^N,b} = 1 - \frac{g^2}{16\pi^2} \left(\alpha + \frac{1}{2} \gamma_0^N \ln P^2/\mu^2\right) + O(g^4) \tag{3.19b}$$

where we have chosen μ^2 to be the point at which the coupling constant is defined. Recalling (3.13) we find that

$$\gamma_{1,a}^N - \gamma_{1,b}^N = -2\alpha\beta_0 \tag{3.20}$$

Of course, physics must be independent of how we choose to renormalize the operators, i.e., the a^N of Eqs. (3.6) and (3.7) must be renormalization prescription independent. We shall see below that this is indeed the case.

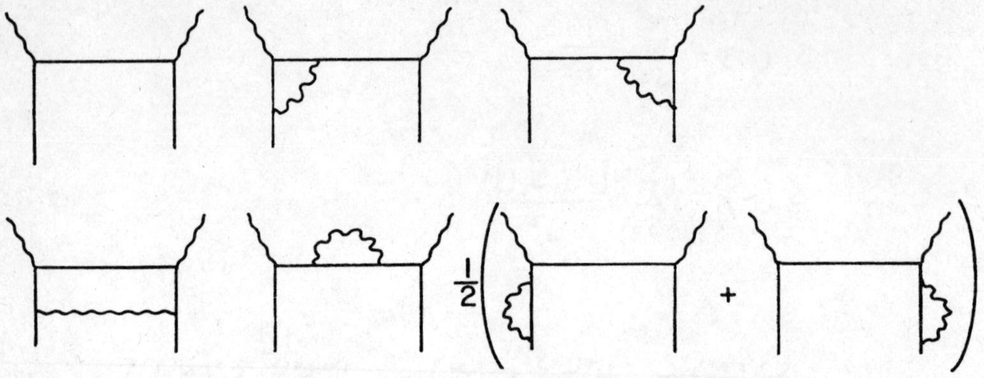

Fig. 9 Diagrams which have to be evaluated in the calculation of the Wilson coefficient functions

c) ε_i^N 23),31)

To calculate the ε_i^N we use the simple fact that the $\bar{F}_i^N$ are functions which depend only on the currents which we are expanding in terms of local operators, and are therefore independent of the states between which we sandwich the currents. Thus, in order to calculate the $\bar{F}_i^N$ we choose the most convenient states, in this case, quark states of momentum p. We consider deep inelastic scattering on a quark of momentum p, (for which Eq. (3.1) is valid) and expand both sides of (3.1) in terms of $g(\mu^2)$ (which we write below as g). For the left-hand side we calculate all the appropriate tree and one loop diagrams (Fig. 9) and for the right-hand side we use Eqs. (3.2) to (3.4) and also Eq. (1.5) in the form

$$g^2(q^2) = g^2 - \frac{g^4}{16\pi^2}\beta_0 \ln q^2/\mu^2 + O(g^6) \tag{3.21}$$

We then find that Eq. (3.1) can be written

$$1 - \frac{g^2}{16\pi^2}\left(\frac{1}{2}\gamma_0^N \ln q^2/p^2 - \delta_i^N\right)$$

$$= \left(1 + \varepsilon_i^N \frac{g^2}{16\pi^2}\right)\left(1 - \frac{1}{2}\frac{g^2}{16\pi^2}\gamma_0^N \ln q^2/\mu^2\right) \cdot \tag{3.22}$$

$$\left(1 + \frac{1}{2}\frac{g^2}{16\pi^2}\gamma_0^N \ln p^2/\mu^2 + \frac{g^2}{16\pi^2}\delta^N\right)$$

where the three factors on the right-hand side correspond to $\bar{F}_i^N$, $\exp \int_g^{g(q^2)} (\gamma_0 N/\beta) dg'$ and $\langle p|O^N|p\rangle$ respectively. The only unknown quantity in (3.22) is ε_i^N and therefore we can obtain it from this equation.

As we have already mentioned above, δ^N depends on the way we choose to renormalize the operators, and hence, physics (in this case the q^2 behaviour of the moments i.e., a_i^N) should be independent of δ^N. δ^N, as we have seen in Eq. (3.20) corresponds to a contribution

$$2\delta^N \beta_0 \tag{3.23}$$

to γ_1^N, and to a contribution of

$$-\delta^N \tag{3.24}$$

to ε_i^N. Hence, from Eq. (3.7) we see that, indeed, the a_i^N are independent of δ^N. Nevertheless, it is crucial to remember, that if we calculate the γ_1^N in a particular scheme (like the MS scheme we have been using), then we must use the appropriate δ^N to extract the ε_i^N.

3.3 Results and an ambiguity

The results of the calculation are presented in detail in Ref. 32). Here, in Table 3.1, we only present the results for a_i^N for $i = \nu W_2$ (since we are still considering non-singlet combinations of structure functions, it is really $\nu W_2^{\mu p} - \nu W_2^{\mu n}$) and taking there to be four flavours.

Table 3.1

Results of the calculation of a_i^N (defined in Eq. (3.6)) for the νW_2 structure function. The calculation was performed in the minimal subtraction renormalization scheme.

N	a_i^N	N	a_i^N
1	0	6	30.73
2	9.045	7	34.75
3	16.02	8	38.22
4	21.75	9	41.38
5	26.64	10	44.29

The relative size of the corrections to the leading terms are the entries of Table 3.1 multiplied by $\alpha_s(q^2)/4\pi$, ($\alpha_s(q^2)/4\pi$ is usually taken to be $\sim$ 2%-3% at $q^2 \sim$ $\sim$ 10 GeV^2). Thus it seems that for N = 4 and above, the corrections are large (50% or more).

Does this mean that these higher order corrections should be detectable in the experimental data?

Not necessarily!! This is because when we expand a physically measurable quantity in terms of a power series in $\alpha_s(q^2)$, we must, of course, specify what we mean by $\alpha_s(q^2)$. In quantum electrodynamics (QED), $\alpha \equiv e^2/4\pi$ is universally accepted as being the coupling constant defined through the "Thompson Limit of Compton Scattering". Such a limit does not exist in QCD so we have to specify precisely what we mean by $\alpha_s(q^2)$, and everyone is free to choose his own definition. In particular, one can relate two definitions by a power series in the form

$$\alpha_{s,1}(q^2) = \alpha_{s,2}(q^2)\left(1 + \rho\, \alpha_{s,2}(q^2) + \cdots\right) \tag{3.25}$$

so that, using Eq. (3.6), we see that a_i^N can be written in the form

$$a_i^N = b_i^N + \rho\, d_N \tag{3.26}$$

where the b_i^N are renormalization scheme independent numbers, whereas ρ is an arbitrary (but N independent) number. d_N is equal to $\gamma_0^N/2\beta_0$. Thus we see that the values of the a_i^N in one particular scheme may not be physically significant. Clearly, one would like to choose a scheme in which the corrections which have not been calculated are reasonable small, but it is impossible to be sure of this without calculating still further corrections. Thus, one is forced either to guess what is a reasonable scheme, or to make predictions which are independent of ρ. We come back to this in Section 3.5, but now we digress to demonstrate that

3.4 β_1 is independent of the definition of α_s, (i.e., of ρ)

Because it is always possible to define α_s in many ways, any two of which can be related by (3.25), one may suspect that in any quantity which can be expanded as a power series in α_s, it is only the term independent of α_s and the first term which depends on α_s which are independent of the definition of α_s. This is indeed true for quantities whose definition is independent of the definition of α_s (such as the ratio R in e^+e^- annihilation, γ_0^N, etc.) but not for the β function, which we recall is defined by

$$\beta(g) = \frac{\partial g(\mu)}{\partial \ln \mu} \tag{3.27}$$

and hence depends on the prescription used to define α_s. In addition to β_o it is found that β_1 is also a physically significant, unambiguous number (given by (3.8)). To illustrate this, let us define two schemes, labelled by a and b and related by*)

$$g_a = g_b(1 + \rho\, g_b^2 + \cdots) \tag{3.28}$$

We write the β functions in the two schemes in the form

$$\beta_a(g_a) = -\frac{\beta_o^a g_a^3}{16\pi^2} - \frac{\beta_1^a g_a^5}{(16\pi^2)^2} + \cdots \tag{3.29a}$$

and

$$\beta_b(g_b) = -\frac{\beta_o^b g_b^3}{16\pi^2} - \frac{\beta_1^b g_b^5}{(16\pi^2)^2} + \cdots \tag{3.29b}$$

Using (3.28) we write

$$\beta_a \equiv \frac{\partial g_a}{\partial \ln\mu} = \frac{\partial g_b}{\partial \ln\mu}\left(1 + 3\rho\, g_b^2\right)$$

$$= \beta_b(g_b)\left(1 + 3\rho\, g_b^2\right)$$

$$= (1+3\rho g_b^2)\left(-\frac{\beta_o^b g_b^3}{16\pi^2} - \frac{\beta_1^b g_b^5}{(16\pi^2)^2}\right) \tag{3.30}$$

$$= (1+3\rho g_b^2)\left(-\beta_o^b\,\frac{g_a^3(1-3\rho g_a^2)}{16\pi^2} - \beta_1^b\,\frac{g_b^5}{(16\pi^2)^2}\right)$$

$$= -\beta_o^b\,\frac{g_a^3}{16\pi^2} - \beta_1^b\,\frac{g_a^5}{(16\pi^2)^2}$$

---

*) ρ here is slightly different from that used in (3.25).

Comparing (3.30) and (3.29a) we deduce that

$$\beta_0^a = \beta_0^b \qquad \text{and} \qquad \beta_1^a = \beta_1^b \tag{3.31}$$

we leave it as an exercise for the reader to convince himself that the coefficients of higher order terms in β are not universal, but are ρ dependent.

3.5 Are the effects of the higher order corrections measurable?

We start by looking for the effects of higher order corrections which are independent of the definition of the coupling constant, i.e., of ρ[33]. One such effect is the behaviour with N of the quantity $R^N(q^2,q_o^2)$, which is defined by

$$R^N(q^2,q_o^2) = \left[\frac{M^N(q^2)}{M^N(q_o^2)}\right]^{1/d_N} \tag{3.32}$$

where we have suppressed the index i. The QCD prediction is

$$R^N(q^2,q_o^2) - R^T(q^2,q_o^2) = \frac{\alpha_s(q^2)}{\alpha_s(q_o^2)}\left(\alpha_s(q^2) - \alpha_s(q_o^2)\right)\left[\frac{a^N}{d^N} - \frac{a^T}{d^T}\right] \tag{3.33}$$

where from Eq. (3.26) we see that the expression in square brackets is independent of ρ, it is in fact positive for $N > T$. Thus for fixed q^2,q_o^2, R^N should decrease with N, and this is a consequence of the higher order corrections. In Fig. 10 we show the behaviour of $R^N(45\ \text{GeV}^2, 6.5\ \text{GeV}^2)$ and $R^N(45\ \text{GeV}^2, 10\ \text{GeV}^2)$ with N (defined by moments of the F_3 structure function) and the theoretical predictions corresponding to the three values of Λ, $\Lambda = 0.3$, 0.5 and 0.7. The experimental data are from Ref. 21). We see that the experimental errors are too large, even to check whether R^N falls or rises with N, and the data are certainly not inconsistent with the theoretical predictions. Hence, even though the entries in Table 3.1 are fairly large, we see that they lead to an unmeasurably small effect, at least as far as the N dependence of R^N goes for the q^2 ranges considered in Fig. 10. If we reduce q_o^2, the magnitude of the effect is increased, but then we have to deal with "higher twist" effects, i.e., terms which violate Bjorken scaling by a power of q^2.

There are several other (related) ways of trying to find (ρ independent) effects of higher order corrections in moments of structure functions. One of these is to check whether the value of the parameter Λ, determined from the data using the lowest order formula

$$R^N(q^2,q_o^2) = \frac{\ln q_o^2/\Lambda^2}{\ln q^2/\Lambda^2} \tag{3.34a}$$

has the correct N dependence. Strictly speaking, one should extract Λ using a slightly more complicated formula than (3.34a), that is

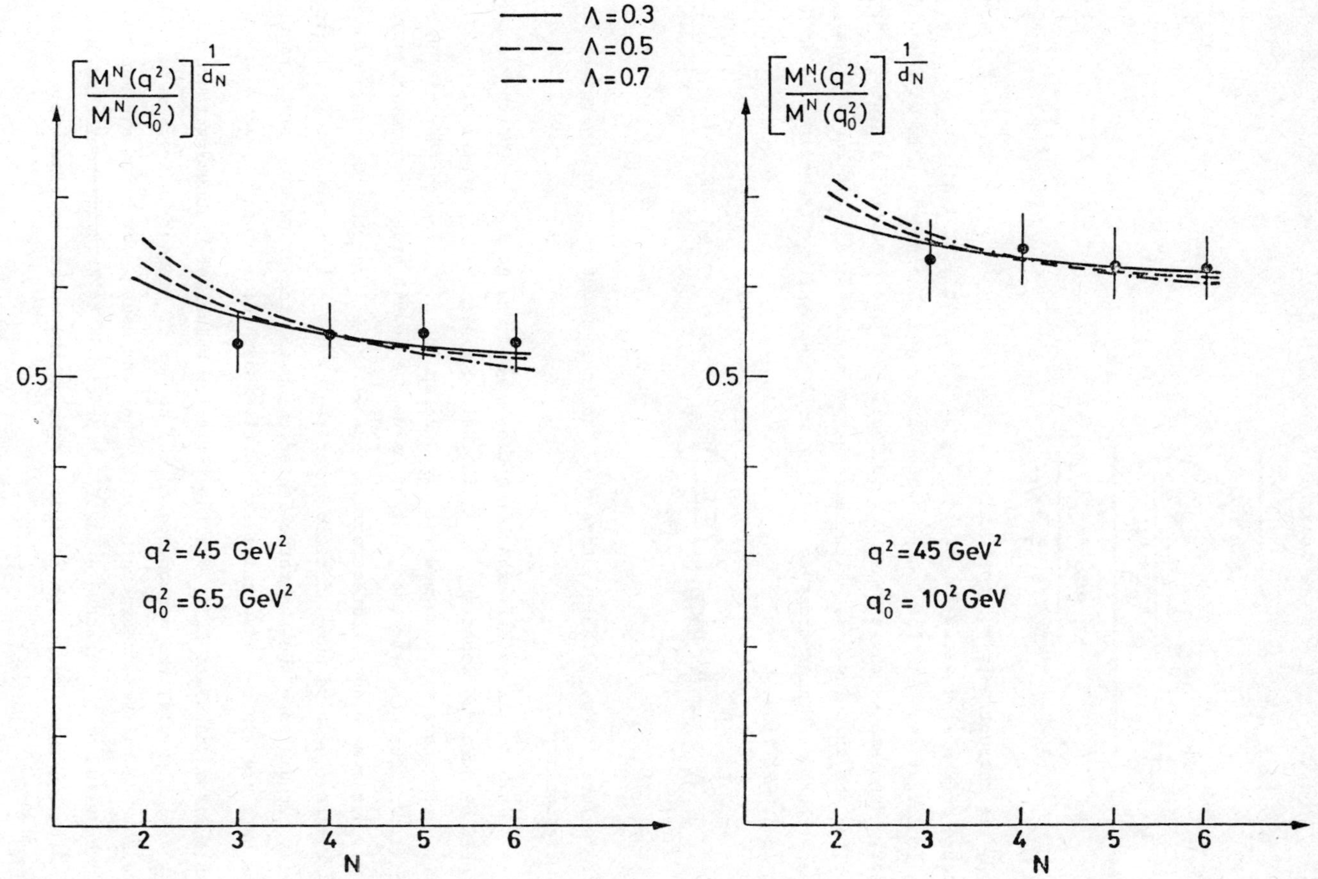

Fig. 10 Values of R^N vs N derived from the CDHS data[21] compared with the expected variation corresponding to different Λ's

$$R^N(q^2,q_0^2) = \frac{\log q_0^2/\Lambda^2}{\log q^2/\Lambda^2} \times \frac{\left(1 - \frac{\beta_1 \log\log q^2/\Lambda^2}{\beta_0^2 \log q^2/\Lambda^2}\right)}{\left(1 - \frac{\beta_1 \log\log q_0^2/\Lambda^2}{\beta_0^2 \log q_0^2/\Lambda^2}\right)} \tag{3.34b}$$

where β_o, β_1 are respectively the coefficients of $-g^3/16\pi^2, -g^5/(16\pi^2)^2$ in the β function. Because of the slow variation of log log q^2/Λ^2 with q^2, at least for q^2 substantially larger than Λ^2, the N dependence of the Λ's obtained in this way should be very nearly the same as those using Eq. (3.34a). We have checked that this, indeed, is the case for the existing data. The fact that we neglect the a_N in Eqs. (3.34a), (3.34b) means that the higher order corrections will now manifest themselves as an N dependence in Λ. We find

$$\Lambda_N = \Lambda \exp\left(\frac{2\pi a^N}{\beta_0 d_N}\right) \tag{3.35}$$

A very nice feature of Eq. (3.35) is that Λ_N/Λ_T is independent of ρ and therefore is a clean prediction of QCD.

This definition of Λ_N is different from that of Bardeen et al.[31] (which was subsequently used by Duke and Roberts[34),*)]); in their definition only a part of the higher order corrections was absorbed into an N dependent Λ. Using the definition of Bardeen et al., Λ_N must be extracted, not by using the lowest order formula (3.34a) (or (3.34b)), but a more complicated formula, one which includes the "unabsorbed" higher order terms; we find our definition preferable, not only because it is simpler, but primarily because both the theoretical prediction and the procedure for extracting Λ_N from the data are (in our case only) explicitly independent of ρ.

In Table 3.2 we tabulate Λ_N/Λ_2 for the first few values of N assuming there to be four (or six) flavours, the dependence on the number of flavours is very weak. The entries are somewhat different depending on whether one uses F_2 or F_3 to extract the Λ's. For N = 2, the value of Λ extracted using F_2 is predicted to be 37% larger than that extracted using F_3; this prediction is independent of the number of flavours.

---

*) From the paper of Duke[34)] one might wrongly conclude that Duke and Roberts have the same definition as ours for Λ_N; in fact, this is not so. Their Λ_N is defined using the definition of Bardeen et al.[31)]. The procedure we are using has previously been proposed by Baĉe[35)].

TABLE 3.2 Predicted values of Λ_N/Λ_2, for four flavours; the values are different if one uses F_2 or F_3 to obtain the Λ's. The values in parentheses are for six flavours.

N	F_2	F_3
2	1.00 (1)	1.00 (1)
3	1.19 (1.17)	1.41 (1.39)
4	1.33 (1.30)	1.67 (1.63)
5	1.45 (1.41)	1.87 (1.82)
6	1.56 (1.51)	2.04 (1.98)
7	1.66 (1.60)	2.18 (2.11)
8	1.74 (1.68)	2.31 (2.23)
9	1.82 (1.75)	2.43 (2.34)
10	1.89 (1.82)	2.53 (2.43)

From this table we see that the parameter Λ_N should change substantially with N. In Fig. 11 we plot the values of Λ_N extracted from existing neutrino[20,21] and muon[22] data. On the same figure we plot the theoretical prediction; this curve may be displaced vertically at will, since the theory only predicts the shape and not its normalization. At first sight, the two sets of neutrino data look inconsistent, but on closed examination[36] it has been found that systematic uncertainties in the analysis (such as the number of flavours used in the formulae, whether or not to include the pseudoelastics and the correction for Fermi motion) account for about half the discrepancy, so that the final discrepancy in the data is not very significant. Moreover, the discrepancy does not stem from the different ranges of Q^2 in the two experiments[36]. Bearing in mind that these systematic uncertainties are not included in the errors, it is clear that the neutrino data are not yet good enough to confirm or disprove the theoretical predictions. The values of Λ_N have also been extracted from the moments of $F_2^{\mu p} - F_2^{\mu n}$[28]; they show an increase of Λ_N with N in remarkable agreement with the predictions of the table (see Fig. 11b). This striking agreement q^2 is somewhat surprising in view of the fact that these values of Λ_N are determined mostly by the low q^2 region. In this region one might expect other (higher twist) non-scaling contributions (especially for the higher moments), still to be relevant, and there is no reason to expect these to reproduce Fig. 11b. We leave the reader to interpret the significance of Fig. 11, and we look forward to the publication of neutrino and muon data which are more accurate, and in the muon case, at higher values of q^2.

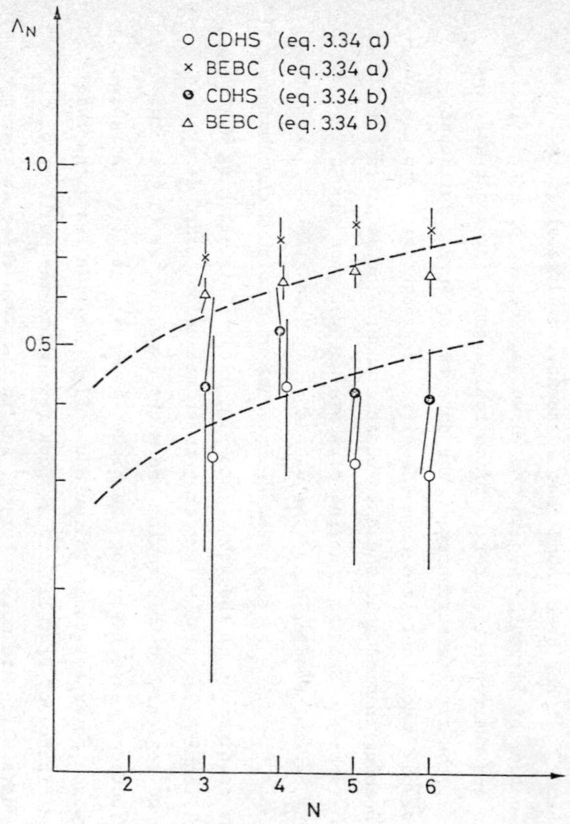

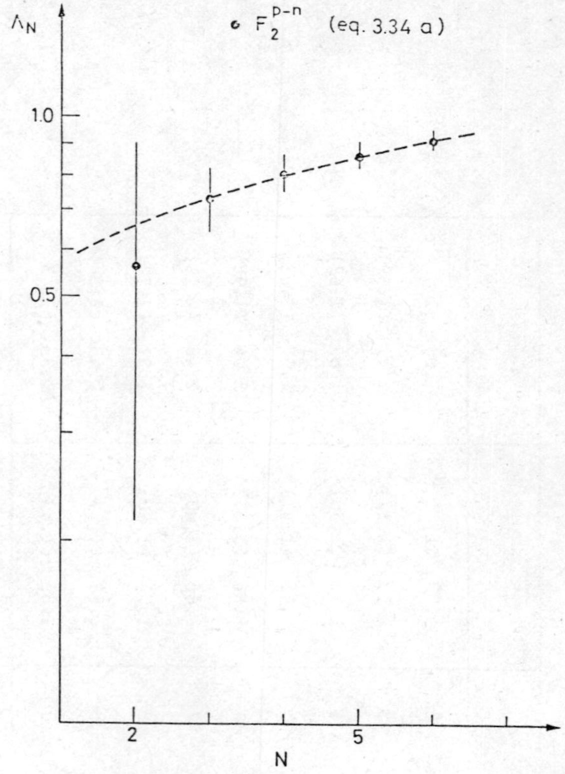

Fig. 11 (a) Values of Λ_N extracted from the F_3 structure function by the BEBC[20] and CDHS[21] neutrino experiments, using the lowest order formula (3.34a) and also formula (3.34b) which include fully the higher order corrections, compared with the QCD prediction. The theoretical curve may be displaced vertically; its normalization is not predicted

(b) Values of Λ_N extracted in Ref. 22) using the moments of $F_2^{\mu p} - F_2^{\mu n}$ and formula (3.34a)

Both the above examples were presented in terms of moments; the same conclusions are obtained when we look at the effects of higher order corrections to the structure functions themselves[37),38),39)]. Even though in a particular renormalization scheme the corrections may seem large, the major part of these corrections can be absorbed into a redefinition of the coupling constant, i.e., of Λ. For example[38)], if one uses the MS scheme, when one compares the ep and μp data to the QCD predictions which include the higher order corrections, one obtains a value of Λ which is about half of that obtained by comparing the lowest order formula to the data. On the other hand, the best fits in the two cases are comparable (there is a slight improvement if one includes the corrections).

3.6 Higher order corrections to singlet combinations of structure functions

So far, we have studied non-singlet combinations of structure functions such as $F_2^{\mu p} - F_2^{\mu n}$ and $F_3^{\nu A}$ (where A is an isoscalar target), where the effects of diagrams such as that in Fig. 12 are zero. For (flavour) singlet combinations of structure functions, for each moment there are two, twist two, operators which contribute, viz.,

$$\bar{\psi}\gamma^{\mu_1}D^{\mu_2}\cdots D^{\mu_N}\psi \tag{3.36}$$

and

$$F_{\mu_1\alpha}D_{\mu_2}\cdots D_{\mu_{N-1}}F_{\alpha\mu_N} \tag{3.37}$$

which mix under renormalization. Thus, the anomalous dimensions now form a 2 × 2 matrix, and its evaluation[26)] is a generalization of the non-singlet case and involves the calculation of the coefficient of the 1/ε term in four sets of two loop diagrams; a sample diagram from each set is shown in Fig. 13. The anomalous dimensions have been calculated[26)]; the only comment we would like to make is that there is a nice consistency check on the calculation. Because the MS scheme preserves the Ward identities, the energy momentum tensor

$$\bar{\psi}\gamma^{\mu_1}D^{\mu_2}\psi + F^{\mu\mu_1}F^{\mu_2\nu}g_{\mu\nu} \tag{3.38}$$

is not renormalized. This means that for N = 2

$$\gamma_{ff}^{N=2} + \gamma_{fg}^{N=2} = 0 \tag{3.39a}$$

and

$$\gamma_{gf}^{N=2} + \gamma_{gg}^{N=2} = 0 \tag{3.39b}$$

for all representations of quarks and gluons; this proves to be a highly non-trivial test of the calculation.

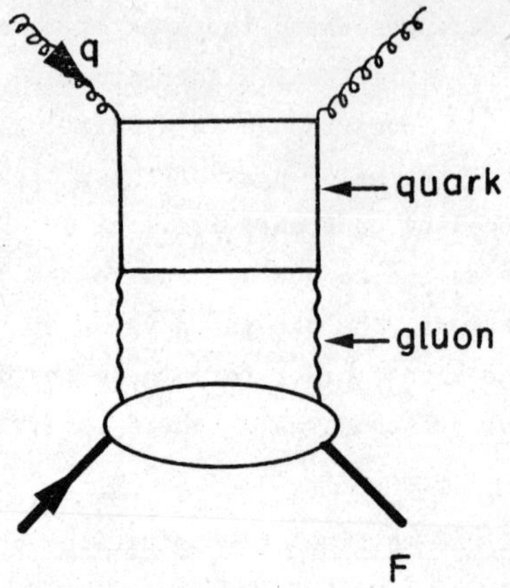

Fig. 12 A "singlet" contribution to deep inelastic lepton-
hadron scattering

$$
\begin{pmatrix} \gamma_{ff} & \gamma_{gf} \\ \\ \gamma_{fg} & \gamma_{gg} \end{pmatrix} = \begin{pmatrix} \quad + \cdots & \quad + \cdots \\ \\ \quad + \cdots & \quad + \cdots \end{pmatrix}
$$

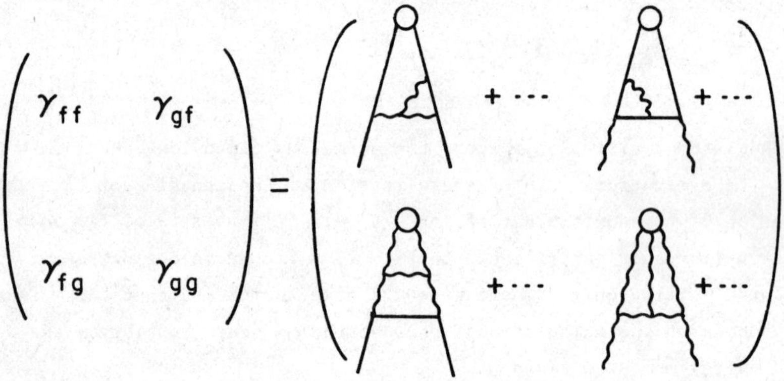

Fig. 13 Sample two-loop diagrams contributing to the anomalous
dimension of the twist-two singlet operators

The ε_i^N corresponding to operator (3.36) is the same as in the non-singlet case; one has to calculate the ε_i^N corresponding to operator (3.37). This has been done in Refs. 26) and 31).

In general, a structure function is a mixture of singlet and non-singlet terms. The phenomenology of the higher order corrections to e.g., $F_2^{\mu p}$ has been studied by D. Ross[40]. The conclusions are similar to those in the non-singlet case; the measurable effects of the higher order corrections are small.

3.7 Comments

We have seen in this section that the higher order corrections to deep inelastic scattering have now been calculated. They turn out to be small, and so they do not invalidate the lowest order picture (discussed in Section 2) to which we have become accustomed. This is certainly a non-trivial result. Because they are small, it is very hard to find any positive sign of these effects unless one is prepared to go to low values of $q^2(\gtrsim 1\ \mathrm{GeV}^2)$ and hope that even here the "higher twist" effects are negligible. It is just possible[33] that future high q^2, ($q^2 > 6.5\ \mathrm{GeV}^2$ [*]) say) data will be able to verify or disprove the existence of the predicted higher order terms.

Many of our parton model ideas, which were still valid in QCD when only the leading terms were included, e.g., the Callan-Gross relation

$$F_2 = 2x F_1 \tag{3.40}$$

and various parton model sum rules, no longer hold when higher order corrections are included. One notable exception is the Adler sum rule

$$\int_0^1 \frac{dx}{x} \left[F_2^{\bar{\nu}p} - F_2^{\nu p} \right] = 2 \tag{3.41}$$

which has no logarithmic corrections, but other parton model sum rules do have $O(\alpha_s(q^2))$ corrections[31),41]. The violation of the Callan-Gross relation leads to a non-zero longitudinal structure function. Since QCD predicts the q^2 behaviour of structure functions, in particular, the longitudinal structure function F_L, and the only parameters to be determined from the data are the operator matrix elements, once these are determined from fits to νW_2 (or W_1 or νW_3) we can predict F_L absolutely (up to corrections of order $\alpha_s(q^2)$, m^2/q^2). F_L is measured very poorly[42]; nevertheless, the theoretical predictions do tend to lie below the data[37]. It would be very interesting (although very difficult) to have better data for this process.

---

*) This value is arbitrarily chosen. It is the smallest value for which moments of structure functions are given by the CDHS group.

4. ASYMPTOTIC PREDICTIONS FOR GENERAL HARD SCATTERING PROCESSES

4.1 Introduction

In this section we will review why one now believes that we can make predictions about the asymptotic behaviour of hard scattering processes. The operator product expansion and renormalization group techniques discussed in the preceding sections are not in general directly applicable*), so we have to search for a new approach. The class of processes which we will consider in this section are those in which there are three mass scales:

a) a large mass scale Q^2

b) a small mass scale (of the order of hadronic masses) p^2, and

c) the renormalization scale μ^2.

We will be interested in the limit $Q^2 \to \infty$, p^2 fixed; physical cross-sections are, of course, independent of μ. In the relevant process there may be more than one variable which is $O(Q^2)$, (e.g., in deep inelastic scattering there is q^2 and $p \cdot q$), but then the ratio of these variables should stay fixed as we take the limit $Q^2 \to \infty$. Similarly, we assume that all hadronic masses are of the same order.

We start by reproducing the results of Section 2 for the deep inelastic structure functions, by using diagrammatic techniques. Later, we will see that the same techniques can be generalized to other hard scattering processes.

4.2 Diagrammatic approach to deep inelastic lepton-hadron scattering

In Sections 2 and 3 we have seen how the use of the operator product expansion and the renormalization group enables us to make predictions for the violation of Bjorken scaling in deep inelastic lepton-hadron scattering. In this subsection we will study the same process with a different approach, one which is less rigorous, but on the other hand, is also applicable to processes which are not light cone dominated. The basic assumption is that asymptotically hard scattering cross-sections can be written as a convolution of soft hadronic wave functions (which are process independent), with the perturbatively calculated cross-section for the "hard subprocess" which involves only quarks and gluons, and which can be studied perturbatively. Thus, for example, in the case of deep inelastic scattering we assume that we can write (see Fig. 14)

$$F(x, q^2) = \sum_{\text{partons}} \int_x^1 d\bar{x} \int d^2k_\perp \, dk^2 \, f(\bar{x}, \underset{\sim}{k}_\perp, k^2) \cdot \sigma_{\text{parton}}\left(\frac{x}{\bar{x}}, k^2, q^2\right) \tag{4.1}$$

---

*) See, however, the recent work of A. Mueller[8].

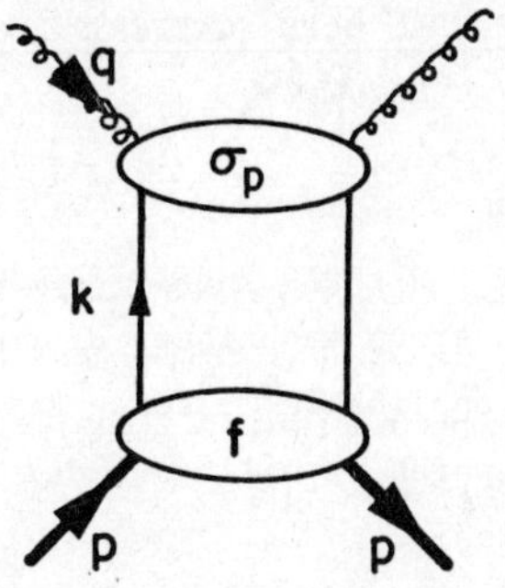

Fig. 14 Schematic representation of diagrammatic approach to deep inelastic lepton-hadron scattering

where f is related to the square of the soft wave function and σ_{parton} is the cross-section for deep inelastic scattering with the parton as the target. Within perturbation theory this assumption can be justified (see below). We will now study σ_{parton} in perturbation theory.

When computing σ_{parton} in perturbation theory, we will find terms of the type $g^2 \log Q^2/\mu^2$ and $g^2 \log Q^2/p^2$ (Q^2, p^2 and μ^2 are defined as in Section 1). We will see below that with each power of g^2 there is, at most, one logarithm.

4.2.1 Equivalence of the light cone result and the summation of leading logarithms

We would now like to demonstrate that the results of Section 2 can be obtained by summing the leading logarithms of perturbation theory. For simplicity, we will consider a non-singlet combination of structure functions; the extension to general combinations of structure functions complicates the technical details, but does not alter the essential features.

We start with Eq. (2.19) for the moment M^N of a structure function

$$M^N(q^2) = \bar{F}^N(1, \bar{g}(q^2)) \exp \int_{g(\mu^2)}^{g(q^2)} \frac{\gamma_{O^N}(g')}{\beta(g')} dg' \qquad (4.2)$$

$$\cdot \langle p | O^N | p \rangle$$

where $|p\rangle$ is now (for example), a quark state of moment p. We now expand the right-hand side of Eq. (4.2) in terms of a power series in $g^2(\mu^2)$; this we can do because we know how to relate $g(q^2)$ to $g(\mu^2)$. This relation is given by the solution of the equation

$$\frac{\partial g(\mu^2)}{\partial \ln \mu^2} = \beta(g) \tag{4.3}$$

In each order of perturbation theory we keep only the terms which are of the type $g^{2n} \log^n X$, where X is a ratio of two of the three mass scales Q^2, p^2 and μ^2, i.e., we work in the leading logarithm approximation. In the leading logarithm approximation the solution of (4.3) is given by

$$g^2(q^2) = g^2(\mu^2) . \frac{1}{1 + \frac{\beta_0}{16\pi^2} g^2(\mu^2) \log q^2/\mu^2} \tag{4.4}$$

We now look at the three factors on the right-hand side of Eq. (4.2) in turn.

a) $\bar{F}^N$: $\bar{F}^N$ is a power series in $g^2(q^2)$ with no logarithms. Therefore, in the leading logarithm approximation, we need to keep only the first term which is 1 *).

b) The exponential factors: here we expand γ_{0^N} and β and use Eq. (4.4) to obtain

$$\exp\left\{\int_{g(\mu^2)}^{g(q^2)} \frac{\gamma_{0^N}(g')}{\beta(g')} dg'\right\} \underset{\substack{\text{Leading}\\ \text{logarithm}\\ \text{Approximation}}}{=} \left\{1 + \frac{\beta_0 g^2(\mu^2)}{16\pi^2} \log q^2/\mu^2\right\}^{-\gamma_0^N/2\beta_0} \tag{4.5}$$

c) $\langle p|0^N|p\rangle$: this can either be calculated directly, or we can use the fact that M^N is μ independent so that the μ dependence of (4.5) must be cancelled by that of $\langle p|0^N|p\rangle$. Since $\langle p|0^N|p\rangle$ is independent of q, we find

$$\langle p|0^N|p\rangle = \left\{1 + \frac{\beta_0 g^2(\mu^2)}{16\pi^2} \log p^2/\mu^2\right\}^{\gamma_0^N/2\beta_0} \tag{4.6}$$

Thus, we find, in the leading logarithm approximation that**)

---

*) For the moments of F_L, the first term in $\bar{F}^N$ is equal to $g^2 \cdot$ constant.

**) If we differentiate the right-hand side of (4.7) with respect to log μ, we find we get 0 in the leading logarithm approximation, i.e., all terms of the form $g^{2n} \log^{n-1} \mu^2$ cancel.

$$M^N(q^2) \sim \left[\frac{1 + \frac{\beta_0 g^2(\mu^2)}{16\pi^2} \log p^2/\mu^2}{1 + \frac{\beta_0 g^2(\mu^2)}{16\pi^2} \log q^2/\mu^2} \right]^{\gamma_0^N/2\beta_0} \tag{4.7}$$

If we did not know anything about the light cone techniques of Section 2, but had instead calculated Feynman diagrams and kept only the leading logarithms, we would have discovered the series generated by (4.7), and thus obtained the result

$$\frac{M^N(q^2)}{M^N(q_0^2)} = \left[\frac{g^2(q^2)}{g^2(q_0^2)} \right]^{\gamma_0^N/2\beta_0} \tag{4.8}$$

which is the result obtained in Section 2. In this sense, the sum of the leading logarithms in perturbation theory is equivalent to the result obtained by light cone techniques.

We will now study some Feynman diagrams to see if we can understand what regions of phase space give the leading logarithms, and how to obtain (4.7).

4.2.2 Which regions of phase space give the leading logarithms?

We choose to write all logarithms as $\log q^2/p^2$ and $\log q^2/\mu^2$, and look for possible sources of $\log \mu^2$ and $\log p^2$. The $\log \mu^2$ terms arise in any renormalizable field theory after renormalization has been performed. In the minimal subtraction (MS) scheme they appear because, although the coupling constant in QCD is dimensionless in the physical number of dimensions, it is not dimensionless in $4 - \varepsilon$ ($\varepsilon \neq 0$) dimensions, and, hence, a mass scale has to be introduced. We will return to the question of the $\log \mu^2$ terms below.

As far as the $\log p^2$ terms go, we can use the machinery which exists for the behaviour of physical processes in field theory in the limit where one or more of the masses vanish. In particular, if the appropriate Feynman integrals are finite in this limit, then there can be no $\log p^2$ term which scales (p^2 is a mass which is being taken to zero). Thus, since we are looking for possible sources of $\log p^2$ terms, we are interested in finding the regions of phase space which yield a divergence as one or more masses vanish. We will discuss the two types of divergence which are often called infra-red divergences and mass singularities. Both of these have been thoroughly studied in field theory.

Infra red divergences[43),44)] arise from the presence of a soft, real or virtual, massless particle (in a frame in which the external particles are not at rest). For example, if we evaluate the diagram of Fig. 15 for the process $e^+e^- \to \mu^+\mu^-$ (or $q\bar{q}$) we find that there is a Feynman integral of the form

$$\int d^4k \; \frac{1}{k^2\,[(p_1+k)^2-m^2]\,[(p_2-k)^2-m^2]} \qquad (4.9)$$

$$\underset{k\to 0}{\sim} \int d^4k \; \frac{1}{k^4} = \infty$$

We see that in Fig. 15, when k is soft, three propagators simultaneously become close to their poles. The theorem of Block and Nordsieck[43),44)] states that in inclusive cross-sections these infra-red divergences cancel; in QED, for example, they cancel between diagrams with real photons and those with virtual photons (see e.g., Fig. 16). In fact, all physical cross-sections are inclusive ones, since in any experiment the energy resolution is not perfect, and hence there may be any number of undetected soft photons. Hence, when calculating predictions for measurable cross-sections, we have to sum over states with these additional photons, and we then arrive at a finite result.

Of course, in deep inelastic scattering this cancellation of infra-red divergences also occurs. The only slightly novel feature occurs at the point x = 1. For example, the diagram of Fig. 17a (evaluated in the Feynman gauge), has a behaviour as $x \to 1$ of the form

$$\text{Diag}\,(17a) \underset{x\to 1}{\sim} \frac{1}{(1-x)^{1+\varepsilon}} \qquad (4.10a)$$

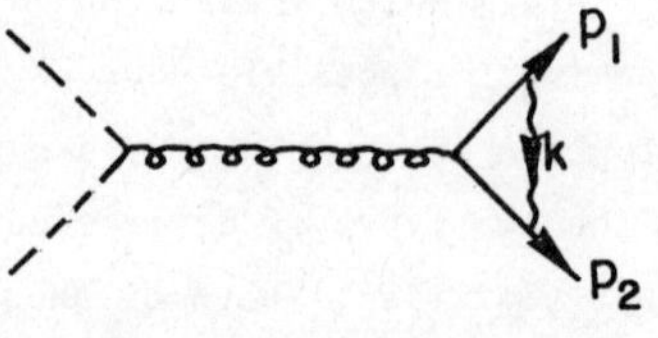

Fig. 15 Example of an infra-red divergent diagram in the process $e^+e^- \to \mu^+\mu^-$ (or $q\bar{q}$)

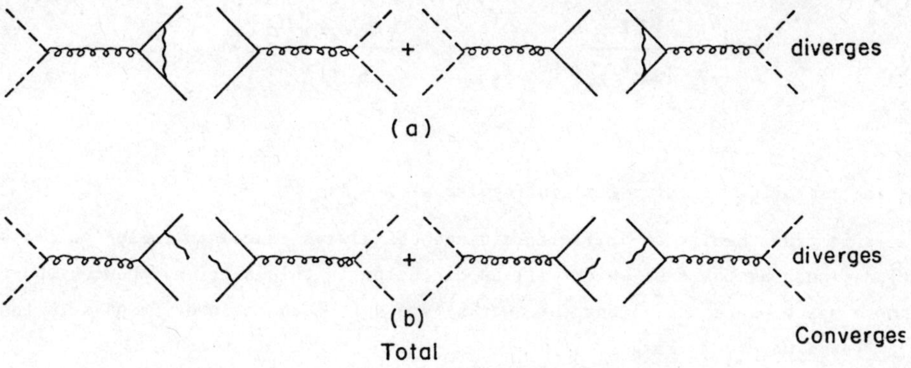

Fig. 16 Example of the Block-Nordsieck mechanism for the cancellation of infra-red divergences in the total e^+e^- annihilation cross-section

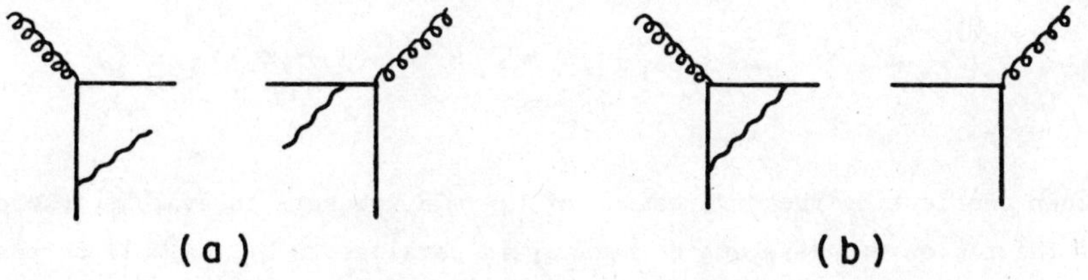

Fig. 17 Infra-red divergences cancel between these two diagrams by the Block-Nordsieck mechanism

where we have chosen to use dimensional regularization to regulate the infra-red divergences[45),46)]. The infra-red divergences occur, of course, at x = 1. The corresponding diagram with a virtual gluon is that of Fig. 17(b) and has a behaviour

$$\mathrm{Diag}\,(17b) \sim \delta(x-1)\,\frac{1}{\varepsilon} \tag{4.10b}$$

The right-hand sides of (4.10a) and (4.10b) are equal and opposite in the sense of distributions[25)], i.e., if we smear $1/(1-x)^{1+\varepsilon}$ with any smooth function of x, which is regular at x = 1, by taking the integral from 0 to 1; we get the opposite result from smearing $\delta(x-1)\,1/\varepsilon$ with the same function. Thus, if we obtain a term of the type 1/1-x, we will be able to rewrite it in terms of $1/(1-x)_+$, which is defined as a distribution through

$$\int_0^1 dz \quad \frac{f(z)}{(1-z)_+} = \int_0^1 dz \quad \frac{f(z)-f(1)}{1-z} \tag{4.11}$$

where the smearing function f(z) is regular at z = 1.

It is true then, that infra-red divergences always cancel in inclusive cross-sections, such as the ones which will be discussed in this section. Hence, the region of phase space where the gluons (or quarks) are soft does not provide us with the $\log p^2$ terms we are looking for.

We come now to mass singularities. These occur in theories with coupled massless particles, and are due to the simple kinematical fact that two massless particles (with momenta k_1, k_2, say) which are moving parallel to each other have a combined invariant mass equal to zero

$$k^2 \equiv (k_1+k_2)^2 = [\omega_1(1,0,0,1) + \omega_2(1,0,0,1)]^2 = 0 \tag{4.12}$$

Thus, when considering the divergences of Fig. 18, we have to evaluate the contribution of the region of phase space where k_1 is parallel to k_2, as well as that where k_1, k_2 are soft.

There is also a theorem, the Kinoshita[47)] -- Lee-Nauenberg[48)], (KLN) theorem, which ensures that for inclusive enough cross-sections, the mass singularities also cancel. For example, the mass singularities of Fig. 16(a), (in the limit where the mass of the muon is zero) cancel those of Fig. 16(b), and so for the process $e^+e^- \to \mu^+\mu^- X$ we can set both the μ and photon masses to zero and still get a finite result. The interpretation of this result is similar to that for infra-red divergences, namely, that in a physically measurable process, the angular resolution is never perfect and therefore we should sum over all indistinguishable states, i.e., all states in which there are some collinear particles. For the exact statement of the KLN theorem we refer the reader to the original papers; here, we will just note that it assures us that all mass singularities coming from final state undetected particles moving parallel to each other cancel.

In deep inelastic scattering, however, we still have mass singularities left over; those coming from regions of phase space in which internal particles have momenta parallel to the momentum of one of the incoming particles. For example, in Fig. we have a factor of $\log p^2$ from the region where k is parallel to p. We will see this explicitly below.

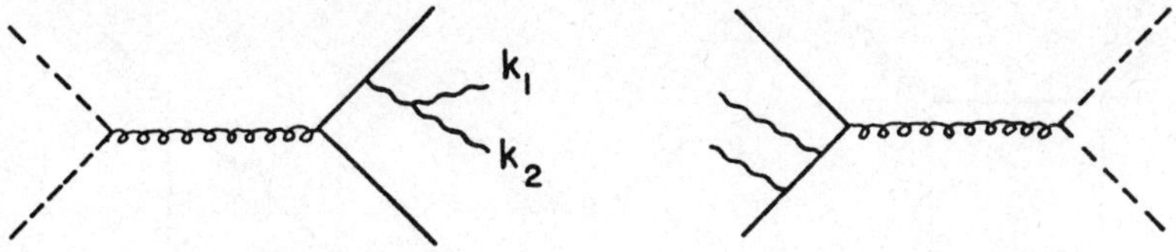

Fig. 18 A QCD diagram contributing to the total e^+e^- annihilation cross-section; this diagram has a mass singularity

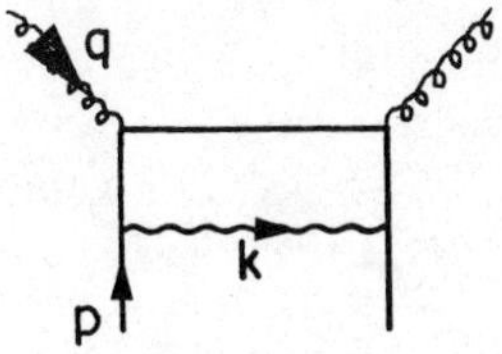

Fig. 19 Sample diagram with a mass singularity which contributes to deep inelastic scattering

Whereas one never gets infra-red divergences if all external particles are neutral, this is not true for mass singularities. For example, if we take deep inelastic scattering on a photon, the lowest order diagrams are shown in Fig. 20, we still get a factor of log p^2.

Thus we see that the required log p^2 terms come from regions of phase space in which (at least some of) the internal particles have momenta parallel to the incoming ones, (or possibly to the trigger particle in an inclusive cross-section).

4.2.3 Some low order diagrams

Let us look at some low order diagrams; this way, we shall see explicitly where the $O(g^2)$ term in the series generated by (4.7) comes from and hopefully get some insight into how the higher orders should work. We consider deep inelastic scattering on a quark and start by looking at the interference term shown in Fig. 21a, in the Feynman gauge. Evaluating this integral, we readily find that it contains an integral of the form

$$I = \int \frac{d^4k}{(p-k)^2} f(k, p, q) \tag{4.13}$$

Evaluating $(p - k)^2$ in terms of the components of the momenta of p and k, we find

$$(p-k)^2 = -2\omega(E - |\mathbf{p}| \cos\theta) \tag{4.14}$$

Fig. 20 Diagrams contributing to deep inelastic scattering on a vector target

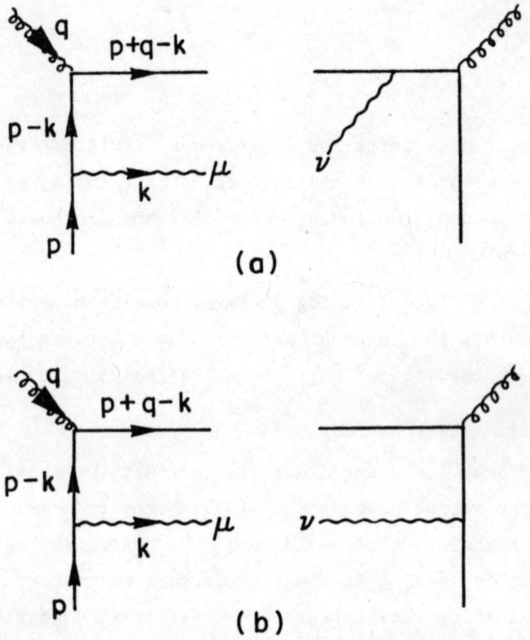

Fig. 21 Low order diagrams contributing to deep inelastic scattering. The Latin letters label the momenty, whereas the Greek letters label the Lorentz indices of the gluons

where $p = (E,0,0,|\underline{p}|)$ in the centre-of-mass frame, for example; ω is the energy of k and θ is the angle between p and k. Thus

$$\underline{I} \propto \int_{-1}^{1} \frac{d(\cos\theta)}{\left(1 - \frac{|\underline{p}|}{E}\cos\theta\right)} \sim \log q^2/p^2 \tag{4.15}$$

where we have used the fact that in our approximation, $E \simeq p$, $(E + p)^2 \simeq q^2$, $E^2 - \underline{p}^2 = p^2$. Thus we have obtained the promised $\log q^2/p^2$. Notice that we could have performed the integral from $1 - \varepsilon$ to 1 for a fixed small ε (so that $\log 1/\varepsilon << \log q^2$) and still have obtained the same result. Thus, the dominant region of integration is that corresponding to small angles as stated above. In terms of k_T, the component of $\underline{k}$ orthogonal to $\underline{p}$, this dominant region corresponds to $m^2/\delta < k^2 < \varepsilon q^2$, where δ, ε are small fixed numbers.

We would like now to concentrate on another feature of the calculation of the diagram of Fig. 21(a). When evaluating the Dirac algebra of this diagram, we will find a term

$$(\not{p} - \not{k})\,\gamma^\mu\,\not{p} \tag{4.16}$$

where μ is the Lorentz index of the gluon. From the form of the integral in (4.15) we see that if we are only interested in the $\log q^2/p^2$ term, we can set $\cos\theta = 1$ everywhere in the numerator, and, in particular, we can write $k = (1 - \rho)p$ (in the numerator we can also consider p to be a null vector). Then in this approximation

$$(\not{p} - \not{k})\,\gamma^\mu\,\not{p} = \frac{\rho}{(1-\rho)^2}\,\not{k}\,\gamma^\mu\,\not{k} = \frac{2\rho}{(1-\rho)^2}\,k^\mu\,\not{k} \tag{4.17}$$

If we now choose a gauge in which we sum only over the physical transverse polarizations diagram by diagram so that

$$k^\mu \sum_{\text{polarisations}} \varepsilon^\mu \varepsilon^\nu = 0 \tag{4.18}$$

we will get no term with a factor of $\log q^2/p^2$ from the diagram of Fig. 21a. Popular choices for the gauge are axial gauges of the type $n \cdot a = 0$ [49] where n is some four vector (n is sometimes chosen to be q [18]).

If instead of setting $\theta = 0$ and writing $k = (1 - \rho)p$, we would have kept θ, we would have found that at each vertex in which the gluon momentum is almost parallel to the incoming quark momentum there is a suppression factor of θ. For small θ,

$$(p-k)^2 \sim 2\omega E(1-\cos\theta)$$

$$\sim 2\omega E\,\theta^2$$

and phase space is proportional to $d(\cos\theta) \simeq \theta d\theta$. Thus, in a transverse gauge the angular integration for $\theta \sim 0$ is of the form

$$\text{Diag}(21a) \sim \int \theta d\theta \cdot \frac{\theta}{\theta^2} = \text{finite} \tag{4.19}$$

as we have already discovered.

For the diagram of Fig. 21b, however, there are two vertices with a suppression factor θ at each, and two factors of $1/(p - k)^2$, so that the angular integration is now of the form

$$\text{Diag}(21b) \sim \int \theta\, d\theta\, \frac{\theta^2}{(\theta^2)^2} = \int \frac{d\theta}{\theta} = \infty \tag{4.20}$$

Thus the diagram of Fig. 21b has a mass singularity, even in a transverse gauge. In fact, it is the only diagram with a real gluon to have a mass singularity.

We now outline the calculation of the diagram of Fig. 21b in the gauge $n \cdot a = 0$. In this gauge the sum over polarizations is given by

$$-\left[g^{\mu\nu} - \frac{k^\mu n^\nu + k^\nu n^\mu}{n.k} + \frac{n^2 k^\mu k^\nu}{(n.k)^2}\right] \tag{4.21}$$

We label the three terms in (4.21) by (a), (b) and (c). It is convenient to use Sudakov variables[50]; we first define q' by $q' \equiv q + xp$ which has the property $q'^2 \simeq 0$. The Sudakov variables ρ, β and k are defined by

$$k = \rho p + \beta q' + k_\perp \tag{4.22}$$

where k_T is orthogonal to p and q. The Jacobian is given by

$$d^4k = p.q\; d\rho\, d\beta\, d^2k_\perp \tag{4.23}$$

and in terms of these variables $(p - k)^2 \simeq p^2 - 2\beta\, p \cdot q$. Performing the Dirac algebra we find that each of the terms (a) and (b) in (4.21) gives one power of $(p - k)^2$ whereas term (c) gives a factor of $(p - k)^4$ and, therefore, does not contribute to the log q^2/p^2 term. The relative contributions of (a) and (b) are

(a) $$C_F\,(1-\rho) \qquad (4.24a)$$

(b) $$C_F\,\frac{2\rho}{(1-\rho)} \qquad (4.24b)$$

where $C_F = 4/3$ is the usual colour factor. Adding these two contributions together we find that we get

$$C_F\,\frac{1+\rho^2}{1-\rho} \qquad (4.25)$$

which is the Altarelli-Parisi[25)] function $P_{q \to q}(\rho)$ for $\rho \neq 1$. If we add the diagrams with the virtual gluons, we would find also the terms proportional to $\delta(\rho - 1)$ in $P_{q \to q}(\rho)$.

The appropriate integral now is of the form

$$I \propto (p\cdot q)\int d\rho\,\frac{1+\rho^2}{1-\rho}\int\frac{d\beta\;d^2k_\perp}{[p^2-2\beta\,p\cdot q]} \qquad (4.26)$$

$$\cdot\;\delta(k^2)\;\delta\left((p+q-k)^2\right)\;\mathrm{Tr}\left[\gamma^\mu(\rho\not p+\not q)\gamma^\nu\not p\right]$$

where the two arguments of the δ functions are given by

$$k^2 = 2\rho\beta\;p\cdot q - k_\perp^2 \qquad (4.27a)$$

and

$$(p+q-k)^2 = 2p\cdot q\,(\rho - x) \qquad (4.27b)$$

Any structure function (F_1, F_2 or F_3) is now given by

$$F \propto g^2 \int d\xi \, \frac{1}{\xi} \, P_{q\to q}(\xi) \, \delta(\xi - x) \int \frac{d^2 k_\perp}{[p^2 + k_\perp^2/\xi]}$$

$$\propto g^2 \int d\xi \, P_{q\to q}(\xi) \, \delta(\xi - x) \int_{p^2}^{q^2} \frac{dt}{t} \qquad (4.28)$$

Although (4.28) can obviously be simplified, we leave it in this form because the higher order diagrams will give us contributions which are simple generalizations of this form. Since the moments of the Altarelli-Parisi function $P_{q \to q}$ are proportional to the anomalous dimensions of the twist two non-singlet operators, we see that (4.28) agrees with the N dependence of the g^2 term in (4.7).

4.2.4 Higher order diagrams and the asymptotic result

Similar features which occurred in the one loop diagrams also appear in the general case*). The general feature is that the dominant diagrams in the transverse gauges are the generalized ladder diagrams (Fig. 22), i.e., ladder diagrams with vertex and self-energy insertions (we denote all these insertions by a cross at the vertex). Gribov and Lipatov[53)] have taught us how to evaluate such diagrams and that the dominant region of integration is where

$$p^2 \ll k_1^2 \ll k_2^2 \ldots \ll k_N^2 \ll q^2 \qquad (4.29)$$

One of the effects of the vertex and self-energy insertions is that at each vertex we should put $g(t_i)$ where $t_i = k_i^2$ is the largest four momentum squared at the vertex. This may seem surprising in view of the fact that the momenta in the different legs are not equal (see Fig. 23). We can choose, however, to associate each self-energy insertion on the fermion lines with the vertex below the insertion, (instead of the usual association of half the self-energy insertion with the vertex above and half with the vertex below). In this way we see that we should write $g(t_i)$ at each vertex.

We now outline the calculation of such a set of diagrams. We start by doing the transverse momentum integrals, which we can rewrite in terms of the t_i analogous to (4.28). The only difference is now that for each t_i we have a factor of $\alpha(t_i) \sim$ $\sim 1/2\beta_0 \log t_i/\Lambda^2$ so that we have

---

*) For comprehensive discussions of this general case, see, for example, References 18), 51), 52) and references therein.

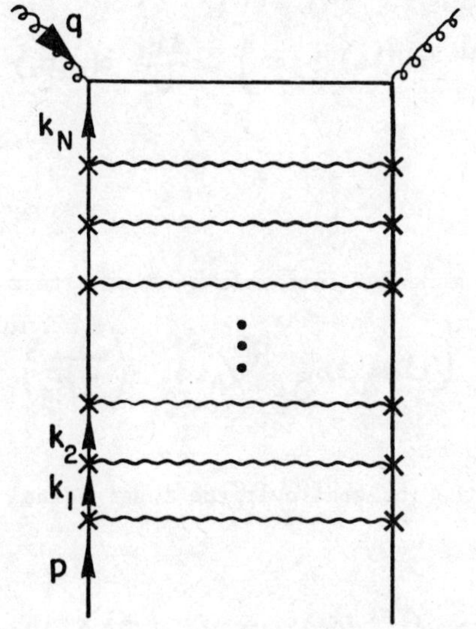

Fig. 22 The dominant structure (in the leading logarithm approximation) for deep inelastic scattering in the Bjorken limit. The x's signify that self-energy and vertex insertions must be added

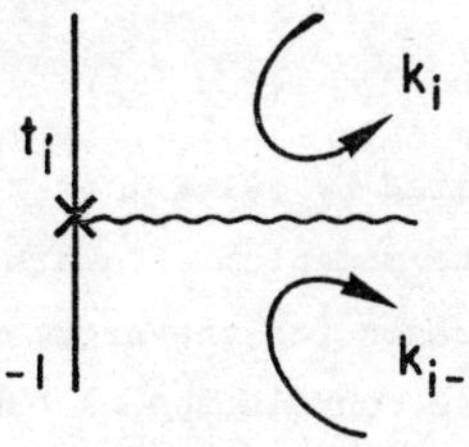

Fig. 23 Auxiliary diagram used to explain Eq. (4.30)

$$I_\perp = \int_{p^2}^{q^2} \frac{dt_N}{t_N} \alpha_s(t_N) \cdots \int_{p^2}^{t_3} \frac{dt_2}{t_2} \alpha_s(t_2) \int_{p^2}^{t_2} \frac{dt_1}{t_1} \alpha_s(t_1) \quad (4.30)$$

Thus, from the transverse momentum integrals we find a term proportional to*)

$$I_\perp \propto \frac{1}{N!} \left(\log\log q^2/\Lambda^2\right)^N \left(\frac{1}{2\beta_0}\right)^N \quad (4.31)$$

Now we have to calculate the integral over the longitudinal component of momentum. This takes the form

$$\int_x^1 dx_1\, P_{q\to q}(x_1) \int_x^{x_1} \frac{dx_2}{x_1} P_{q\to q}\left(\frac{x_2}{x_1}\right) \cdots \int_x^{x_{N-2}} \frac{dx_{N-1}}{x_{N-2}}$$
$$\cdot P_{q\to q}\left(\frac{x_{N-1}}{x_{N-2}}\right) \cdot \int \frac{dx_N}{x_{N-1}} P_{q\to q}\left(\frac{x_N}{x_{N-1}}\right) \delta(x_N - x) \quad (4.32)$$

The fraction of momentum of p carried by quark i is x_i, and that by quark i + 1 is x_{i+1}, so that the fraction of the momentum of quark i which is carried by quark i + 1 in x_{i+1}/x_i; this is the reason for the arguments in the $P_{q \to q}$ functions in (4.32). Since (4.32) is a multiple convolution we take moments,

$$\int_0^1 x^{r-1} I_\perp(x)\, dx = \left[\int_0^1 dx\, P_{q\to q}(x)\, x^{r-1}\right]^N$$
$$\propto \left(-\gamma_0^r\right)^N \quad (4.33)$$

Thus we find that for the rth moment of the diagram of Fig. 22 the asymptotic behaviour is proportional to

---

*) This result is obtained by performing the integrals in (4.30) and keeping the top limit of the integration range. Below we will come back to the terms obtained by keeping one or more of the bottom limits.

$$\frac{1}{N!}\left(\log\log q^2/\Lambda^2\right)^N\left(-\frac{\gamma_0^\Gamma}{2\beta_0}\right)^N \tag{4.34}$$

whence on summing over N we reproduce the standard light cone result

$$M^\Gamma(q^2) \propto \left(\log q^2\right)^{-\gamma_0^\Gamma/2\beta_0} \tag{4.35}$$

We would like to make a few comments concerning this calculation. First of all, we notice that the corrections to (4.31) (obtained by keeping a term which involves at least one of the lower limits) are of the form

$$\frac{1}{(N-1)!}\left(\log\log q^2/\Lambda^2\right)^{N-1} + \ldots \tag{4.36}$$

and, therefore, it may seem that the corrections to (4.35) should be supressed by one or more powers of log log q^2/Λ^2 (rather than powers of log q^2/Λ^2, which we know to be correct, see Section 3). This is not so, however, since if we keep the lower limit in one of the integrals of (4.30), the t_i integral, say, then we keep a term from the first j integrals which is q^2 independent. If for $j \leq i \leq N$, we now keep the upper limit in the t_i integrations, we generate a term proportional to

$$\frac{1}{(N-j)!}\left(\log\log q^2/\Lambda^2\right)^{N-j}$$

whence, summing over N from j to infinity, after having performed the integrals over the longitudinal components of momenta, we obtain a new contribution of the form (4.35). This just means that we are in effect including the part of the diagram in which the transverse momenta are finite in the soft wave function f and not in σ parton. Of course, we are not able to calculate the constant of proportionality in (4.35); this is analogous to not being able to calculate the operator matrix elements.

Another relevant question is, why are we entitled to substitute the asymptotic form for $\alpha(t_i)$ in (4.30), when we perform the integral from a finite mass p^2? This is so because in the leading logarithm approximation, we could equally well have taken the integrals from Xp^2, where X is a large fixed number, such that $\alpha(Xp^2)$ is well approximated by its asymptotic form.

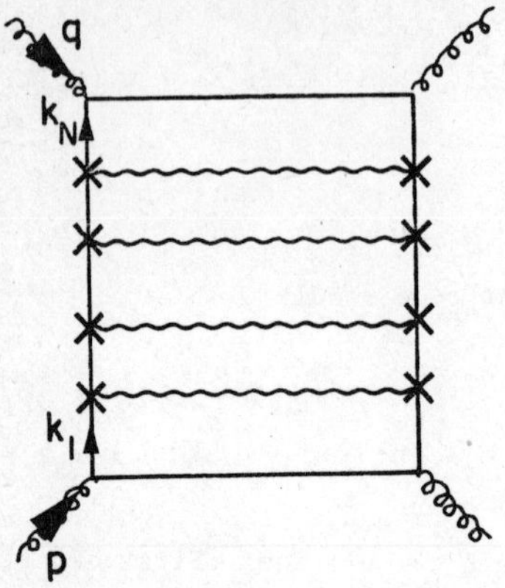

Fig. 24 The dominant structure for deep inelastic scattering on a photon target in the Bjorken limit

Finally, we would like to comment about the fact that, although the dominant region of phase space is where the transverse momenta k_i^2 are << q^2, nevertheless, the k_i do grow with q^2. For example, the average transverse momentum of a quark jet is predicted to behave like

$$\langle k_\perp^2 \rangle \sim \alpha_s(q^2)\, q^2 \tag{4.37}$$

So we have to abandon the notion of fixed k_T jets, not only in deep inelastic lepton-hadron scattering, but also in other processes. Perhaps the most conclusive evidence that this is so experimentally comes from e^+e^- annihilation at PETRA[17)].

We are now in a position to understand the interesting result of Witten[54)] about the structure function of a photon. In the transverse gauges the dominant diagrams are again the generalized ladder diagrams[55),56)] (see Fig. 24). The only difference is at the bottom of the diagram where we now have two electromagnetic vertices. These vertices are hard vertices, so that we can have large values of t_1, but we have no factor of $\alpha_s(t_1)$. Thus, the transverse momentum integral, the analogue to (4.30), is now

$$I_\perp^\gamma = \int_{p^2}^{q^2} \frac{dt_N}{t_N}\, \alpha_s(t_N) \ldots \int_{p^2}^{t_3} \frac{dt_2}{t_2}\, \alpha_s(t_2) \int_{p^2}^{t_2} \frac{dt_1}{t_1} \tag{4.38}$$

which can be evaluated to give

$$I_\perp^\gamma \propto \log Q^2 \quad \frac{1}{(2\beta_0)^N} \tag{4.39}$$

We obtain this result when we keep the upper limit in all the integrals of (4.38). If somewhere we keep a lower limit, we get terms of the form which we encountered in the hadronic case, i.e., (4.31). The rth moment of the integral over longitudinal momenta is now equal to (compare with (4.33))

$$(\gamma_0^r)^{N-1} f^r \tag{4.40}$$

where f^r is the rth moment of $P_{\gamma \to q}(x)$ [57], the probability density of finding a quark in a photon with fraction x of its momentum. We now sum over N, as in the hadronic case, but now we have a geometric series which we sum to find that the rth moment of the structure function of a photon is given by

$$\log q^2 \cdot \frac{f^r}{1 + \frac{\gamma_0^r}{2\beta_0}} \tag{4.41}$$

Witten first obtained this result by using the operator product expansion[54]. In addition to the usual twist two operators, we now have operators based on $F^{\mu\nu}$, the electromagnetic field strength tensor. It is these operators which lead to the behaviour (4.41). These predictions will be tested in future e^+e^- experiments at PEP and PETRA.

In a fixed point theory with a small ultra-violet stable fixed point, the structure functions of hadrons are still dominated by the diagrams of Fig. 22, but now in the analogous integral to (4.30) we have no $\alpha_s(t_i)$ factors. This means that we can immediately deduce the result; we just make the substitution $\log \log q^2/\Lambda^2 \to \log q^2/\Lambda^2$ and hence the rth moment of a structure function in such a theory has q^2 dependence given by

$$(q^2)^{-\gamma_0^r/2\beta_0} \tag{4.42}$$

Thus, in such theories the violations of Bjorken scaling are given by powers of q^2.

4.3 Asymptotic predictions for other hard scattering processes

Once we understand the ideas and calculations of Section 4.2 for the deep inelastic case, it is conceptually fairly straightforward to generalize the ideas to other hard scattering processes. We start by giving a few examples

4.3.1 The Drell-Yan process

We now consider the inclusive production of a pair of leptons with large invariant mass in hadronic collisions, the Drell-Yan process[14]. Sample diagrams which contribute to this process are shown in Fig. 1. Let us denote the lowest order contributions to the cross-section for the subprocess $q\bar{q} \to \ell^+\ell^- x$ (that corresponding to Fig. 1a) by σ_o. Then when all the radiative corrections of order g^2 are calculated (including those of Fig. 1b and 1c) in the leading logarithm approximation, one finds[58] that these corrections are equal to

$$a(\tau)\, g^2 \left[\log Q^2/p_1^2 + \log Q^2/p_2^2\right] \tag{4.43}$$

where $\tau \equiv Q^2/s$. It turns out that $a(\tau)$ is a very significant function; it is exactly the same function which appears in the order g^2 corrections to deep inelastic scattering on a quark. The structure function of a quark is proportional to

$$\delta(x-1) + a(x)\, g^2 \log q^2/p^2 + O\left(g^4 \log^2 q^2/p^2\right) \tag{4.44}$$

Thus, to this order, at least, it seems that the deviations from the naive Drell-Yan picture are intimately related to the violations of Bjorken scaling in deep inelastic scattering. In other words, the diagrams which spoil the simple Drell-Yan picture (such as those of Figs. 1b and 1c) are related to those which are responsible for the violation of Bjorken scaling (such as those of Figs. 25a and 25b). In particular, to this order we can write a modified Drell-Yan formula

$$\frac{d\sigma}{dQ^2} = \frac{4\pi\alpha^2}{3nQ^4} \sum_{\substack{\text{quark}\\ \text{flavours}\\ a}} Q_a^2 \int_0^1 dx_1\, dx_2\, x_1 x_2\, \delta(x_1 x_2 - \tau)$$

$$\cdot \left[G_{q^a/h_1}(x_1, Q^2)\, G_{\bar{q}^a/h_2}(x_2, Q^2) + G_{\bar{q}^a/h_1}(x_1, Q^2)\, G_{q^a/h_2}(x_2, Q^2)\right] \tag{4.45}$$

where the G distribution functions are the appropriate linear combinations of experimentally determined deep inelastic structure functions.

This result is easy to understand in transverse gauges, because in these gauges it is again the generalized ladder diagrams (Fig. 26) which dominate. In particular, in the planar gauge[18] $(p_1 + \alpha p_2) \cdot A = 0$, and in the leading logarithm approximation, the top of the diagram decouples from the bottom, and so we get (4.45) naturally. In other axial gauges we also have to consider the mass singularities in the vertex corrections at the electromagnetic vertex. Equation (4.45) is true to all orders of perturbation theory in the leading logarithm approximation[18),59),60)]. The proof is

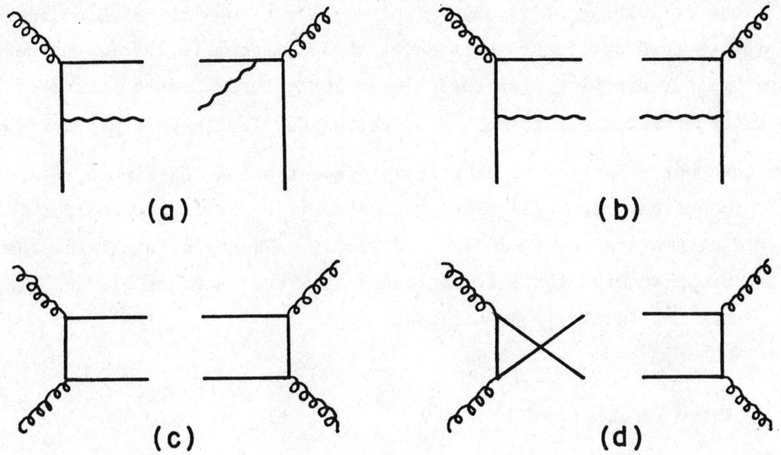

Fig. 25 Sample diagrams contributing non-scaling terms to deep inelastic structure functions

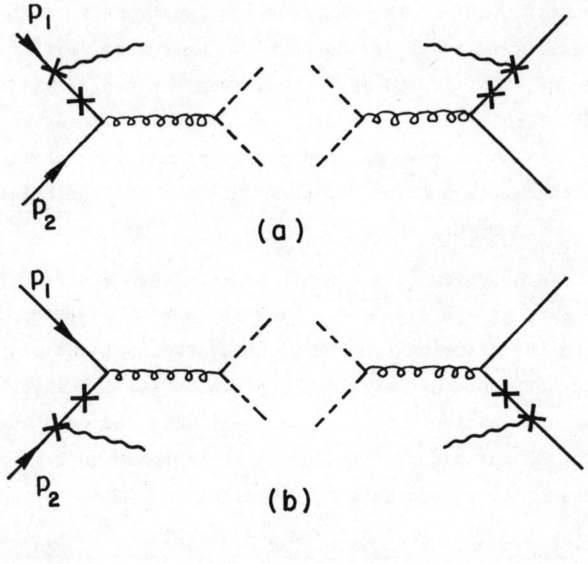

Fig. 26 Sample diagrams which contribute terms which violate Drell-Yan scaling

analogous to the derivation of (4.35). Thus the large logarithms which appear in the Drell-Yan process turn out to be the same as those in deep inelastic scattering. So once we have "measured" the sum of these large logarithms in deep inelastic scattering we can use these measurements to make predictions for lepton-pair production.

We now turn our attention to other subprocesses which contribute to massive lepton pair production. As an example, let us consider qg scattering for which some lowest order diagrams are shown in Figs. 1d and 1e. When one calculates the contribution to the cross-section for the subprocess from the diagrams of Figs. 1d and 1e one finds[61)] that the result is proportional to

$$g^2 \left(1 - 2\tau(1-\tau)\right) \log Q^2/p_2^2 \ \sigma_0 \tag{4.46}$$

Again the coefficient of $g^2\sigma_0$ is highly significant. It is the probability of finding an antiquark in a gluon with a fraction τ of its longitudinal momentum[*)]. When the distribution functions are measured in deep inelastic scattering, they include the effects of gluons through diagrams such as those of Figs. 25c and 25d. Equation (4.46) implies that the contribution to the distribution functions from these diagrams is related to the contribution to the cross-section for lepton pair production from diagrams such as those of Figs. 1d and 1e. Thus, even these diagrams are included in the right-hand-side of Eq. (4.45). In other words, the dominant contribution from the subprocess $qg \to \ell^+\ell^- x$ can be expressed as the probability of finding an antiquark in the gluon times the cross-section for this antiquark to annihilate the incident quark.

A similar feature has been found in all other subprocesses which have been studied (e.g., $qq \to \ell^+\ell^- x$, $gg \to \ell^+\ell^- x$ [61)]). In each case the leading logarithmic terms can be absorbed into the distribution functions of the incident hadrons, so that the asymptotic behaviour of the cross-section is given by (4.45), where the G's are extracted from linear combinations of experimentally measured deep inelastic structure functions. Thus the dominant contribution can be represented by Fig. 27 where the shaded blocks represent these non-scaling distribution functions.

4.3.2 Inclusive production of particles and jets with large transverse momenta

Deep inelastic scattering and massive lepton pair production both involve an off-shell photon; here we study a purely strong interaction process, the production of particles and jets with large transverse momenta in hadronic collisions. The mechanism

---

*) The moments of $1 - 2\tau(1 - \tau)$ with respect to τ are thus proportional to one of the off-diagonal elements in the anomalous dimension matrix of the lowest twist quark and gluon operators.

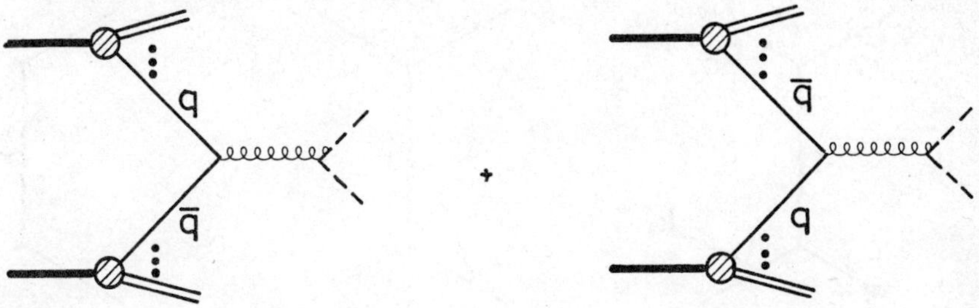

Fig. 27 The dominant contribution to massive lepton pair production in QCD. The shaded blobs represent experimentally determined, non-scaling distribution functions

usually believed to be responsible for this process has been described in Section 1 and can be summarized by Eq. (14) and Fig. 2. To see what, if anything, QCD can teach us about this process, let us study a concrete example quark (A) + quark (B) → → quark (C) + anything, up to order g^6 in the cross-section and, as before, in the leading logarithm approximation[62]. For simplicity we start with the case $x < 1$, where $x = (2E_c/\sqrt{s})$, in which case only the inelastic diagrams of Fig. 28 contribute. We notice that even in this order there are diagrams with a three-gluon vertex.

There are a number of kinematic regions which contribute to the leading logarithmic behaviour. All these contributions can be interpreted in an elegant way. From the region where k is parallel to $p_A(p_B)$ we obtain a contribution to $E_c(d\sigma/d^3p_c)$ of the form $\log s/p_A^2$, $(\log s/p_B^2)$ which can be interpreted as being the convolution of the probability of finding a quark in quark A(B), and the Born term for quark-quark elastic fixed angle scattering (Figs. 29a and 29b). In other words, the coefficient of these logarithms is just the function "a" of Eqs. (4.43) and (4.44) (in spite of the presence of the three-gluon vertex). From the region where k is parallel to p_C, we obtain a contribution which can be interpreted as the convolution of the Born term for q-q elastic fixed angle scattering, and the probability for one of the resulting quarks to fragment into the observed one + anything (Fig. 29c). The fragmentation function is just that which would be calculated from $e^+e^- \to qX$ and is related to the distribution function (in this order, at least) by the Gribov-Lipatov reciprocity relation[63]. The final dominant contribution comes from the region in which k balances the transverse momentum of p_C, and this can be interpreted as the probability of finding a gluon in quark B with a fraction of the longitudinal momentum of B[*)] convoluted with the Born term for gq elastic fixed angle scattering (Figs. 29d, 29e and 29f).

---

*) The moments of this probability are, of course, proportional to an off-diagonal element in the anomalous dimension matrix for the lowest twist quark and gluon operators.

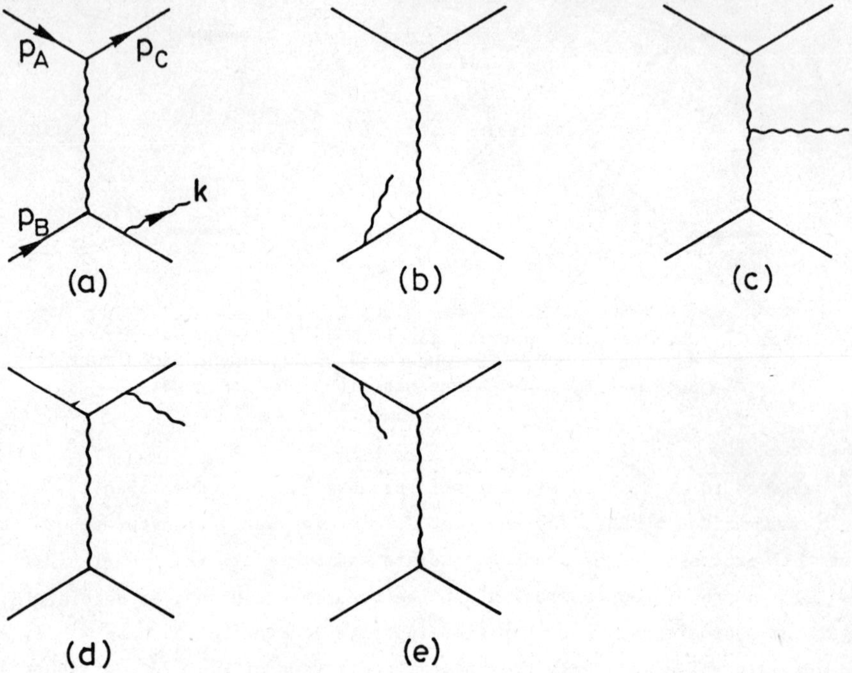

Fig. 28 Lowest order inelastic diagrams which contribute to the process $qq \to qx$

At $x = 1$, in addition to the contributions from the diagrams of Fig. 28, one must also include the diagrams which contribute to the elastic qq scattering amplitude. The leading $\log s/p^2$ contributions can also be absorbed into the initial distribution and final fragmentation functions. In addition to these $\log s/p^2$ terms, after renormalization there will also be $g^6(\mu^2) \log s/\mu^2$ terms. These terms are exactly those which, when combined with the $g^4(\mu^2)$ contribution from the Born term give the first two terms in the expansion of $g^4(s)$, the running coupling constant.

Thus we conclude that the cross-section for $qq \to qX$ to order g^6 in the cross-section and in the leading logarithmic approximation can be written in the form (Fig. 30)

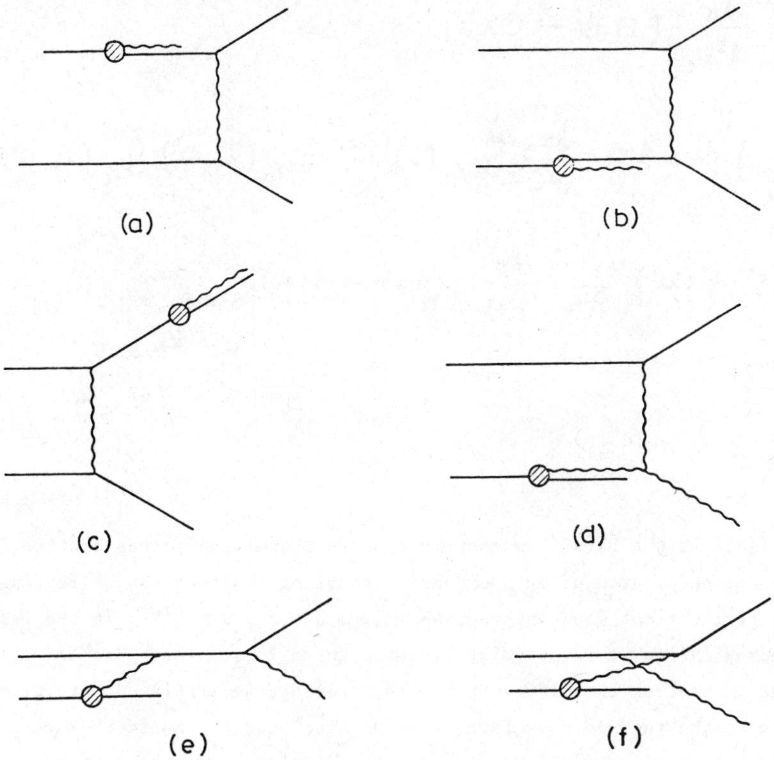

Fig. 29 Symbolic representation of the dominant contribution from the diagrams up to order g^6 for the process $qq \to qx$. The shaded blobs represent the non-scaling distribution and fragmentation functions

$$E_c \frac{d\sigma}{d^3p_c} (A + B \to CX) = \sum_{\substack{\text{quarks} \\ \text{gluons}}}$$

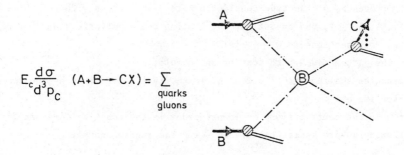

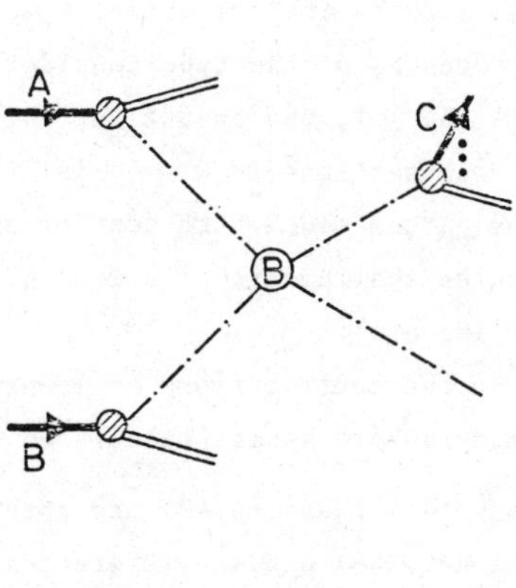

Fig. 30 Dominant contribution to $E_c\, dv/d^3p_c$. The -· -· -· line represents quarks and gluons, the shaded blobs represent non-scaling distribution and fragmentation functions and the "B" implies that we take the Born term contribution calculated using the running coupling constant

$$E_c \frac{d\sigma}{d^3p_c}(AB \to CX)$$

$$= \sum_{a,b:c,d} \int_0^1 dx_a\, dx_b\, \frac{dx_c}{x_c^2}\, G_{a/A}(x_a,Q^2)\, G_{b/B}(x_b,Q^2)\, \tilde{G}_{c/C}(x_c,Q^2)$$

$$\cdot\, \delta(s'+t'+u')\, \frac{s'}{\pi}\, \frac{d\sigma^{(B)}}{dt'}(a+b \to c+d)\Bigg|_{\substack{s'=x_a x_b s \\ t'=x_a/x_c\, t \\ u'=x_b/x_c\, u}} \tag{4.47}$$

where $d\sigma^{(\beta)}/dt'$ is the Born term contribution to the elastic cross-section for qg or qq scattering, calculated using the running coupling constant and Q^2 represents a typical large invariant mass squared (we assume $s \sim t \sim u \sim Q^2$). In the leading logarithm approximation we cannot distinguish between log s, log t and log u. It is interesting to see how the different subprocesses are interrelated, starting with the quark-quark scattering, we have found that we also need to consider quark-gluon scattering. If we had started off by looking at radiative corrections to quark-gluon scattering, we would have found that we also need to consider the third subprocess, gluon-gluon scattering. Again, general arguments such as those in Ref. 59) ensure the validity of (4.47) to all orders of perturbation theory in the leading logarithm approximation.

Studies, such as the two examples we have discussed above, show that for all hard scattering processes of the type considered here (i.e., those with the three mass scales Q^2, p^2 and μ^2), the ansatz for calculating the asymptotic behaviour of hard scattering cross-sections as $Q^2 \to \infty$ is:

a) Take the parton model hard scattering formula;
b) Replace the scaling distribution and fragmentation functions by the appropriate non-scaling ones;
c) Keep only the contributions of lowest order in the coupling constant for the hard scattering subprocesses, calculated using the running coupling constant.

Equations (4.45) and (4.47) are specific examples of this ansatz. For other examples, see Ref. 64) and the references therein.

We now understand how to calculate the leading asymptotic prediction for these hard scattering processes. In the next section we will see that all the logarithmic corrections are also calculable.

5. BEYOND LEADING LOGARITHMS

5.1 Introduction

In Section 3 we saw how to calculate logarithmic corrections to the predictions for the scaling violations in deep inelastic structure functions. In this section we will outline how one would approach the problem of calculating these logarithmic corrections to lowest order formulae [such as Eqs (4.45) and (4.47)] for other processes, processes in which we relate the cross-section in question to the deep inelastic structure functions or e^+e^- fragmentation functions.

5.2 A simple example: The longitudinal structure function

As a simple example of a hard scattering process, let us take the longitudinal structure function F_L. In the parton model (with spin ½ quarks) this structure function decreases with q^2 like a power of q^2, e.g., under certain assumptions the ratio F_L/F_T is given by[65)]

$$\frac{F_L}{F_T} = \frac{4\langle k_\perp^2\rangle}{(-q^2)} \tag{5.1}$$

where $\langle k_T\rangle$ is the average transverse momentum squared of the quarks in the hadron. To see that $F_L \sim 1/q^2$ in the parton model one has to calculate the contribution from Fig. 31(a).

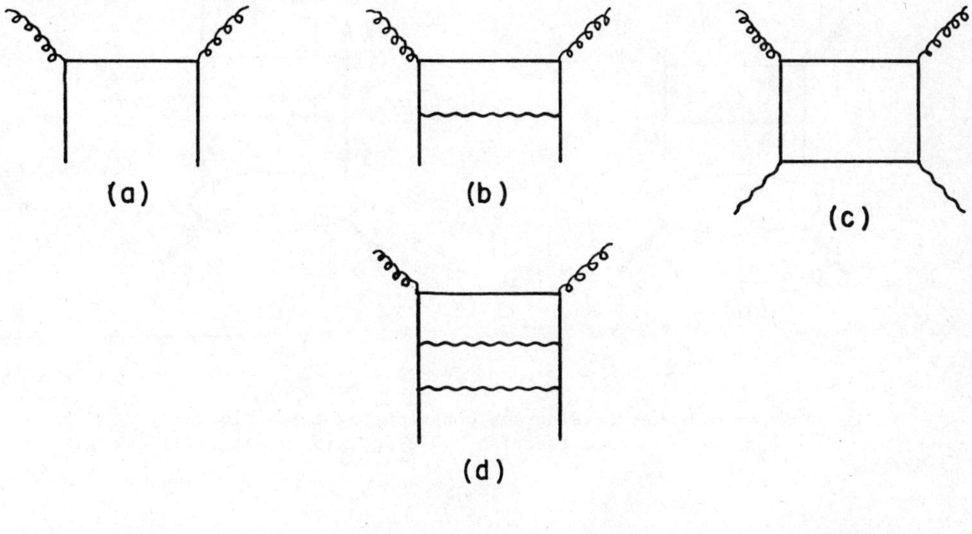

Fig. 31 Sample diagrams contributing to deep inelastic scattering

In QCD, however, there are contributions from higher order diagrams, such as those of Figs 31(b) and (c) which scale (up to the logarithmic corrections). Thus, using the ansatz at the end of Section 4 we would write for the leading term asymptotically

$$F_L(q^2,x) = \sum_{f=q,\bar{q},g} \int_x^1 \frac{dy}{y} f(y,q^2)\, F_{L,f}^{(0)}\left(q^2, \frac{x}{y}, g(q^2)\right) \tag{5.2}$$

where f runs over all quarks and gluons and is defined by means of F_T, for example, and $F_{L,f}^{(0)}$ is the lowest order contribution to the longitudinal structure function of f, evaluated using the running coupling constant. This can easily be checked to be the same answer as would be obtained by the light cone techniques of Section 2. Even though in QCD, F_L/F_T and $\langle k_T^2\rangle/q^2$ are now both only logarithmically suppressed, relation (5.1) is no longer valid.

Now we would like to calculate the next-to-leading corrections to Eq. (5.2); in order to do this we start by calculating both F_L and F_T in terms of perturbation theory*). We start by calculating F_L, and for simplicity we will restrict ourselves to the non-singlet quantity $F_L^{\mu p} - F_L^{\mu n}$, the generalization to $F_L^{\mu p}$ itself is straightforward. The lowest order diagram is shown in Fig. 32a and its contribution can be written

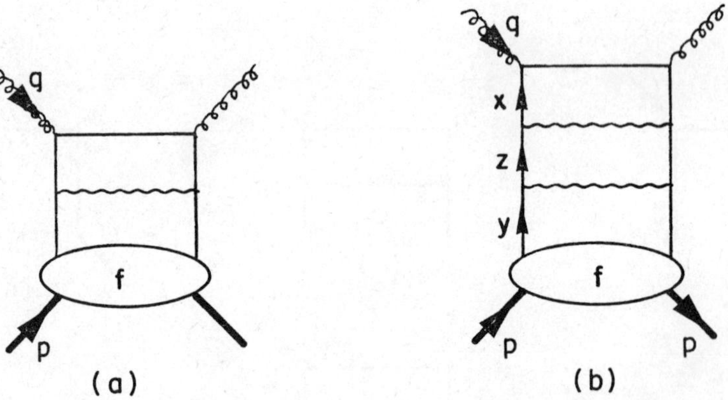

Fig. 32 Sample low order diagrams contributing to the longitudinal structure function. See Eqs. (5.3) and (5.4)

*) For a light-cone derivation of the relation between F_L and F_T beyond leading logarithm see, e.g., Ref. 64).

$$\alpha_s \int dk^2\, d^2k_\perp \int_x^1 \frac{dy}{y}\, f(y,k^2,k_\perp^2)\, F_L^{(0)}\left(\frac{x}{y}\right) \tag{5.3}$$

where f is defined as in Section 4 and $F_L^{(0)}$ is the contribution from the lowest order diagram (Fig. 32a) to the longitudinal structure function of a quark. We now have to calculate the two-loop diagrams, for example, that of Fig. 32b. Now the extra loop integration will give us a logarithmic term of the type log q^2/k^2 and a constant term. The contribution from the two-loop diagrams can be written in the form

$$\frac{\alpha_s^2}{2\pi} \int dk^2 d^2k_\perp \int_x^1 \frac{dy}{y} f(y,k^2,k_\perp^2) \int_x^y \frac{dz}{z} \left[P_{q\to q}\left(\frac{z}{y}\right) F_L^{(0)}\left(\frac{x}{z}\right)\right.$$
$$\left. \cdot \log q^2/k^2 \right] \tag{5.4}$$
$$+ \frac{\alpha_s^2}{2\pi} \int dk^2\, d^2k_\perp \int_x^1 \frac{dy}{y}\, f(y,k^2,k_\perp^2)\, F_L^{(1)}\left(\frac{x}{y}\right)$$

where $F_L^{(1)}$ is the constant term (the term without the log q^2) in the contribution to the longitudinal structure function of a quark from all the two-loop diagrams such as that of Fig. 31d. The first term on the right-hand side of (5.4) has, of course, precisely the parton model-like interpretation of the leading logarithms discussed in Section 3.

We now have to calculate F_T to the same accuracy. We start by calculating the lowest order diagram, that of Fig. 33a and we find for the contribution to F_T:

$$\int dk^2\, d^2k_\perp\, f(x,k^2,k_\perp^2) \tag{5.5}$$

From the one-loop diagrams, such as those of Fig. 33b we find for the contribution to F_T:

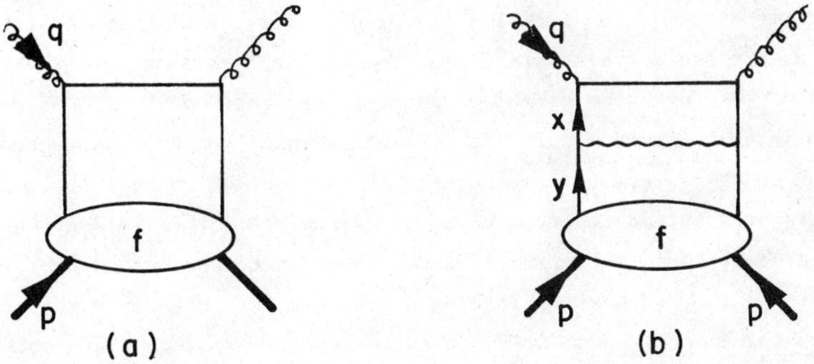

Fig. 33 Sample low order diagrams which contribute to the transverse structure function. See Eqs. (5.5) and (5.6)

$$\frac{\alpha_s}{2\pi}\int dk^2\, d^2k_\perp \int_x^1 \frac{dy}{y}\, f(y,k^2,k_\perp^2)\, P_{q\to q}\left(\frac{x}{y}\right) \log q^2/k^2$$

$$+ \frac{\alpha_s}{2\pi}\int dk^2\, d^2k_\perp \int_x^1 \frac{dy}{y}\, f(y,k^2,k_\perp^2)\, F_T^{(1)}\left(\frac{x}{y}\right) \qquad (5.6)$$

where $F_T^{(1)}$ is the constant term in the contribution from the one-loop diagrams to the transverse structure function of a quark (such as that of Fig. 31b).

Since Eqs (5.3) to (5.6) contain convolutions it is convenient to take the moments with respect to x, and we denote the Nth moment of the longitudinal and transverse structure functions by F_L^N and F_T^N, respectively. It is possible now to eliminate f (of which we are ignorant) and we find

$$\frac{F_L^N(q^2)}{F_T^N(q^2)} = \alpha_s(q^2)\, F_L^{N(0)}(q^2) + \frac{\alpha_s^2(q^2)}{2\pi} F_L^{N(1)}(q^2)$$

$$- \frac{\alpha_s^2(q^2)}{2\pi} F_L^{N(0)}(q^2)\, F_T^{N(1)}(q^2) \qquad (5.7)$$

which is the final result. It can also be obtained by light cone techniques[64].

5.3 General discussion

For (5.7) to be valid, the "next-to-leading" logarithms in all orders of perturbation theory must arrange themselves in a particular way, again generating the series (4.7) in both F_L and F_T. That this is the case for general hard scattering processes has been proved in (for example) Refs 52), 59) and 60). In view of the discussion of Section 4 this is not so surprising, since, whereas we saw that the leading logarithms in the axial gauge come from "ladder" diagrams, the next-to-leading logarithms come from modified ladder diagrams in which two of the rungs have been crossed (these are common to all hard processes and therefore do not change the leading order formulae) or from non "collinear" regions of integration in the innermost loop. This last effect is exactly taken into account by calculating the constant term (i.e., the term independent of log q^2) in the lowest but one order of perturbation theory, in the way discussed for F_L in Section 5.2. Thus for a general hard scattering process, in order to obtain the expression for the cross-section, σ, to next-to-leading order, one must calculate the constant terms in the lowest order but one for the cross-section for the hard scattering subprocesses ($\Delta\tilde{\sigma}$), and also the corrections to the appropriate quark and gluon distributions (Δf, $f = q, g, \bar{q}$). The correction is then the appropriate difference of terms involving the $\Delta\tilde{\sigma}$ and Δf (analogously to (5.7), except that in general there are more than one $\Delta\tilde{\sigma}$ and Δf) at the level of moments. Notice that because the quarks and gluons are not on mass-shell $\Delta\tilde{\sigma}$ and Δf separately are not gauge invariant, it is only the difference which is physically meaningful.

In addition to including non-leading terms from dominant subprocesses, for consistency one must also include subprocesses which start in the next-to-leading order in α_s. The discussion of Sections 5.2 and 5.3 can be easily generalized to still less leading logarithms.

Recently the $O(\alpha_s)$ corrections to the Drell-Yan process have been calculated[66)-69)], they turn out to be very significant.

5.4 Interactions involving "spectator" partons[70)]

In the previous sections (with the exception of Section 2.3), we calculated predictions for hard scattering processes by evaluating low order Feynman diagrams for quark and gluon subprocesses. We have not considered diagrams such as those of Fig. 34, which involve the interactions of spectator partons. By assuming that the hadronic wave function f is soft, one can argue that these diagrams are logarithmically suppressed (the softness of the wave function prevents the loop integral diverging, thus there is no mass singularity logarithm), compared to the leading ones discussed in Section 4, but that's all; it we want to calculate the logarithmic corrections we must study these diagrams some more. Moreover, the contributions of these diagrams are gauge dependent. In the case of deep inelastic scattering the light cone analysis guarantees that Eq. (2.19) is correct, irrespective of the contributions of diagrams such as Fig. 34a. The effects of such diagrams for deep inelastic scattering (in a general gauge) are included in the operator matrix element, and the higher order radiative corrections again generate the series (4.7). However, a priori, we have no such guarantee that diagrams involving spectator quarks for other hard scattering processes (e.g., Fig. 34b) are negligible [i.e., $O(m^2/Q^2)$]. We shall show in this section that this is, nevertheless, the case.

For definiteness let us take a simple ϕ^3 model for the wave function. We take an interaction term of the type $\lambda\phi\chi^+\chi$, where ϕ and χ are scalar fields, with ϕ being both electrically neutral and a colour singlet, whereas χ is a charged and coloured (belonging to the fundamental representation of colour SU(3)) field. Thus ϕ in our model for the "hadron" and the χ's are our "quarks". The quarks interact in the usual way with the coloured gluons, which belong to the adjoint representation. λ has dimensions of mass and therefore the $\phi\chi^+\chi$ vertex is soft. We believe that this is the only relevant assumption, the further details of the model (such as whether the quarks have spin 0 as in this case, or spin ½) are not relevant. The advantage of working in a specific simple model is that we can obtain explicit answers for the

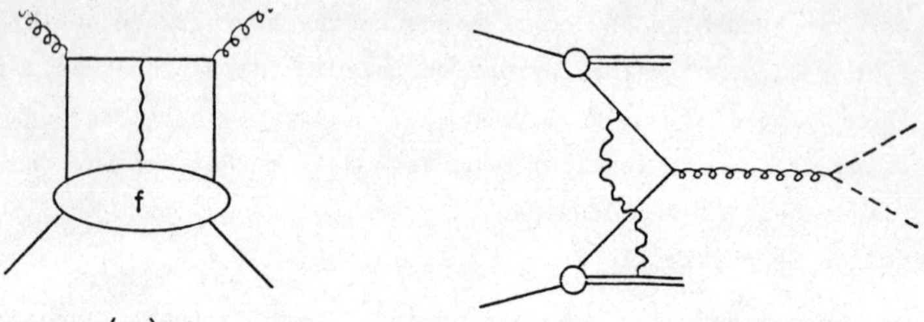

(a) (b)

Fig. 34 Some diagrams relevant for (a) deep inelastic scattering and (b) massive lepton pair production in which "spectator partons" interact

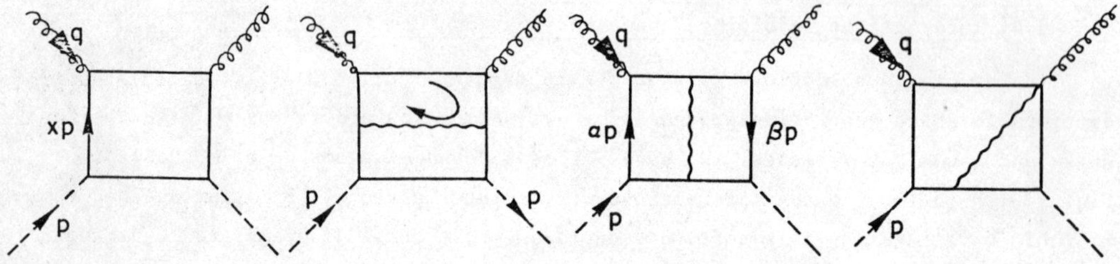

(a) (b) (c) (d)

Fig. 35 Some low order diagrams which need to be calculated in the evaluation of deep inelastic structure functions

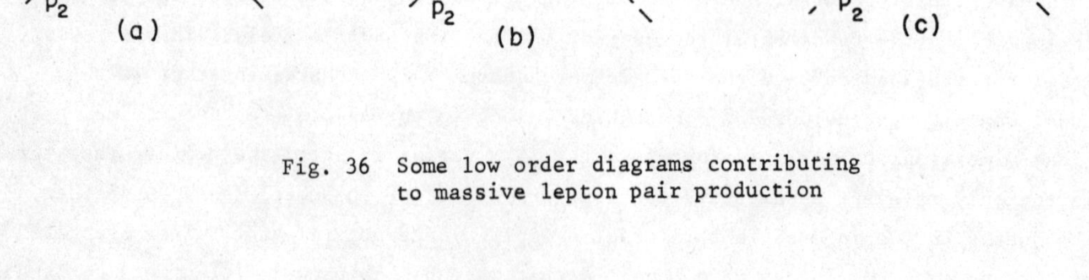

(a) (b) (c)

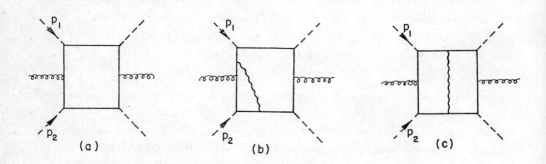

Fig. 36 Some low order diagrams contributing to massive lepton pair production

Feynman diagrams, in particular, we can see how gauge invariance is restored between diagrams involving only the "participating" quarks and those which also involve "spectator" quarks.

In order to try and get some insight into the problem, we start by calculating some low order diagrams. For the moment we will work in the Feynman gauge. The lowest order diagram is that of Fig. 35a and gives an exactly scaling contribution. Some sample diagrams contributing to the next order are shown in Figs 35b, c and d. The dominant contribution from Fig. 35b has a log q^2/p^2 term (analogously to the diagram of Fig. 19), this comes from the usual "collinear" region of integration corresponding to $p^2 \ll k_T^2 < \varepsilon q^2$. The transverse momentum is allowed to get as large as εq because in the loop marked with the arrow in Fig. 35b there are no soft vertices. This is not the case in Figs 35c and d, and we find that for these diagrams we get a scaling contribution, i.e., no logarithms. This is because there is no loop in which the transverse momentum could get large without flowing through a soft vertex. So, in this order of perturbation theory, the contributions from diagrams which include interactions involving spectator quarks is suppressed by only one logarithm.

Let us now study the diagram of Fig. 35c in some detail. An analysis of the dominant momentum flow in the diagram (using, for example, the techniques of Refs. 71) and 72)) shows that is as marked on the diagram, the momentum flowing in the vertical propagators is proportional to p, whereas the two top horizontal propagators are hard. Comparing this to the lowest order scaling diagram of Fig. 35a we now see that we have:

a) 1 extra hard propagator giving a factor $\sim 1/p\cdot q$;

b) 1 factor from the numerator which is $\sim p\cdot q$.

Thus the diagram of Fig. 35c gives a scaling contribution. We notice that to obtain the factor of $p\cdot q$ in the numerator, we have had to take a factor of p^λ from the correction current corresponding to the gluon vertex at the bottom of the diagram (the only large momentum available at the bottom is proportional to p) and therefore a q^λ from the convection current corresponding to the gluon vertex at the top of the diagram. If we now choose to work in the gauge

$$q.A = 0 \tag{5.8}$$

[or indeed $(q + \alpha p)$ A = 0, where α is an arbitrary constant], it is easy to see that we do not get the power of $p\cdot q$ in the numerator. Thus the diagrams involving spectator quark interactions are gauge dependent and in some gauges contribute to the "leading twist" terms and in other gauges do not.

Before trying to understand this feature, let us look at the Drell-Yan process, $h_1h_2 \to \ell^+\ell^-X$ and study some of the diagrams involving interactions of spectator quarks, e.g., those of Figs 36b and 36c. We leave the diagram of Fig. 36c till later, and start by looking at that of Fig. 36b, which has many similar features to that of Fig. 35c in the deep inelastic case. In the dominant region of phase space, and in the Feynman gauge this diagram also has

i) one more hard propagator and hence a factor of $\sim 1/s$; and

ii) one extra numerator and hence a factor of $\sim s$

compared to the lowest order diagram, that of Fig. 36a. So, in the Feynman gauge, the diagram of Fig. 36b gives a (Drell-Yan) scaling contribution, and must be considered. However, also in this Drell-Yan case, we can define an axial gauge

$$(p_1 + \beta p_2) \cdot A = 0 \tag{5.9}$$

where β is an arbitrary constant, in which the diagram of Fig. 36b is suppressed by a factor of s.

We notice that in both the deep inelastic and Drell-Yan cases, the gauge in which the diagrams involving spectator parton interactions are negligible (e.g., Figs 35c and 36b) depends on the external momenta. Can we find a single gauge in which both the diagrams of Fig. 35c and 36b are negligible? To this end we define

$$q_1 = \frac{1}{x_1} y_1 \left(p_2 - x_1 p_1\right) \tag{5.10a}$$

and

$$q_2 = \frac{1}{x_2} y_2 \left(p_1 - x_2 p_2\right) \tag{5.10b}$$

so that

$$\frac{-q_1^2}{2p_1 \cdot q_1} = x_1 \quad , \quad \frac{-q_2^2}{2p_2 \cdot q_2} = x_2 \tag{5.11}$$

and y_1, y_2 are arbitrary constants. Now, if we calculate the deep inelastic structure functions in the cases $q = q_1$, $p = p_1$ and $q = q_2$, $p = p_2$ in the gauge (4.9), we see that there is no contribution from the diagram of Fig. 35c. Hence in the gauge (5.9) we can simultaneously eliminate the "unwanted" diagrams of Fig. 35c in the deep inelastic scattering case, and that of Fig. 36b in the Drell-Yan case. We also notice from (5.10) that although $q_1^2 (q_2^2)$ is proportional to s, the constant of proportionality depends on $y_1 (y_2)$ and hence is not determined; this is a statement that these diagrams do not really contribute to the next-to-leading logarithms.

In the Feynman gauge, as we have seen above, both the diagrams of Figs 35c and 35b contribute to the next-to-leading logarithms. These diagrams have many similar features and closer examination reveals that they are related by the Drell-Yan formula, so that when we write down this formula, and insert the experimentally extracted parton distribution functions, we are including the contributions from diagrams such as Fig. 35c for these distribution functions and hence the contributions of diagrams such as 36b for the Drell-Yan cross-section.

We are not finished yet, there is a class of diagrams which have not been included, for example, the diagram of Fig. 36c. This diagram has no analogue in the deep inelastic case. The dominant region of phase space is that in which k is orthogonal to p_1 and p_2 and is therefore spacelike. This means that if we evaluate the contribution to the lepton-pair cross-section from this diagram by taking all the unitarity cuts, we can neglect any cuts through the gluon. This diagram is very reminiscent of that studied by De Tar et al.[73)], in which the gluon is replaced by a Pomeron. The conclusion of Ref. 73) was that the Pomeron diagram was suppressed by a factor $\sim 1/s$. When evaluating the diagram, since we are dealing with an inclusive cross-section, we have to take care with the $i\varepsilon$'s in the propagators (we are taking the Mueller discontinuity[74)]), and, having done this, it turns out that there is always an integration which can be shown to be zero by using Cauchy's theorem. In our case we have an extra singularity due to the pole in the gluon propagator, however, as stated above, the contribution from this pole is also suppressed by a factor of s. It is amusing to see how one obtains this answer of zero by taking unitarity cuts. In the Pomeron case, the Pomeron is imaginary and there are three cuts (see Fig. 37a) giving the relative contributions 1, -2, 1. In the gluon case, the gluon is real so there are just two cuts (see Fig. 37b) (remember the cut through the gluon is zero) giving the relative contributions 1, -1.

Thus we find that in lowest order perturbation theory, when we follow the prescription for calculating hard scattering cross-sections outlined in Sections 3 and 4 above, there is no additional contribution from diagrams involving spectator partons (these diagrams are suppressed by a power of Q^2). This feature can be generalized to higher orders of perturbation theory[70)]. It is straightforward to see that (in the Feynman gauge, for instance), there will be many contributions to the Drell-Yan cross-sections which have corresponding contributions to the deep inelastic structure functions and (somewhat less straightforwardly) it can be shown that interactions between "spectators" of different hadrons cancel out. We have used the Drell-Yan process as an example, but the above argument can easily be generalized to other hard scattering processes.

The arguments given in this section lead us to believe that we now know how to calculate, not only the asymptotically leading contribution to the cross-section for hard scattering processes, but also all the logarithmic corrections.

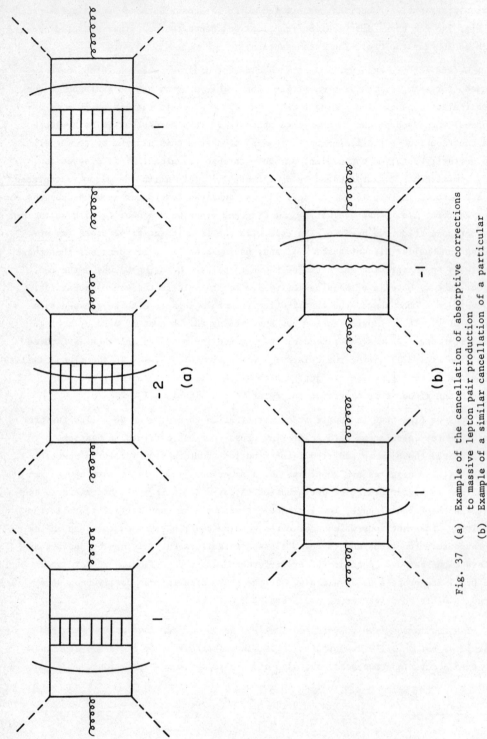

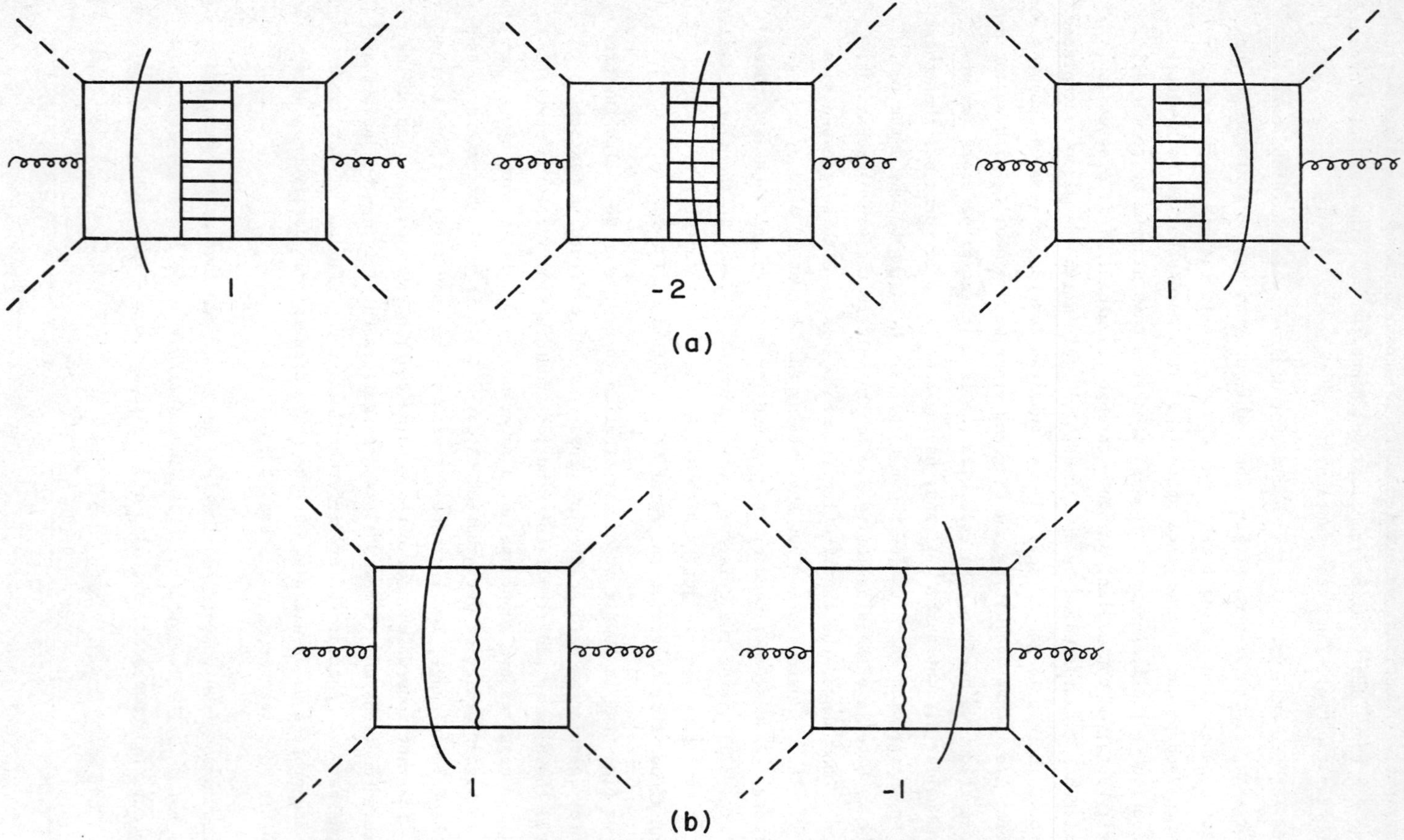

Fig. 37 (a) Example of the cancellation of absorptive corrections to massive lepton pair production
(b) Example of a similar cancellation of a particular type of gluonic corrections (see text)

6. APPLICATIONS OF QCD TO (HARD) EXCLUSIVE PROCESSES

6.1 Introduction

All the discussion of the previous sections has concerned inclusive reactions; but recently, in the past few months, there has been some very interesting progress in the application of QCD to exclusive processes[75], and in this lecture we will briefly review these developments. Before doing this, however, let us recall how one approached exclusive processes within the context of the parton model.

It is popular, within the parton model, to think of hard exclusive processes in terms of "dimensional counting rules"[12),13)]. For example, for the spin averaged form factor of a hadron H, these give

$$F_H(t) \sim \frac{1}{t^{N-1}} \tag{6.1}$$

where $t = q^2$ and N is the number of fundamental constituents in H. Thus the form factor of a pion is predicted to behave asymptotically as $1/t$, and that of a proton as $1/t^2$, both of which agree with the data [which exists for $|t|$ up to 4(6) GeV^2 for the pion (proton)]. We can see heuristically that (6.1) is correct by counting the "hard fermion propagators" (see Fig. 38). We iterate the hadronic wave function, so that the large transverse momentum is routed through the wiggly lines (which we would now call gluons). In deriving the dimensional rules, it is assumed that the four quark interaction is scale invariant. We see now that in the meson form factor there is (at least) one hard fermion propagator ($\sim 1/t$), and in the proton case there are two such propagators ($\sim 1/t^2$). The generalization to arbitrary N is trivial.

For elastic scattering at fixed angle the dimensional counting rules are

$$\frac{d\sigma}{dt}(AB \to CD) \sim \frac{1}{t^{N-2}} f(\theta) \tag{6.2}$$

where N is the total number of constituents in A, B, C, and D, and θ is the scattering angle. Thus $d\sigma/dt$ is predicted asymptotically to behave like $1/t^6$, $1/t^8$, and $1/t^{10}$ for meson-meson, meson-baryon and baryon-baryon scattering, respectively. These predictions seem to be consistent with the experimental data. To see that (6.2) is reasonable we repeat the "hard fermion" counting arguments which we used in the form factor case. For example, in Fig. 39 we present some sample diagrams which contribute to meson-meson scattering at large angle. The lines marked with a cross represent hard quark propagators, and we see that in each diagram there are two such propagators. Thus the amplitude $\sim 1/t^2$ and hence $d\sigma/dt \sim 1/t^2(1/t^2)^2 \sim 1/t^6$ which satisfies (6.2). The generalization to higher N is straightforward.

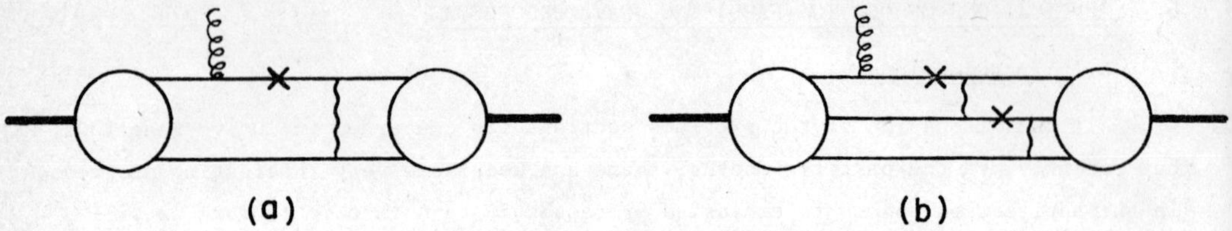

Fig. 38 Diagrams calculated in the derivation of the dimensional counting rules for the form factor of (a) a meson and (b) a baryon

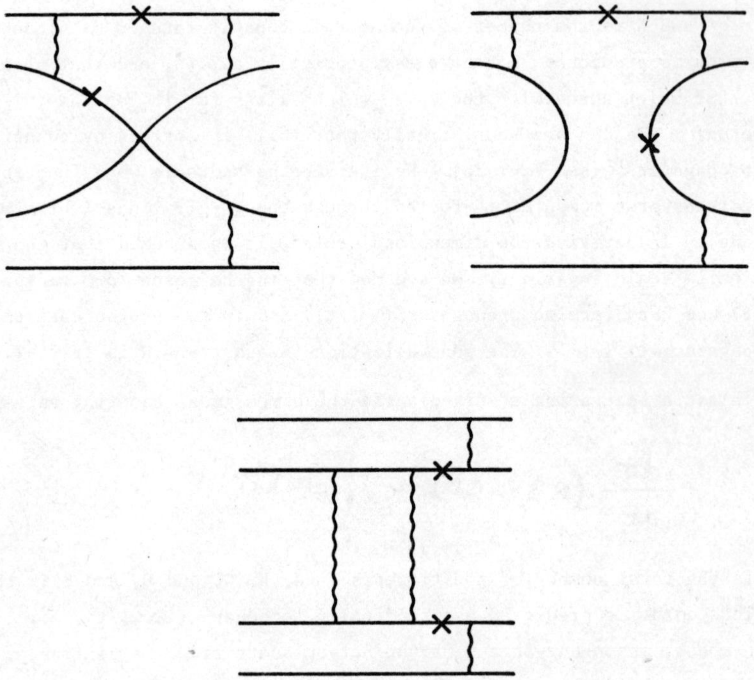

Fig. 39 Diagrams calculated in the derivation of the dimensional counting rules for meson-meson fixed angle scattering

There is one further complication. Landshoff[76)] proposed that instead of diagrams such as those in Fig. 39 in which there are "hard fermion" propagators, the relevant mechanism for fixed angle scattering may contain no such propagators at all. For example, he suggested that meson-meson scattering may be dominated by two separate scatterings of the two valence quarks through the same angle (see Fig. 40a). Although there are no hard fermion propagators to suppress the amplitude, there is little phase space for both quarks to scatter through the same angle. Calculation of the diagram of Fig. 40 reveals that it contributes to the amplitude $M_{\pi\pi}$ a term

$$M_{\pi\pi} \sim \frac{1}{t^{3/2}} \tag{6.3a}$$

which leads to

$$\left.\frac{d\sigma}{dt}\right|_{\pi\pi} \sim \frac{1}{t^5} \tag{6.3b}$$

which we notice should dominate asymptotically over the dimensional counting contribution which behaves like $1/t^6$. Similarly the analogous "multiple scattering" contribution in the proton-proton case gives a contribution

$$\left.\frac{d\sigma}{dt}\right|_{pp} \sim \frac{1}{t^8} \tag{6.4}$$

which dominates over the dimensional counting contribution which gives $1/t^{10}$. The meson-baryon case is not so clear[77)].

In any renormalizable field theory we expect there to be at least some logarithmic modifications to Eqs (6.1) to (6.4). We shall see that this is indeed the case in QCD.

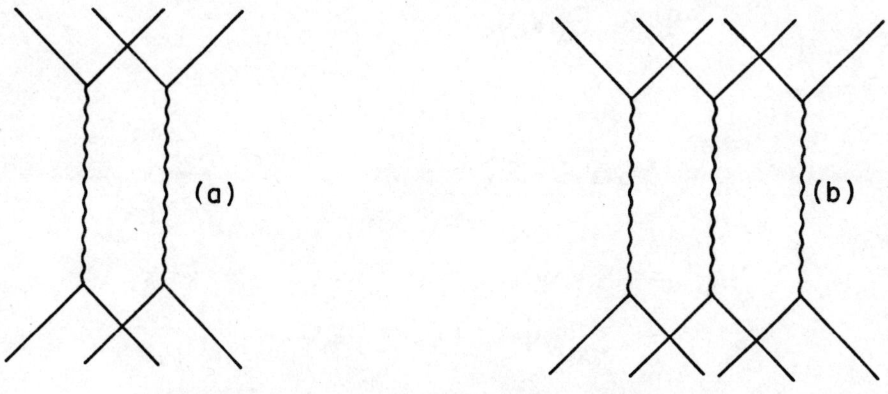

Fig. 40 Some "multiple-scattering" diagrams for (a) meson-meson and (b) baryon-baryon fixed angle scattering

6.2 Pion form factor in QCD

In this section we review the approach and results of Brodsky and Lepage[75),78)] concerning the pion form factor in QCD which is the most complete discussion of the problem. Other contributions to this problem can be found in Refs 79) to 82). We will study the form factor in time ordered perturbation theory in the infinite momentum frame. The meson Fock state ψ can be represented as a column vector with infinitely many components, corresponding to $q\bar{q}$, $q\bar{q}g$, $q\bar{q}q\bar{q}$, etc., states ($q \equiv$ quark, $g \equiv$ gluon). It satisfies the bound state equation

$$\psi = SK\psi \tag{6.5}$$

which is represented in Fig. 41. S is a diagonal, infinitely dimensional, matrix whose elements are the multiparticle propagators. K is the irreducible kernel.

The first step is to separate the hard and soft components in the wave function. To this end we introduce a mass cut-off λ, and define S_λ by

$$S_\lambda = \begin{cases} 0 & \text{if } \left| M^2 - \sum \frac{k_\perp^2 + m^2}{x} \right| > \lambda^2 \\ S & \text{otherwise} \end{cases} \tag{6.6}$$

where

$$M^2 - \sum_{i=1}^{n} \frac{k_{\perp,i}^2 + m_i^2}{x}$$

is the inverse n particle propagator. $S_\lambda = 0$ for virtual states which are far off the energy shell, in particular, it is zero if any of the constituents has transverse momentum $> \lambda$. Defining the soft component in the wave function ψ_λ by

$$\psi_\lambda \equiv S_\lambda K \psi \tag{6.7}$$

and

$$\Delta S \equiv S - S_\lambda \tag{6.8}$$

we find

$$\begin{aligned} \psi_\lambda &= SK\psi \\ &= S_\lambda K\psi + (S - S_\lambda) K\psi \\ &\equiv \psi_\lambda + \Delta S K \psi \\ &= \psi_\lambda + \Delta S K \psi_\lambda + \Delta S K \Delta S K \psi_\lambda + \cdots \end{aligned} \tag{6.9}$$

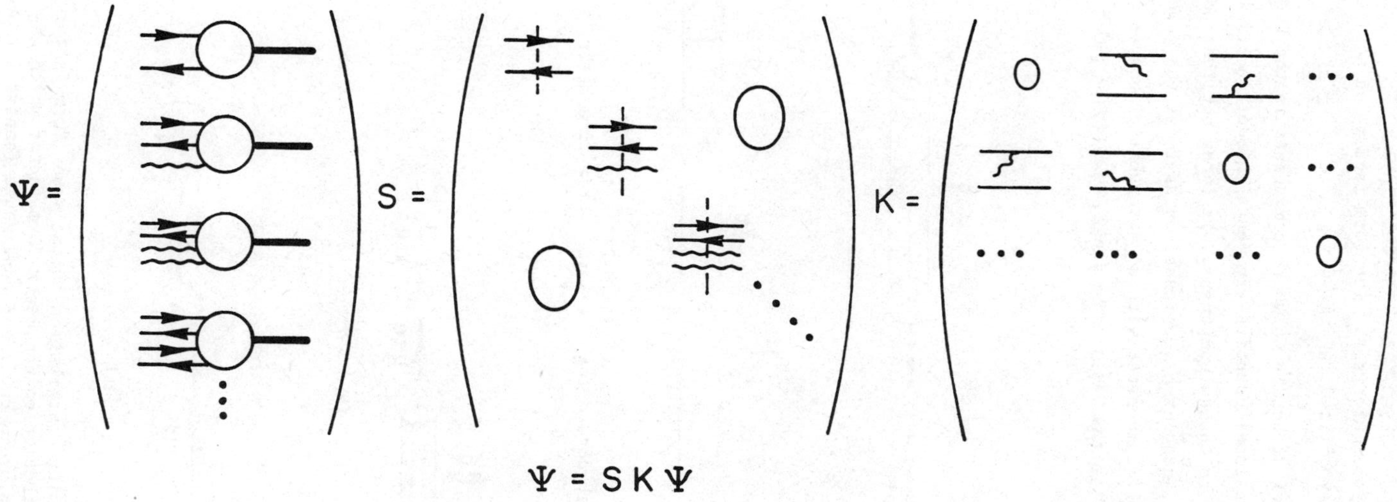

Fig. 41 Diagrammatic representation of the elements in the bound state equation of a meson

so that we can obtain the full wave function by perturbation theory from the soft wave function ψ_λ. We notice also that only off-shell propagation occurs. From now on we need only consider ψ_λ in Feynman diagrams. The form factor can now be represented by the set of diagrams symbolized in Fig. 42.

We would like to get some sort of handle on which components in the Fock space we need to consider. In the dimensional counting arguments in Section 6.1 we assumed we could restrict ourselves to the $q\bar{q}$ component. It turns out that components with more than two fermions (e.g., $q\bar{q}q\bar{q}$) are suppressed by at least one power of Q^2 (see, e.g., Fig. 43a). In general, components with $q\bar{q}$ + several gluons (e.g., Fig. 43b) are not suppressed, but in the light cone gauge, viz. $(p_0 + p_3) \cdot A = 0$,

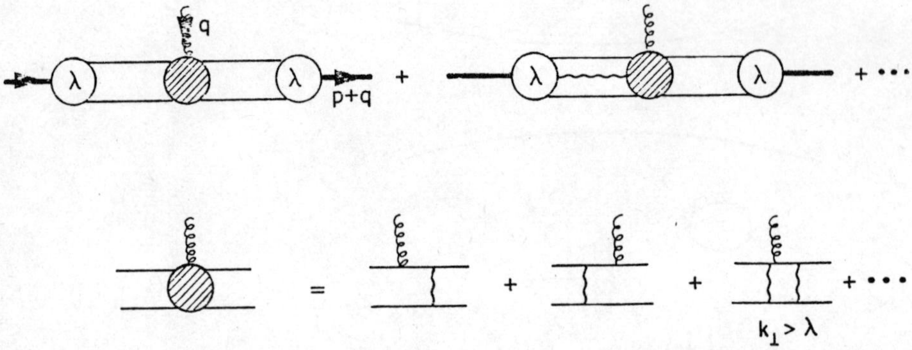

Fig. 42 Diagrammatic structure for the form factor of a meson

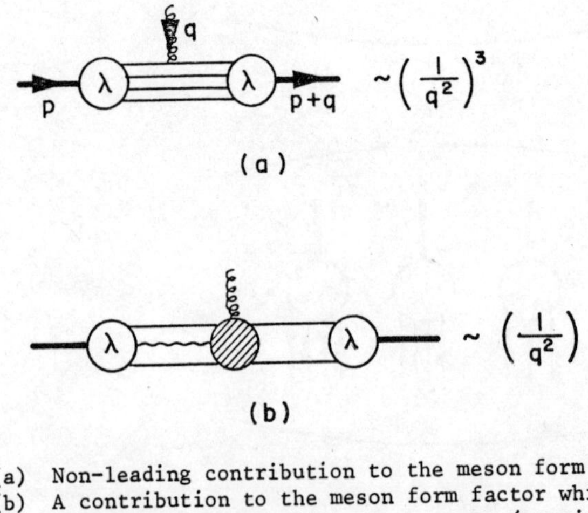

Fig. 43 (a) Non-leading contribution to the meson form factor
(b) A contribution to the meson form factor which is non-leading in the light cone gauge (considered in the text)

these components are also down by powers of Q^2. Thus in this gauge we need only to consider the $q\bar{q}$ component in the Fock space, i.e., the diagrams of Fig. 44.

We now turn our attention to the leading logarithms. Analogously to the deep inelastic structure functions we find that in the light-cone gauge the dominant diagrams are the generalized ladder diagrams of Fig. 45, where the dots represent vertex and self-energy corrections. The dominant region of integration is (again analogously to the deep inelastic case):

$$k_{1,\perp}^2 \ll k_{2\perp}^2 \ll k_{3\perp}^2 \ll \cdots \ll Q^2 \tag{6.10}$$

Because of the structure of Fig. 45 we can now write

$$F_\pi(Q^2) = \int_{-1}^{1} dx \int_{-1}^{1} dz \; \phi^+(z,Q^2)\, T_B(z,x,Q^2)\, \phi(x,Q^2) \tag{6.11}$$

where T_B can be readily evaluated and one finds*)

$$T_B = \frac{16\pi\, \alpha_s(Q^2)\, C_F}{Q^2} \qquad \frac{1}{(1+x)(1+z)} \tag{6.12}$$

and $\phi(x,Q^2)$ is defined by

$$\phi(x,Q^2) = \int_0^{Q^2} \frac{dk_\perp^2}{16\pi^2}\, \psi(x,k_\perp^2) \left(\log Q^2/\Lambda^2\right)^{-C_F/\beta_0} \tag{6.13}$$

Explicit evaluation of T_B for a vector meson shows that the behaviour (6.12) comes from the case where the helicities of the quark and antiquark are opposite; if we calculate the contribution from the equal helicity case (the form factor of a transversely polarized ρ meson, e.g.), we find it is suppressed by one additional power of Q^2.

Because of the ladder-like structure of the dominant diagrams in the wave function (Fig. 45) we can write down an integral equation for the function $\tilde{\phi}$, where

$$\tilde{\phi} = \phi \left(\log Q^2/\Lambda^2\right)^{-C_F/\beta_0} \tag{6.14}$$

---

*) Strictly speaking, because of the way T_B is defined in Fig. 45, there should be a factor of $(\log Q^2/\Lambda^2)^{-2C_F/\beta_0}$ on the right-hand side of (6.12) from the vertex and self-energy corrections. We include this factor in the ϕ^+ and ϕ.

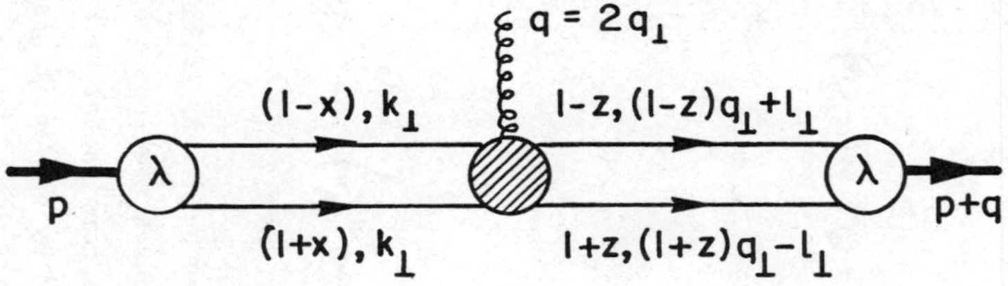

Fig. 44 Only diagrams of this form contribute in the leading logarithm approximation (in the light cone gauge) to the meson form factor

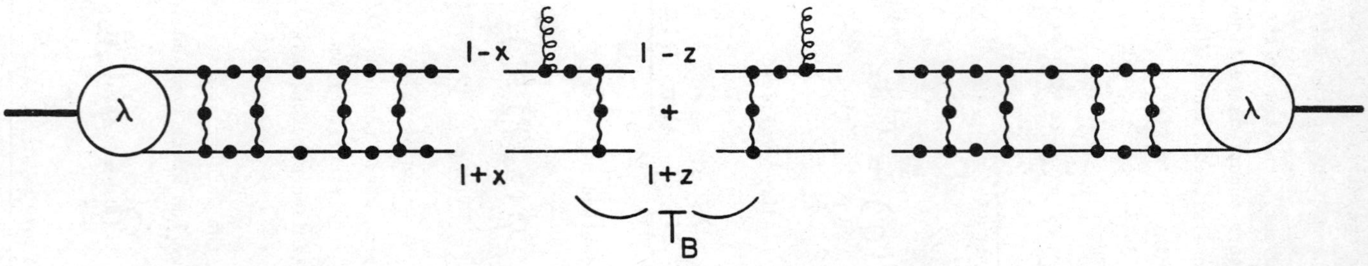

Fig. 45 Dominant structure for the meson form factor. The dots signify vertex and self-energy corrections

which is

$$\tilde{\varphi}(x,Q^2) = \tilde{\varphi}(x,\lambda^2) + \int_{\lambda^2}^{Q^2} \frac{dk_\perp^2}{k_\perp^2} \frac{C_F}{4\pi} \alpha_s(k_\perp^2) \int_{-1}^{1} dy\, V(x,y)\, \tilde{\varphi}(y,k_\perp^2) \tag{6.15}$$

where the kernel V(x,y), is the one-gluon potential

$$\begin{aligned} V(x,y) &= \frac{1-x}{1-y}\left(1 + \frac{\Delta}{x-y}\right)\theta(x-y) \\ &= \frac{1+x}{1+y}\left(1 - \frac{\Delta}{y-x}\right)\theta(y-x) \end{aligned} \tag{6.16}$$

where Δ is defined by

$$\Delta\tilde{\varphi} = \tilde{\varphi}(y,Q^2) - \tilde{\varphi}(x,Q^2) \tag{6.17}$$

Equation (6.15) can be solved explicitly, the solution is given by

$$\phi(x,Q^2) \underset{Q^2\to\infty}{\simeq} (1-x^2) \sum_{\substack{\text{even}\\ N}} a_N\, C_N^{3/2}(x) \left(\log Q^2/\Lambda^2\right)^{-d_N} \tag{6.18}$$

where the d_N are the familiar anomalous dimensions of the twist two non-singlet operators (see Sec. 2), $C_N^{3/2}$ are the Gegenbauer polynomials [complete orthonormal polynomials on measure $\int_{-1}^{1} dx\ (1-x^2)$]. a_N are in general uncalculable constants. Since the d_N are monotonically increasing ($d_0 = 0$, $d_2 = 0.62$, $d_4 = 0.90$, ...) the asymptotic behaviour of the pion form factor is given by

$$F_\pi(Q^2) \underset{Q^2\to\infty}{\sim} \frac{\alpha(Q^2)}{Q^2} \tag{6.19}$$

We can go even further and calculate the normalization factor in (6.19), since the same wave function appears in the decay $\pi \to \mu\nu$. We find

$$F_\pi(Q^2) \underset{Q^2\to\infty}{\longrightarrow} \frac{36\pi\,\alpha_s(Q^2/4)}{Q^2} \frac{f_\pi^2 C_F}{3} \tag{6.20}$$

which is the final answer. The pion decay constant f_π is equal to 0.94 MeV. This result (6.20) is much smaller than the values of F_π measured in the range

$2 < Q^2 < 4\ \text{GeV}^2$ [75], however, Eq. (6.20) is an asymptotic prediction, the expected value at finite Q^2 depends substantially on the input form of the driving term in (6.15), i.e., $\tilde{\phi}(x,\lambda^2)$. A reasonable assumption which is made in the above derivation is that $\phi(x,\lambda^2)$ falls at least like $(1 - x^2)^\varepsilon$ at $x^2 \to 1$ (where $\varepsilon > 0$). In this case the singularity in T_B as $x \to \pm 1$ is integrable, so that up to logarithmic corrections (which in this case asymptotically happen to be negligible because $d_0 = 0$) the dimensional counting rules are valid.

The results above (6.18), (6.20) can be understood in terms of the operator product expansion and the renormalization group [82]. In the wave functions ψ in Fig. 45, the outermost legs are hard, so that we want

$$\langle 0 | \psi(z_1) \bar{\psi}(z_2) | \pi \rangle \tag{6.21}$$

in the limit $z_1^+ = z_2^+$. We expand $\psi\bar{\psi}$ in terms of local operators (we are still working in the light cone gauge and in this gauge the path ordered exponential of the gauge potential just gives a factor 1, so we can neglect it), and again the non-singlet twist two operators are the dominant ones. There is a slight difference however, since we now have to consider operators of the type

$$\partial_{\mu_{k+1}} \cdots \partial_{\mu_m} \bar{\psi} \gamma_\mu D_{\mu_1} \cdots D_{\mu_k} \psi \tag{6.22}$$

In the deep inelastic case we take the forward matrix element of the operators, in that case it is only the operator which has $k = m$ which has a non-zero matrix element. Thus, now for each value of m, we have m operators, and therefore, in principle, m eigenvalues of the anomalous dimension matrix. However, the anomalous dimension matrix is a triangular one, renormalization of the operator with $m - k$ external derivative, gives a linear combination of the operators (6.22) with $m - k$, $m - k + 1$, ... etc., external derivatives. Thus the eigenvalues of this matrix are just the diagonal elements, and these are just the γ_0^N for $N < m$. In this way we understand Eq. (6.12).

6.3 Form factors of other hadrons

In the previous section we saw that the QCD prediction for the asymptotic behaviour of the form factor was very similar to the dimensional counting prediction, the only difference is an extra factor of $\alpha_s(Q^2)$ in the QCD case. In general, however, the lowest anomalous dimension is not zero, and there are additional logarithmic modifications to Eq. (6.1).

Perhaps the simplest example is the form factor of the vector meson. For the helicity zero form factor of the ρ meson for example, the arguments of the previous section go through exactly so that

$$F_\rho(Q^2, \lambda_\rho = 0) \sim \frac{\alpha(Q^2)}{Q^2} \tag{6.23}$$

In this case the normalization constant can also be determined, it is the same as for the pion (6.20) with f_π replaced by the ρ decay constant g_ρ defined by

$$\langle 0|\bar{\psi}\gamma_\alpha(\tfrac{1}{2}\tau^+)\psi|\rho^+\rangle = \varepsilon_\alpha g_\rho m_\rho \tag{6.24}$$

In fact $g_\rho \approx f_\pi$ due to the KSFR relation[83], so that $F_\pi \approx F_\rho$ asymptotically. For the transverse ρ (helicity = ±1) there are two differences:

i) T_B is suppressed by an additional power of Q^2 and
ii) the structure of the potential is different.

The bound state equation can be solved[78] giving

$$F_\rho(Q^2, \lambda = \pm 1) \sim \frac{m^2(Q^2)}{Q^2} \frac{\alpha_s(Q^2)}{Q^2} \left[\alpha_s(Q^2)\right]^{2C_F/\beta_0} \tag{6.25}$$

The exponent $2C_F/\beta_0$ is the anomalous dimension of

$$\bar{\psi}\, \sigma_{\alpha\beta}\, (\tfrac{1}{2}\tau^+)\, \psi \tag{6.26}$$

The subdominant contributions to this form factor have exponents which are just the anomalous dimensions of the tower of operators based on (6.26).

For the baryons we can carry out the same procedure as for mesons[84]. The integral equation for the wave function is more complicated, nevertheless it can be solved (at least numerically). For example, for the magnetic form factor of a nucleon, the lowest anomalous dimension is not zero, and one finds

$$G_M(Q^2) \sim \left(\frac{\alpha_s(Q^2)}{Q^2}\right)^2 \left(\alpha_s(Q^2)\right)^{C_F/\beta_0} \tag{6.27}$$

so that the nucleon form factor satisfies the dimensional counting rules, up to logarithmic modifications. As in the meson case, form factors in which the baryon's helicity is changed, or in which its helicity is > 1, are suppressed by powers of Q^2.

C_F/β_0 is just the anomalous dimension of the fermion field ($\equiv \gamma_F$) in the Feynman gauge. One can readily see why the exponents in (6.25) and (6.27) are $2\gamma_F$ and γ_F, respectively. The lowest anomalous dimension of the familiar non-singlet twist two operator, $\bar{\psi}\gamma_\mu\psi$ is proportional to the ultra-violet divergence in the diagrams of Figs 46a and 46b. These cancel by the Ward identity $Z_1 = Z_2$, thus there is zero

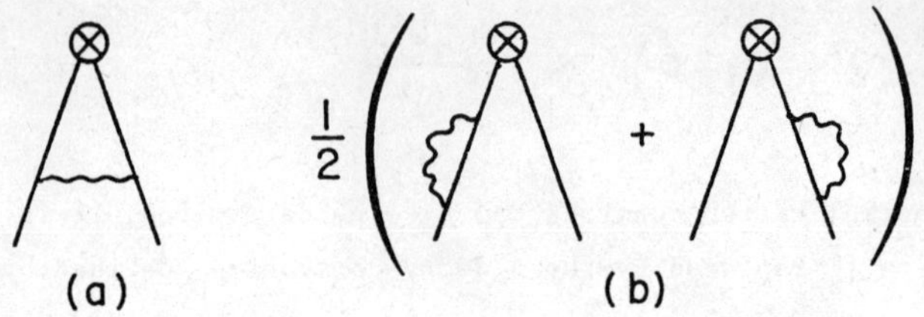

Fig. 46 Diagrams which have to be calculated when evaluating the anomalous dimension of the operator $(\bar{\psi}\sigma_{\alpha\beta}\,\tfrac{1}{2}\tau^{+}\psi)$ which dominates the form factor of a vector meson with helicity equal to ± 1

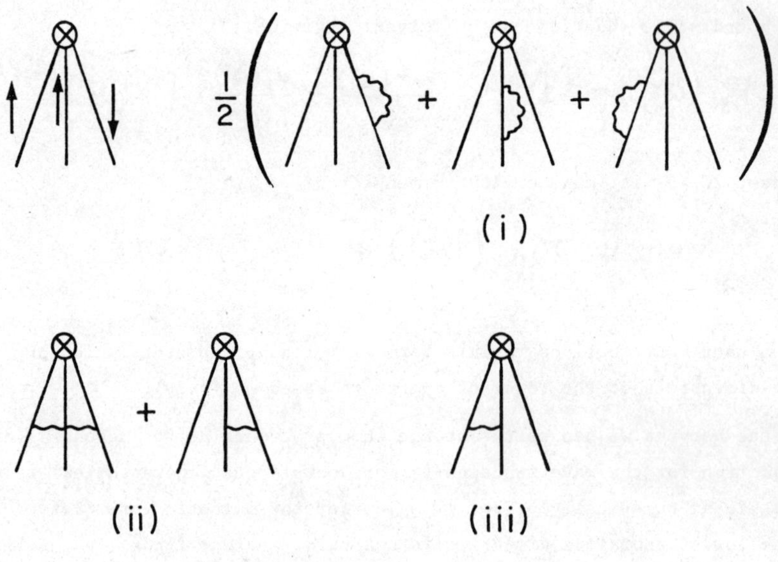

Fig. 47 Diagrams which have to be calculated when evaluating the anomalous dimension of the three quark operator which dominates the form factor of a baryon

anomalous dimension in (6.9). If we next take the operator (6.26), and calculate the same diagrams, we find (by simple Dirac algebra) that the diagram of Fig. 46a vanishes in the Feynman gauge, leaving us with a contribution equal to $2\gamma_F$ from the two diagrams of Fig. 46b. The anomalous dimension in the case the form factor of a helicity ½ baryon can be deduced from the helicity 0 and 1 meson form factors discussed above. When we calculate the lowest anomalous dimension of the three quark operator (Fig. 47) we find we get the following contributions:

i) $3\gamma_F$ from the self-energy diagrams;

ii) $-2\gamma_F$ from the two vertex diagrams where the gluon links quarks of opposite helicity; and

iii) 0 from the vertex where the gluon links quarks of the same helicity.

Hence we end up with an anomalous dimension equal to γ_F in (6.27). Similarly, for a helicity 3/2 baryon we deduce that the relevant anomalous dimension is equal to $3\gamma_F$ (only the first contribution (i) above is not zero).

Before proceeding with a discussion of elastic scattering at large angle, we review very briefly the properties of the Sudakov form factor; this form factor plays an important role for these large angle processes.

6.4 Sudakov form factor[85)]

It is well known that in quantum electrodynamics (QED) the form factor of the electron (of momentum p) at large momentum transfer (q) in the leading logarithmic approximation falls very rapidly with q^2, in fact like

$$\exp\left[-\text{const}\cdot e^2 \cdot \log^2 q^2/p^2\right] \tag{6.28}$$

We notice that in this case every loop gives two logarithms, in contrast to the inclusive processes where there was only one logarithm per loop, corresponding to the mass singularities. The second logarithm corresponds to an infra-red divergence, these divergences now survive, since there are no diagrams with real photons to cancel them.

Equation (6.28) is interpreted as signifying that it is difficult to bend a charged particle through a finite angle, without it radiating many photons. Since the result of summing leading logarithms is much smaller than the neglected non-leading terms, it is relevant to ask whether these terms also sum to something small. Calculations by Korthales-Altes and de Rafael[86)] seemed to indicate that this may be so, and recently Mueller[87)] has argued that all the non-leading logarithms, in massive QED, sum up to give an asymptotic form

$$\exp\left[-c(g^2)\log^2 q^2/M^2\right] \tag{6.29}$$

where M is a finite mass. It is still just possible that the terms, which are suppressed by powers of q^2 in perturbation theory, sum up to give something larger than (6.29).

Similarly, when one studies electron-electron scattering through a fixed angle in perturbation theory, we find factors such as (6.28) appearing. For example, to order α^2 in the amplitude the dominant diagrams (in the Feynman gauge) are those in Fig. 48. The relative contributions of Figs 48a, b, and c are $\log^2 t$, $-\log^2 t$, and $-y^2 t$, respectively (in our approxiamtion we do not distinguish log s from log n or log t). We interpret this result as being that the diagrams of Figs 48a and b cancel, leaving us with the diagrams of Fig. 48c, which is just the Born diagram modified by the Sudakov form factor. This is a general feature to all orders of perturbation theory[72].

Although the calculations are more complicated in QCD, it seems that the same results hold (up to a trivial colour factor), when we consider the form factors of coloured fields or fixed angle scattering of coloured fields[18),88),89)].

6.5 Elastic scattering at fixed angle

The problem of elastic scattering at fixed angle has not been studied very thoroughly yet in QCD, but certain features seem to have emerged[75]. In particular in time ordered perturbation theory in the light cone gauge, the dominant radiative corrections to diagrams such as those of Fig. 39 which obey the dimensional counting rules, lead to a ladder like structure of Fig. 49, in which each rung joins together two quarks moving almost parallel to each other. These radiative corrections are identical to those in the form factors (Sections 6.2 and 6.3) so that we can write

$$\frac{d\sigma}{dt}(AB \to CD) = \frac{\alpha_s^2(t)}{t} F_A(t) F_B(t) F_C(t) F_D(t) f(\theta) \qquad (6.30)$$

In principle, the function $f(\theta)$ and the normalization are also calculable, but quite clearly this is a horrendous task. Thus we see that the diagrams which obeyed the dimensional counting rules, now in QCD are only modified by calculable logarithmic corrections.

What about the Landshoff diagrams, e.g., those of Fig. 40 which were the dominant ones in the parton model. When we evaluate the radiative corrections to these diagrams, we find that the leading contributions are exactly those corresponding to the Sudakov form factor, thus each quark-quark scattering amplitude should be multiplied by $F_q^2(t)\,\alpha(t)$ (where F_q is the Sudakov form factor). Since asymptotically the Sudakov form factor falls faster than any power of t, the "multiple scattering" diagrams will be suppressed relative to those obeying the dimensional counting rules. Moreover, even at low values of t (~ 0 (10 GeV^2)) they can be estimated to be less important.

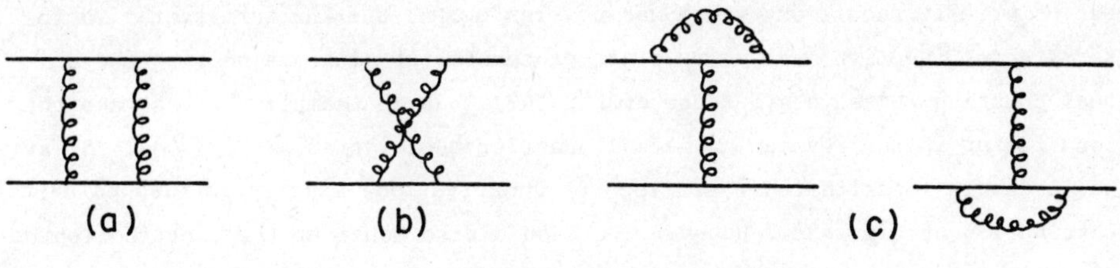

Fig. 48 Some order α^2 diagrams which contribute to fixed angle electron-electron scattering in QED

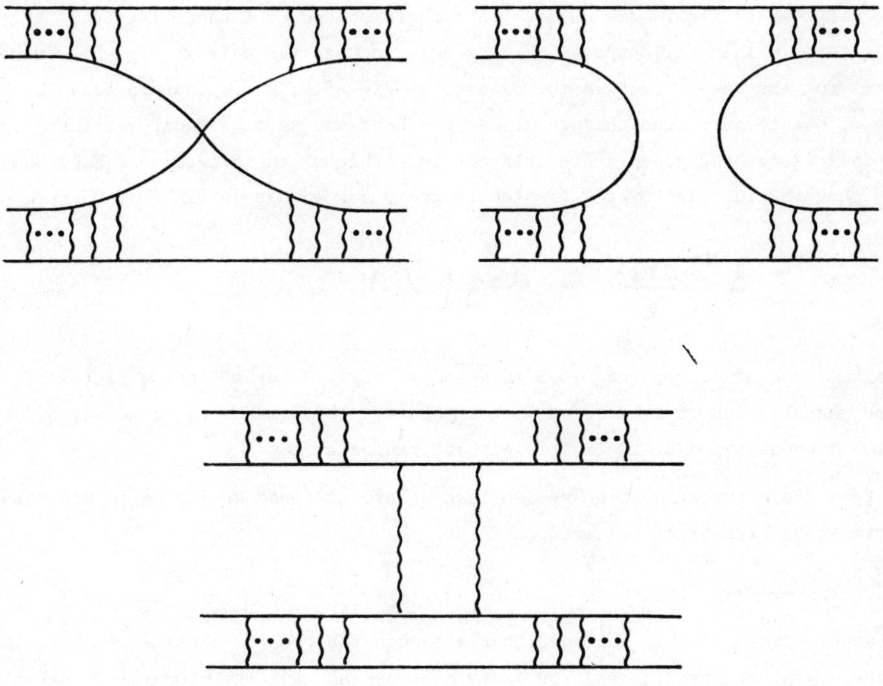

Fig. 49 Dominant gluonic corrections to the diagrams of Fig. 39

Let us briefly recall why there was no sign of the Sudakov form factor in the pion form factor (Section 6.2) or in other processes for diagrams obeying the dimensional counting rules (e.g., those of Fig. 49). As an example, let us take the pion form factor in the Feynman gauge and consider the diagram of Fig. 50a. Clearly, this has a double logarithmic (i.e., $\log^2 t$) contribution, exactly corresponding to the contribution of Fig. 48c. However, we should also consider the contribution of Fig. 50b which also has a $\log^2 t$ term, which cancels that of Fig. 50a. It is crucial that the external particle is a colour singlet, otherwise the colour factors of the two diagrams of Figs 50a and 50b would not be equal and the cancellation would not take place. This cancellation corresponds to the cancellation of the infra-red divergences when we consider radiation from a (colour) neutral particle. This cancellation occurs in all diagrams which do not have nearly on-shell partons scattering through a finite angle.

Why does such a cancellation not place in the multiple scattering diagrams? In these diagrams it might seem natural that there should be a cancellation of the $\log^2 t$ terms within various sets of diagrams, such as the pair in Fig. 51. However, this is not the case, since in the diagram of Fig. 51b, k_T is bounded above by λ, since it has to be routed through at least one of the soft hadronic vertices. Notice that this is not the case in Fig. 51a nor in either of the diagrams of Fig. 50. This means that the integral over k_T, which in the Sudakov case is

$$\int_{p^2}^{t} \frac{dk_\perp^2}{k_\perp^2} \sim \log\left(t/p^2\right) \tag{6.31}$$

is in the case of diagram Fig. 51b independent of t. Thus we lose a factor of log t and no cancellation of the $\log^2 t$ factor occurs. There seems to be no way of avoiding a Sudakov suppression of the multiple scattering diagrams[89].

From the above discussion we see that we are at least beginning to understand elastic scattering processes in QCD.

6.6 Conclusion

We have seen in this section, that although the results obtained so far concerning the asymptotic predictions for elastic processes at large momentum transfer in QCD are not phenomenologically very useful, very significant progress has been made towards the understanding of the nature of these processes.

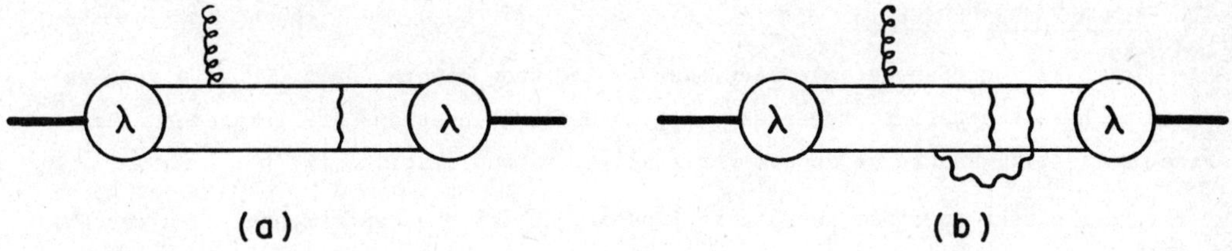

Fig. 50 Example of the cancellation of the "Sudakov double logarithm". Both diagrams (a) and (b) have such a $\log^2 q^2$ term; however, their sum does not

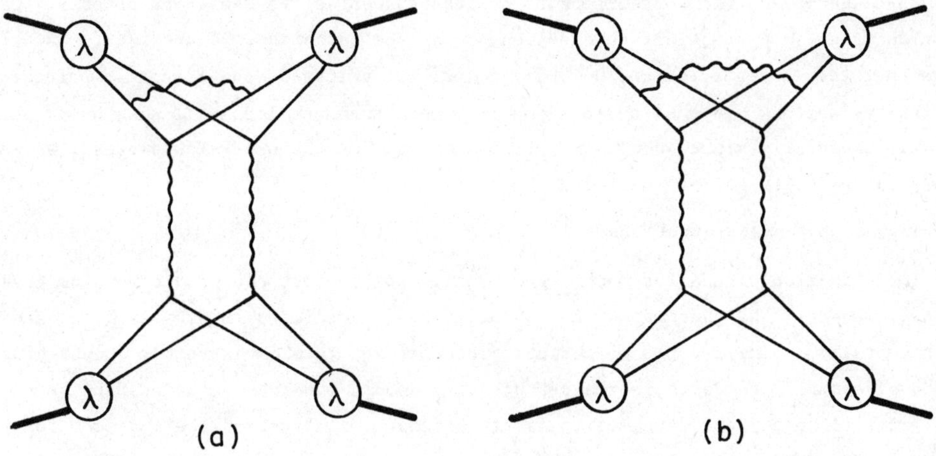

Fig. 51 Examples of radiative corrections to the multiple scattering diagram of Fig. 40a. The diagram (a) has a $\log^2 q^2$ factor whereas that of (b) does not

7. SUMMARY AND PROSPECTS

In these lectures we have seen that we now know how to apply QCD to a wide variety of hard scattering processes. Although the foundations are laid, much theoretical work still has to be done. For example, we would still like to:

i) bring closer together the predictions of QCD and the experimental data for the hard exclusive processes (those discussed in Section 6). At the moment, the predictions are for $Q^2 \to \infty$ for the form factors, whereas the data are at relatively low Q^2 so that it is not clear whether the QCD predictions can be directly applied. Almost nothing has so far been calculated for fixed angle scattering.

ii) understand much better processes with more than one independent large variable (see, for example, Ref. 18)). These processes are much more complicated than the ones discussed in Section 4. Experimental data for processes such as $hh \to \ell^+\ell^-X$ (where h = hadron, ℓ = lepton and the $\ell^+\ell^-$ pair have large invariant mass ($= \sqrt{Q^2}$) and large transverse momentum Q_T, where $Q_T^2 << Q^2$), and $hh \to hX$ (where the transverse momentum of the trigger hadron p_T is large but $p_T^2 << s$) does exist, and would provide interesting tests of QCD if our predictions were reliable.

iii) get a handle on the corrections to processes such as the Drell-Yan process. Calculation of these corrections to one-loop (using the techniques of Sections 4 and 5) shows that they are very large (O(100%)) at present values of the kinematic parameters and the indications are that corrections from more than one loop will also be large. Clearly, especially since such a lot of data exists for the Drell-Yan process, we want to understand these corrections, and

iv) answer many other questions.

There is also a need for much experimental work to test the predictions we have so far. Recently the phenomenon of "Jet Broadening" predicted in QCD was discovered experimentally[17)] with all the predicted properties. This is extremely encouraging. If QCD passes all its tests in the perturbative domain (the one we understand best of all), it will be a tremendous motivation to attempt to understand the more complex properties of QCD, hopefully eventually leading to an understanding of confinement and the hadron spectrum. Are we beginning to understand the strong interactions?

REFERENCES

1) H.D. Politzer, Phys. Rep. 14C (1974) 129.

2) D.J. Gross, in Methods in Field Theory, Les Houches (1975) ed. R. Balian and J. Zinn Justin (North Holland, Amsterdam).

3) C. Nash, Relativistic Quantum Fields (London, Academic Press, 1978).

4) J. Ellis, in Weak and Electromagnetic Interactions at High Energy, Les Houches (1976) ed. R. Balian and C.H. Llewellyn Smith (North Holland, Amsterdam).

5) A.J. Buras, Fermilab preprint PUB-79/17-THY (1979).

6) P. Pascual, Lectures on Deep Inelastic Scattering and Asymptotic Freedom, GIFT publication (1979).

7) A. Peterman, Phys. Rep. 53 (1979) 158.

8) A. Mueller, Phys. Rev. D18 (1978) 3705.

9) S. Gupta and A. Mueller, Columbia University preprint CU-TP-139 (1979).

10) P.V. Landshoff and J.C. Polkinghorne, Phys. Rep. 5C (1972) 1.

11) D. Sivers, R. Blankenbecler and S.J. Brodsky, Phys. Rep. 23C (1976) 1.

12) S.J. Brodsky and G.R. Farrar, Phys. Rev. Lett. 31 (1973) 1153; Phys. Rev. D11 (1975) 1309.

13) V.A. Matveev, R.M. Muradyan and A.V. Tavkhelidze, Lett. Nuovo Cimento 7 (1973) 719.

14) S.D. Drell and T.M. Yan, Phys. Rev. Lett. 25 (1970) 316; Ann. Phys. 66 (1971) 578.

15) S. Berman, D. Levy and T. Neff, Phys. Rev. Lett. 23 (1972) 1363.

16) A. De Rújula, J. Ellis, E.G. Floratos and M.K. Gaillard, Nucl. Phys. B138 (1978) 387.

17) Talks of N. Newman, Ch. Barger, G. Wolf and S. Orito, in the 1979 Lepto-Photon Conference, Fermilab (to be published in the proceedings).

18) Yu. Dokshitzer, D.I. Dyakonov and S.I. Troyan, SLAC-TRANS -183 (1978); Phys. Rep. (to be published).

19) E. de Rafael, these proceedings.

20) P.C. Bosetti et al., Nucl. Phys. B142 (1978) 1.

21) J. de Groot et al., Z. Phys. C Particles and Fields 1 (1979) 143.

22) T.W. Quirk, J.B. Mac Allister, W.S.C. Williams and C. Tao, Oxford University preprint (1979).

23) E.G. Floratos, D.A. Ross and C.T. Sachrajda, Nucl. Phys. B129 (1977) 66; E. B139 (1978) 545.

24) D. Gross and F. Wilczek, Phys. Rev. D8 (1973) 3633 and D9 (1974) 980.

25) G. Altarelli and G. Parisi, Nucl. Phys. B123 (1977) 298.

26) E.G. Floratos, D.A. Ross and C.T. Sachrajda, Nucl. Phys. B152 (1979) 493.

27) W. Caswell, Phys. Rev. Lett. 33 (1974) 224.

28) D.R.T. Jones, Nucl. Phys. B75 (1974) 531.

29) G. 't Hooft and M. Veltman, Nucl. Phys. B44 (1972) 189.

30) A. González-Arroyo, C. López and F.J. Ynduráin, Nucl. Phys. B153 (1979) 161.

31) W.A. Bardeen, A.J. Buras, D.W. Duke and T. Muta, Phys. Rev. D18 (1978) 3998.

32) E.G. Floratos, D.A. Ross and C.T. Sachrajda, Phys. Lett. 80B (1979) 269.

33) A. Para and C.T. Sachrajda, CERN preprint TH. 2702 (1979) (to be published in Phys. Lett.).

34) D.W. Duke and R.G. Roberts, Rutherford Laboratory preprint RL-79-025 (1979); D.W. Duke, Rutherford Laboratory preprint RL-79-044 (1979).

35) M. Baće, Phys. Lett. 78B (1978) 132.

36) J. Wotschack, Talk at the 1979 EPS Conference on High Energy Physics, Geneva (to be published in the proceedings).

37) A.J. Buras, E.G. Floratos, D.A. Ross and C.T. Sachrajda, Nucl. Phys. B131 (1977) 308.

38) A. González-Arroyo, C. López and F.J. Ynduráin, Madrid University preprint FTUAM/79-3 (to be published in Nucl. Phys. B).

39) L. Abbott and M. Barnett, SLAC prepring SLAC-PUB-2325 (1979).

40) D.A. Ross, Caltech preprint 68-699 (1979).

41) D.A. Ross and C.T. Sachrajda, Nucl. Phys. B149 (1979) 497.

42) E.M. Riordan et al., SLAC-PUB-1634 (1975) (unpublished).

43) F. Block and A. Nordsieck, Phys. Rev. 52 (1937) 54.

44) D.R. Yennie, S.C. Frautschi and H. Suura, Ann. Phys. 13 (1961) 379.

45) R. Gastmans and R. Meuldermans, Nucl. Phys. B63 (1973) 277.

46) W.J. Marciano and A. Sirlin, Nucl. Phys. B88 (1975) 86.

47) T. Kinoshita, J. Math. Phys. 3 (1962) 650.

48) T.D. Lee and M. Nauenberg, Phys. Rev. 133 (1964) 1549.

49) See Ref. 18) and also D.J. Pritchard and W.J. Stirling, Cambridge preprint DAMTP 79-1 (1979) and references therein.

50) V. Sudakov, JETP 3 (1956) 65.

51) C.H. Llewellyn Smith, Acta Physica Austriaca, Suppl. XIX (1978) 331.

52) D. Amati, R. Petronzio and G. Veneziano, Nucl. Phys. B140 (1978) 54; B146 (1978) 29.

53) V.N. Gribov and I.N. Lipatov, Soviet Journal of Nucl. Phys. 15 (1972) 438.

54) E. Witten, Nucl. Phys. B120 (1977) 89.

55) C.H. Llewellyn Smith, Phys. Lett. B79 (1978) 83.

56) W.R. Frazer and J.F. Gunion, Phys. Rev. D20 (1979) 147.

57) C.F. von Weizacker, Z. Phys. 88 (1934) 612;
E.J. Williams, Phys. Rev. 45 (1934) 729.

58) H.D. Politzer, Nucl. Phys. B129 (1977) 301.

59) R.K. Ellis, H. Georgi, M. Machacek, H.D. Politzer and G.G. Ross, Phys. Lett. 78B (1978) 281; Nucl. Phys. B152 (1979) 285.

60) G. Sterman and S. Libby, Phys. Rev. D18 (1978) 3252, 4737.

61) C.T. Sachrajda, Phys. Lett. 73B (1978) 185.

62) C.T. Sachrajda, Phys. Lett. 76B (1978) 100.

63) V.N. Gribov and I.N. Lipatov, Soviet Journal of Nucl. Phys. 15 (1972) 675.

64) C.T. Sachrajda, Physica Scripta 19 (1979) 85.

65) R.P. Feynman, Photo Hadron Interactions, W.A. Benjamin, Reading, Mass. 1972.

66) J. Kubar André and F.E. Paige, Phys. Rev. D19 (1978) 221.

67) G. Altarelli, R.K. Ellis and G. Martinelli, MIT preprint 776 (1979).

68) J. Abad and B. Humpert, Phys. Lett. 78B (1978) 627; 80B (1979) 433.

69) K. Harada, T. Kaneko and N. Sakai, CERN preprint TH. 2619 (1979).

70) C.T. Sachrajda and S. Yankielowicz, in preparation.

71) N. Nakanishi, Graph Theory and Feynman Integrals, (Gordon and Breach 1971).

72) I.G. Halliday, J. Huskins and C.T. Sachrajda, Nucl. Phys. B83 (1974) 189.

73) C.E. De Tar, S.D. Ellis and P.V. Landshoff, Nucl. Phys. B87 (1975) 176.

74) A.H. Mueller, Phys. Rev. D2 (1970) 2963.

75) S.J. Brodsky and G.P. Lepage, SLAC preprint SLAC-PUB-2294 (1979).

76) P.V. Landshoff, Phys. Rev. D10 (1974) 1024.

77) A. Donnachie and P.V. Landshoff, Z. für Physik C2 (1979) 55.

78) S.J. Brodsky and G.P. Lepage, SLAC preprint SLAC-PUB-2343 (1979).

80) A.V. Efremov and A.V. Radyushin, Dubna preprint JINR-E2-11983 (1979).

81) G. Parisi, Ecole Normale preprint LPTENS 79-7 (1979).

82) S.J. Brodsky, Y. Frishman, G.P. Lepage and C.T. Sachrajda, CERN preprint TH. 2731 (1979).

83) K. Kawarabayashi and M. Suzuki, Phys. Rev. Lett. 16 (1966) 255; Fayyazuddin and Riazuddin, Phys. Rev. 147 (1966) 1071.

84) S.J. Brodsky and G.P. Lepage, SLAC preprint SLAC-PUB-2348 (1979).

85) V. Sudakov, Sov. Phys. JETP 3 (1956) 65.

86) C.P. Korthals-Altes and E. de Rafael, Nucl. Phys. B106 (1976) 237.

87) A.H. Mueller, Columbia University preprint CU-TP-158 (1979).

88) J.J. Carazzone, E.R. Poggio and H.R. Quinn, Phys. Rev. D11 (1975) 2286 and Phys. Rev. D12 (1975) 3368 (E).

89) J. Cornwall and G. Tiktopoulos, Phys. Rev. D13 (1976) 3370 and Phys. Rev. D15 (1977) 2937.

Experimental Aspects of Quantum Chromodynamics

Harvey L. Lynch
DESY
2000 Hamburg 52
W. Germany

These three lectures will concentrate almost exclusively on the application of e^+e^--annihilation to QCD. Since the intent is purely pedagogic for a primarily theoretically oriented audience, no attempt will be made to give a comprehensive review of all possible relevant experimental contributions. Rather experiments have been selected mainly on the basis of how easily they may be used to discuss a given point.

The lectures may be outlined as follows:

Lecture I: Basic Facts and the Beginning of Charm

A. Characteristics of e^+e^- Storage Rings

B. $R = \sigma_{had} / \sigma_{\mu\mu}$

C. ψ Family Characteristics: Foundation of Charm Physics

D. Related $(c\bar{c})$ Systems (not $J^{PC} = 1^{--}$)

Lecture II: Open Charm and a Heavy Lepton

A. D-States' Characteristics

B. F-States' Characteristics

C. τ: the Other Side of QCD, and an Indication of More to Come

Lecture III: Beyond Charm

A. The Υ Family Characteristics

B. Recent PETRA Results

1. R
2. Jet Structure and Relation to New Physics
3. Inclusive Distributions
4. What is to come?

I A. Characteristics of Storage Rings

Let us begin with a brief discussion of why e^+e^- storage rings are such a valuable tool and what basic characteristics are needed to make use of them: One of the most important facts to recognize is that the center of mass is in the laboratory; this greatly simplifies numerous kinematic problems. That is to say that the 4 vector of the initial state is given by $(E, \vec{p}) = (w, \vec{0})$. The center of mass energy w is known to very high relative precision (of the order of 10^{-4}) and also high absolute precision (of the order of 10^{-3}). This fact will prove to be very valuable. If we now believe that the e^+e^- annihilation proceeds via a single intermediate photon, then we know the initial state has the quantum numbers $J^{PC} = 1^{--}$. The fact that the initial state is so well defined is one of the major reasons one goes to all the trouble to build and use e^+e^- storage rings (and they are a lot of trouble!). There is, however, one more advantage brought by the intermediate photon: it provides a natural mechanism to couple to new quark flavours in a clean way. This single fact alone would make rings worthwhile.

To perform an experiment, and this generally means to measure some sort of cross section, one needs the ring, some measure of beam intensity, and some sort of detector. Historically the rings have grown in physical dimensions in large steps: The very first storage ring ADA, built in early 1960's at Frascati, had a bending radius of about 30 cm. This was only a tool to study how to make a storage ring work, but it was very valuable in this regard. The rings SPEAR at SLAC and DORIS at DESY, built in the early 1970's, have a bending radius of about 30 m. These are very much larger than ADA and have been extraordinarily fruitful in physics output. The next generation machines, e.g. PETRA at DESY, is another order of magnitude larger, having a bending radius of about 350 m. This ring has only very recently come into operation, and naturally we hope that it will also be very fruitful in producing new physics.

In order to measure a cross section it is necessary to have some measure of the beam intensity, called the "luminosity". Operationally one can define the luminosity by reaction rate = cross section x luminosity. (As an exercise, compute the reaction rate for a cross section σ under the condition that bunches of N_+ positrons and N_- electrons each uniformly distributed in bunches of cross sectional area A and length l collide with a frequency of ν.) Typically one measures the luminosity by measuring the reaction rate for some known cross section, such as small angle e^+e^- elastic scattering. As a rule of thumb such luminosity measurements are accurate to 5 - 10%, unless extraordinary care is exercised to reach higher precision.

Naturally, to do an experiment, some kind of detector is required. The reaction rates are usually rather small; as a consequence a detector must cover as large a solid angle as possible. In addition the large solid angle is essential to study many body final states, which cannot be analyzed if some of the particles are not observed. Here is the only disadvantage to being in the center of mass: covering $\approx 4\pi$ solid angle in the center of mass implies covering the same solid angle in the lab. Other constraints set minimum dimensions resulting in a detector which is very large. Figure 1.1 show what is now the father of today's detectors. The so called MK-I detector used at SPEAR. This detector's output will be central to much of the experimental results discussed in the first two lectures. This detector exhibits the basic characteristics sought: (a) There must be some means of analyzing the momenta of charged particles. In practice this means measuring the curvature of a trajectory in a magnetic field. In the MK-I this meant a nearly uniform field parallel to the e^+e^- beam direction and a set of proportional and spark chambers to measure positions at different points. Today the technology is slightly different, using drift chambers instead of spark chambers, but the principle is the same. (b) Outside the tracking system (spark chambers) there is a set of scintillation counters. These serve not only as part of a trigger system but also as a time-of-flight system used to identify charged particles: Having both the momentum and velocity of a particle, the mass may be computed. (c) In addition there is a system used to identify electrons and photons, which cause a large signal due to the electromagnetic shower; on the other hand pions, kaons, protons, or muons produce small signals. (d) Naturally to make a magnetic field a coil is required. Similarly, to keep Mr Gauss happy there must be a magnetic flux return path; this is made of iron, and it also serves as a hadron filter to separate muons from hadrons. (Electrons and photons were stopped in the shower counter, and hadrons are usually absorbed or suffer large scattering in the iron). (e) Finally a tracking device outside the iron (here muon spark chambers) provide identification of muons.

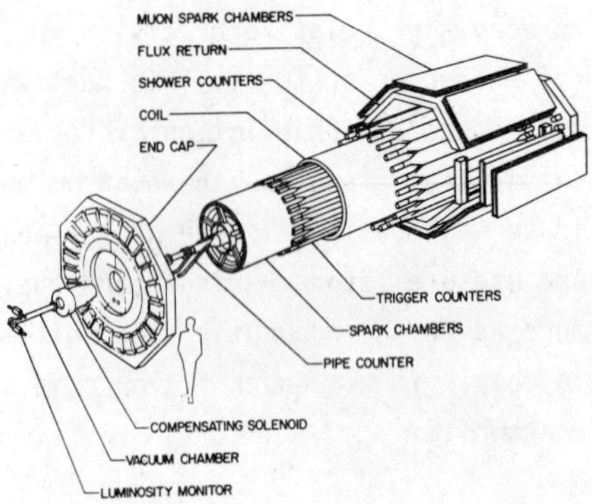

Fig. 1.1 Basic elements of a detector for e^+e^- storage ring experiments, as illustrated by the SLAC-LBL MARK I (now retired).

I B. $R = \sigma_{Hadron} / \sigma_{\mu\mu}$

One of the most fundamental measurements to be made is the total cross section for $e^+e^- \rightarrow$ hadrons. Instead of this cross section, however, it is more convenient to consider the ratio of the total cross section for hadron production to the simple QED cross section for μ-pair production

$$R = \sigma_{hadron} / \sigma_{\mu\mu} \quad . \tag{1.1}$$

This is a particularly convenient quantity because within the framework of quark-parton model (or light-cone algebra) R is expected to be a constant, independent of w. The constant is just the sum of the squares of the quark charges. Clearly, if a new threshold is crossed, then new quarks can enter into the sum with a resulting increase in the value of R. Equally clearly the photon can couple to a $q\bar{q}$ bound state manifested as a resonance. Figure 1.2 shows the state of knowledge of R as of 1975: Below w = 3 GeV one has a somewhat complex region which is related just to u, d and s quarks.

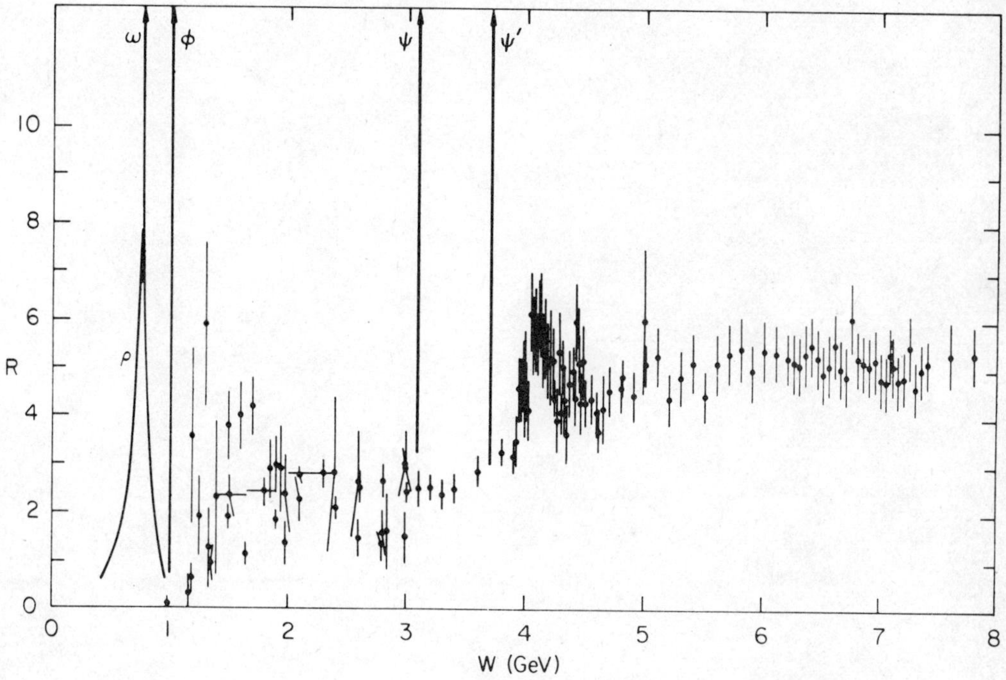

Fig. 1.2 The ratio R of the total cross section for e^+e^- annihilation to hadrons to the QED μ-pair cross section.

One sees clearly the ρ, ω, and ϕ resonances as well as a continuum. There is evidence that other resonances also exist in this region. Starting at 3.095 GeV the world of physics took a great leap forward with what turned out to be the next quark flavor, charm. This is manifested by the two large resonances at 3.095 and 3.684 GeV, and then a rather complex region starting at about 4 GeV. The ψ (3095) and ψ (3684) are bound states of $c\bar{c}$, but the region above 4 GeV is above threshold for pairs of states having net charm $\neq 0$, as we shall see in the next lecture.

I C. The ψ Family Characteristics

Since the ψ's form the foundation on which much will be built, it is worthwhile to study in detail what is really known about them. In particular, what are the masses, widths, and branching ratios;, what is the spin, parity, charge conjugation, G-parity, and isospin of these states?

Fig. 1.3 shows the cross sections [1] for $e^+e^- \rightarrow$ hadrons, μ-pairs, and e-pairs as a function of w near the ψ. The three signals are all very large. Fig. 1.4 shows the same quantities [2] for $\psi(3684)$. Here the lepton signals are much smaller than for $\psi(3095)$.

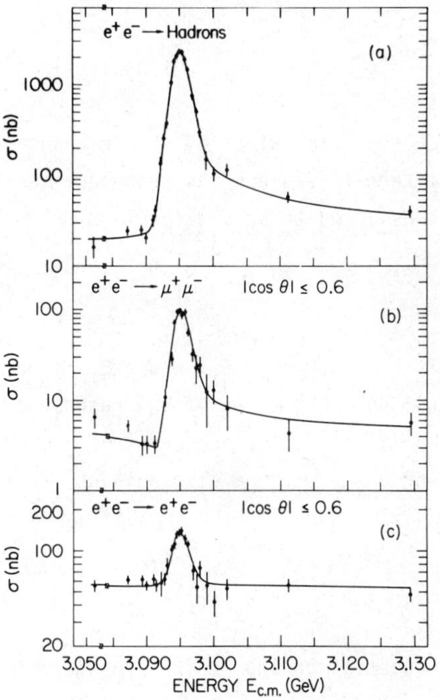

Fig. 1.3 Excitation curves for $\psi(3095)$ in the channels (a) hadrons, (b) μ-pairs, and (c) e-pairs.

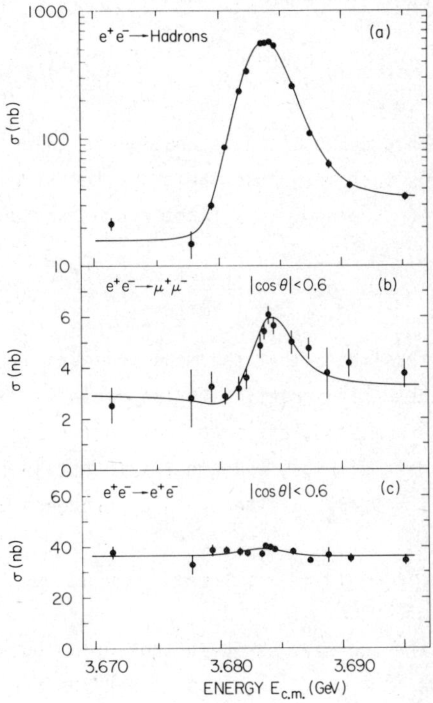

Fig. 1.4 Excitation curves for $\psi(3684)$ in the channels (a) hadrons, (b) μ-pairs, and (c) e-pairs.

Clearly since we have an initial state of e^+e^- the "law of least amazement" says that the ψ is formed by a single photon annihilation of the e^+ and e^-, meaning that the ψ must share the quantum numbers of the photon $J^{PC} = 1^{--}$. For simplicity let us begin with this premise to find the mass and width; then we shall take up the evidence for this assignment.

On very general grounds the cross section for the production of a resonance from e^+e^- and its subsequent decay into a given channel "f" may be described by the Breit-Wigner relation:

$$\sigma_f(w) = \frac{(2J+1)\pi}{w^2} \frac{\Gamma_e \, \Gamma_f}{(w-m)^2 + \Gamma^2/4} , \qquad (1.2)$$

where J is the spin of the resonance, m is the mass, Γ is the total width, Γ_e is the partial width for the resonance to decay into e^+e^-, and Γ_f is the partial width to decay into any channel f. In practice one does not measure $\sigma_f(w)$ directly, but rather a convolution with the inherent energy spread of the machine and somewhat complicated radiative effects,

$$\sigma_f^{\text{observed}}(w) = \int \sigma_f(w') \, K(w', w) \, dw' . \qquad (1.3)$$

This convolution is crucial to understand, because the total width of the resonance is much less than the characteristic width of the kernel K. Since the resonance is very narrow, to a first approximation it may be considered to be a δ-function,

$$\sigma_f(w) \rightarrow \frac{2(2J+1)\pi^2}{m^2} \Gamma_e \frac{\Gamma_f}{\Gamma} \delta(w-m) ; \qquad (1.4)$$

this means that the measured quantity is really $\Gamma_e \cdot \Gamma_f/\Gamma = \Gamma_e B_f$, where B_f is the branching fraction to the state f. By assuming that no decay modes of the resonance are missed, after correction for obvious detection efficiencies, one can thus use the data of Figs. 1.3 and 1.4 to obtain Γ_e, B_e, B_μ, and B_{hadrons}. One also obtains $\Gamma = \Gamma_e/B_e$. If the $\psi \rightarrow \mu^+\mu^-$ proceeds via

$$e^+e^- \rightarrow \gamma \rightarrow \psi \rightarrow \gamma \rightarrow \mu^+\mu^- \qquad (1.5)$$

then we have observed the vacuum polarization enhancement of a pure QED process $e^+e^- \rightarrow \gamma \rightarrow \mu^+\mu^-$. If so, then the process

$$e^+e^- \rightarrow \gamma \rightarrow \psi \rightarrow \gamma \rightarrow \text{hadrons} \qquad (1.6)$$

must also exist, as a vacuum polarization enhancement of $e^+e^- \rightarrow \gamma \rightarrow$ hadrons. One can thus use the reaction (1.5) to measure reaction (1.6), which is otherwise indistinguishable from direct decays $\psi \rightarrow$ hadrons. Thus one calculates $\Gamma_{\gamma h}$, which is the partial width for reaction (1.6). Table 1.1 summarizes these results. Note that $\Gamma_{\gamma h}$ is a significant, but by no means overwhelming, part of Γ. We shall return to this vacuum polarization enhancement again later.

Why are the ψ's so different from what was known before? The mystery is that one has such a massive resonance with such a very small width; without invoking some new almost conserved quantity it is extremely hard to understand why this comes about. This fact is what really started the big rush into the field of "charm" and as a direct result why QCD has become so successful.

_If_ we wish to build the whole structure of charm etc. using the ψ's as a foundation, we must be quite sure of the J^{PC} assignment, and that we are not deceived by some sort of unexpected weak interaction effect. (Note, weak interactions would very nicely explain the very small width of the ψ's). To establish J^{PC} we can use the decay $\psi \to \mu^+\mu^-$ and look for an interference with the QED production of μ-pairs. Since the quantum numbers of the photon are known, the quantum numbers of the ψ's can be inferred. Figure 1.5 shows the ratio of the cross section [1,2] for μ-pair to e-pair production as a function of energy. The solid line shows the expectation in the case of maximum interference between the ψ and the QED amplitudes; the dashed line shows the expectation if there is no interference. For the ψ(3095), the no interference case may be excluded by 3 standard deviations; for ψ(3684) the effect is 5 standard deviations. From the clear observation of interference and the symmetry of the detector one can immediately conclude that the ψ's must be mainly P = C = -1 decays. (We allow for the possibility that the resonances are not eigenstates of P or C). The spin may not so quickly concluded because not all the solid angle is covered. Stated differently, spin 1 and spin 2 states, for example, integrated only over the solid angle of the detector are not orthogonal. Detailed consideration, however, shows that only spins 2 and 3 can possibly show any interference with spin 1, and these will have wrong sign of interference. Thus we conclude that J = 1. We have learned that the ψ's _primarily_ decay into a final state having P = C = -1, but the question is still open whether there is a parity violation, i.e. ψ is not an eigenstate of P and C, as would be expected for weak interactions. To study this question one can study the front-back asymmetry of μ-pair production. If ψ has an admixture of P = +1, then there will be an asymmetry near the resonance. This has been measured, and the conclusion is that any possible wrong-parity admixture must be very

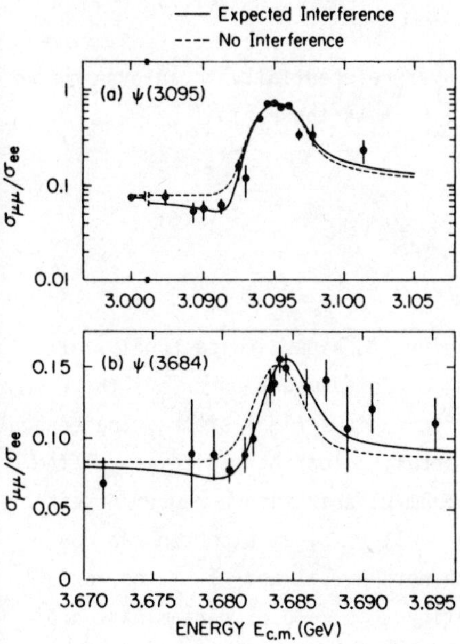

Fig. 1.5 Test for interference between the resonance (a) ψ(3095) and (b) ψ(3684) by us of the ratio of cross sections for μ-pairs to e-pairs for $|\cos\Theta| \leq 0.6$. The full line represents no interference.

small, so we can make the final assignment $J^{PC} = 1^{--}$.

To study the G-parity and isospin of the ψ's one must study the hadronic decays. The list of all observed decays is staggering; there are about 50 distinct modes which have been measured for $\psi(3095)$ (not counting trivial isospin or charge conjugate variations) and about 20 upper limits. For our purpose we shall use only a small subset to study the questions at hand. It is easiest to concentrate on just two general types to answer our question, (a) $\psi \to$ multipions (b) $\psi \to$ baryon pairs. This will unambigously establish $I^G = 0^-$ for $\psi(3095)$. For the ψ's we shall use I^G for $\psi(3095)$ plus the observed decays $\psi' \to \pi\,\pi\,\psi$ and $\psi' \to \eta\,\psi$.

Recall that a final state having n pions has G parity

$$G = (-1)^n. \tag{1.7}$$

Thus it is a natural question to ask whether ψ decays preferentially to an even number of pions (G = +1) or an odd number (G = -1). Fig. 1.6 shows the ratio[3)]

$$\alpha = \frac{R_{on}}{R_{off}} = \frac{\sigma^F_{on}}{\sigma^\mu_{on}} \frac{\sigma^\mu_{off}}{\sigma^F_{off}} \tag{1.8}$$

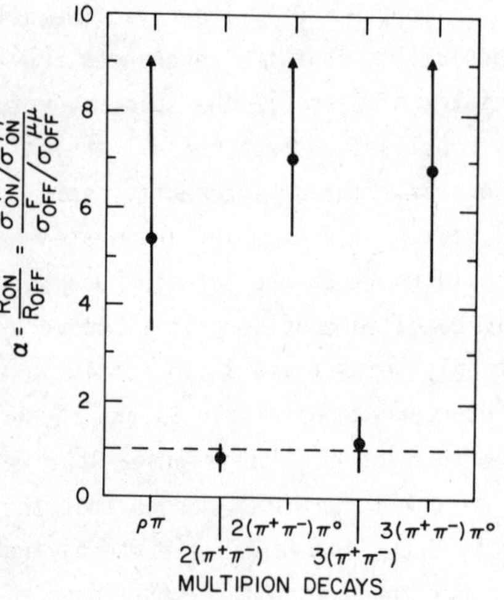

Fig. 1.6 The ratio of the quantities $\sigma^F/\sigma_{\mu\mu}$ on the $\psi(3095)$ resonance to the value off the resonance for the channels F shown. The dashed line represents the expectation if the final state F does not come from a direct ψ decay.

where F refers to a particular final state F and "on" or "off" means on or off the resonance, for various final states. The reason for this peculiar looking choice is related to the vacuum polarization enhancement effect discussed earlier: Since a photon can contribute to both G = +1 and -1 states, one must identify this kind of contamination of the selection one seeks. The ratio α allows one to do this. Those states proceeding purely via a vacuum polarization enhancement will have $\alpha = 1$; those states resulting from vacuum polarization + direct decays will have $\alpha > 1$. The evidence is very clear that 4 and 6 pion states have little or no direct component, while the 3, 5, and 7 pion states have a substantial direct component. Since one sees this selection so clearly the assumption of a normal hadronic-like decay seems reasonable, leading to G = -1.

The relation

$$G = C\ (-1)^I \qquad (1.9)$$

plus G = -1 and C = -1 immediately implies I = even. Since the photon communicates to I = 0 or 1 states, one is tempted to assign I = 0 for the ψ. The assignment I = 0 is a crucial prediction of a $c\bar{c}$ model, and therefore one wants better evidence than prejudice. The 3-pion state has been observed to be predominantly $\rho\pi$. Furthermore the relative mixture of $\rho^0\pi^0$ vs. $\rho^+\pi^-$ depends on the isospin:

$$\frac{\sigma\ (\psi \rightarrow \rho^0\pi^0)}{\sigma\ (\psi \rightarrow \rho^+\pi^- \text{ or } \rho^-\pi^+)} = \underset{(I=0)}{0.5} \quad \text{or} \quad \underset{(I=2)}{2} \qquad (1.10)$$

The measured value of this ratio is 0.59 ± 0.17, so that I = 0 is strongly favored. Further evidence for I = 0 are the direct decays [4] $\psi \rightarrow p\bar{p}$ and $\psi \rightarrow \Lambda\bar{\Lambda}$. The $p\bar{p}$ state can only have I = 0, 1 and $\Lambda\bar{\Lambda}$ is only I = 0. Thus the case for I = 0 for $\psi(3095)$ is closed.

It is perhaps worthwhile to mention that from this α plot we obtain corroborative evidence that the ψ couples to e^+e^- via a photon. This is because both the even number of pion states agree well with a simple vacuum polarization enhancement, rather than being an arbitrary number. Thus if the original 3 standard deviation evidence for J = 1 seemed somewhat meagre for so important a number there is strong circumstantial evidence here to support this assignment. It is furthermore evidence that the e and μ pair coupling to the ψ is not direct but proceeds rather via an intermediate photon.

For $\psi(3684)$ as somewhat different direction must be taken, because of the very different character of its decay. The largest single branching fraction [5] of $\psi(3684)$ is $B(\psi(3684) \rightarrow \psi(3095) + \text{anything}) = 0.57 \pm 0.08$. This alone suggests a very strong relation between the two ψ's. The branching ratio [5] $B(\psi(3684) \rightarrow \psi(3095)\ \pi^+\pi^-) = 0.32 \pm 0.04$ is also quite large. This immediately allows us to assign G = -1 and therefore I = even. We can get information on I itself by the ratio

$$\frac{B(\psi(3684) \rightarrow \psi(3095)\ \pi^0\pi^0)}{B(\psi(3684) \rightarrow \psi(3095)\ \pi^+\pi^-)} = \underset{(I=0)}{0.5} \quad \underset{(I=1)}{0} \quad \underset{(I=2)}{2} \qquad (1.11)$$

(Note there are some phase space corrections to the simple isospin ratios: 0.5 → 0.52 and 2 → 2.12). The measured value for $\psi' \rightarrow \psi$ + neutrals assuming all neutrals are π^0's is [6] 0.78 ± 0.10. The value I = 0 is clearly favored, but the observed value is about 2 standard deviations above the "expected value". This is because [6] $B(\psi(3684) \rightarrow \psi(3095) + \eta) = 0.043 \pm 0.008$. This branching ratio alone immediately implies I = 0.

Correcting the measured value to be used in reaction 1.11 for this η decay mode, one finally obtains 0.53 ± 0.06 for the ratio, in excellent agreement with I = 0 assignment.

Table 1.1 ψ Parameters

	$\psi(3095)$	$\psi(3684)$	
m	3.095 ± 0.004	3.684 ± 0.005	GeV/c^2
Γ_e	4.8 ± 0.6	2.1 ± 0.3	keV
$B_e \approx B_\mu$	0.069 ± 0.009	0.0093 ± 0.0016	
B_h	0.86 ± 0.02	0.981 ± 0.003	
Γ	69 ±15	228 ±56	keV
$\Gamma_{\gamma h}$	12 ± 2	6.6 ± 0.9	keV

I D. Related $c\bar{c}$ states, not $J^{PC} = 1^{--}$

Having the ψ's alone is hardly a compelling case for the existence of a $c\bar{c}$ system, even though the observed properties are consistent with this picture. Recall that 2 fermions having a relative angular momentum L and total spin S should be able to make states having $P = -(-1)^L$ and $C = (-1)^L\,(-1)^S$. Therefore besides $J^{PC} = 1^{--}$ states there should be other hydrogen atom-like states. Fig. 1.7 shows the expected levels, using spectroscopic notation. There should be two pseudoscalar states, each one slightly below the $\psi(3095)$ and $\psi(3684)$. There should also be a set of 3 P-states between the ψ's. If the $c\bar{c}$ model is to make any sense one must be able to observe these "other" states. Naturally, since these states do not have the quantum numbers of the photon, they are not expected to be seen in e^+e^- annihilation. One should, however, be able to couple these states via a photon to the ψ's.

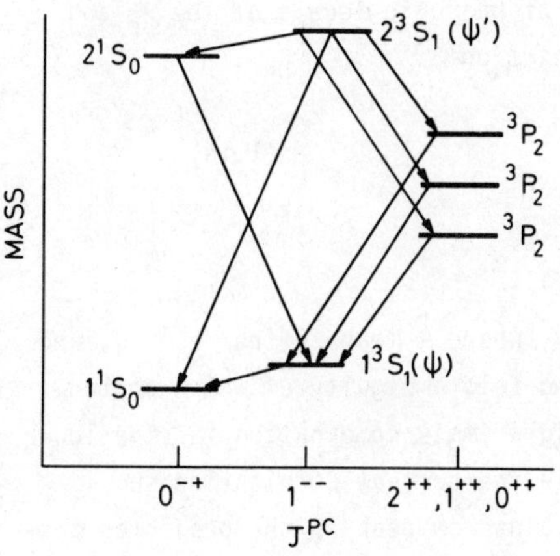

Fig. 1.7 Theoretical level scheme for a $q\bar{q}$ system. Spectroscopic notation is used for the states.

The lowest such member should be the pseudoscalar, generically called the η_c. Evidence has been presented [7] for the existence of such a state called the X(2830), seen in the decay scheme

$$\psi(3095) \to \gamma X(2830) . \qquad (1.12)$$
$$\hookrightarrow \gamma\,\gamma$$

This observation certainly fits into the broad outline of the $c\bar{c}$ model, but there was one thing which causes great trouble theoretically. The mass of 2.83 GeV/c^2 is very much further below the $\psi(3095)$ than expected, and this causes some troubles with the expected width of such a state. Another curious fact is that no hadronic decay mode has ever been seen. There is recent conflicting evidence from another experiment [7] which <u>does</u> <u>not</u> see reaction 1.12 at a level significantly below the original observation. This contradiction between two experiments, both of which taken alone look quite convincing, is an embarassing situation. It may be some time until this dichotomy is adequately resolved. Still, if the latter experiment is correct then one must ask, "where is the η_c?". The answer may well be that it lies very close to the $\psi(3095)$ so that the emitted photon is too soft to have been observed so far.

Evidence for the other pseudoscalar state has been rather weak and also contradictory. Evidence has been reported [9] for a $\psi' \to \gamma(\eta_c' \to \gamma\psi)$ cascade sequence for a mass of about 3.45 GeV/c^2, however higher statistics data [10] have not supported this suggestion.

Let us now turn our attention to the P-states. Historically the first evidence for such states was seen [11] in the reaction

$$\psi(3864) \rightarrow \gamma\ P_c(3505) \quad . \qquad (1.13)$$
$$\qquad\qquad \hookrightarrow \gamma\ \psi\ (3095)$$

This was very quickly followed [12] by observation of hadronic decays of the P_c and also the existence of 2 more states seen in the reaction

$$\psi(3684) \rightarrow \gamma\ \chi \quad . \qquad (1.14)$$
$$\qquad\qquad \hookrightarrow \text{hadrons}$$

To measure reaction 1.13 one observes the $\psi \rightarrow \ell^+\ell^-$, where ℓ means either e or μ, and at least one of the two photons. Then one has a two-fold ambiguity of which photon to pair with the ψ. Therefore one plots [10] the higher mass combination vs. the lower mass combination as shown in Fig. 1.8. In principle the correct combination should produce a narrow peak in the projected distribution, while the wrong combination should be wider because of the doppler-shifting of the second photon. Two clear peaks are seen in the high mass projection near 3.50 and 3.55 GeV. The hadronic decays provide a clearer signal for these two peaks at 3.50 and 3.55 as well as revealing a third state near 3.40 GeV. Fig. 1.9 and 1.10 show 5 different hadronic decay modes sought. Table 1.2 summarizes the known branching ratios of these intermediate states. (Note, the peak near 3.7 GeV arises from events where $\psi(3684) \rightarrow$ hadrons, but a missing photon was assumed). These distributions provide precise mass determinations of 3.414 ± 0.003, 3.502 ± 0.006, and 3.555 ± 0.005 GeV/c^2 for the three states.

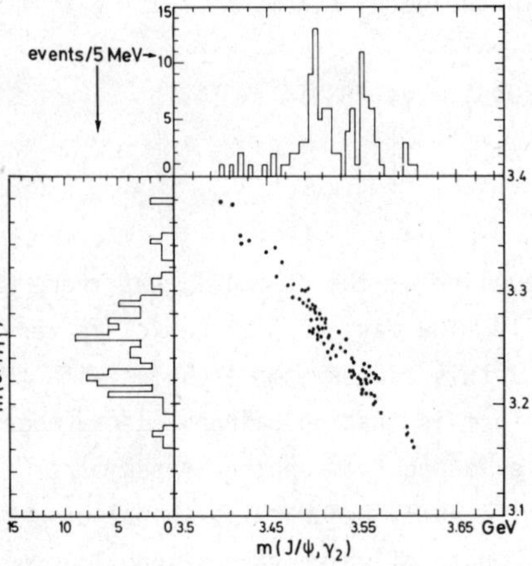

Fig. 1.8 Plot of the higher vs. the lower $\gamma\psi(3095)$, mass combinations from $\psi(3684) \rightarrow \gamma\ \gamma\psi(3095)$.

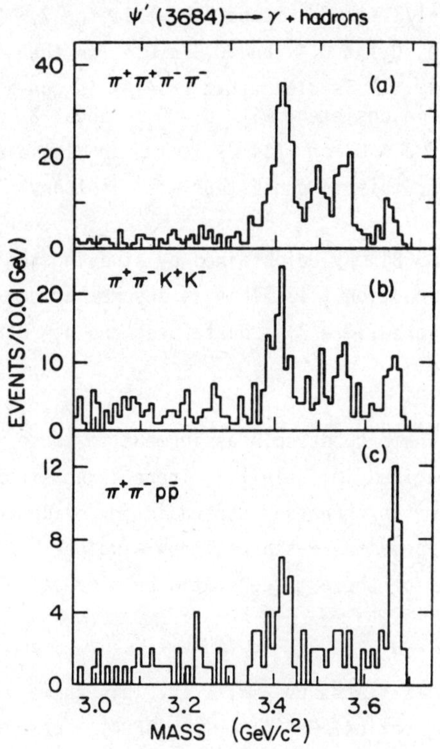

Fig. 1.9 Hadron decays of C = +1 states between ψ(3684) and ψ(3095). The peak near 3.7 GeV is due to events which have no photon but were assumed to have a missing photon.

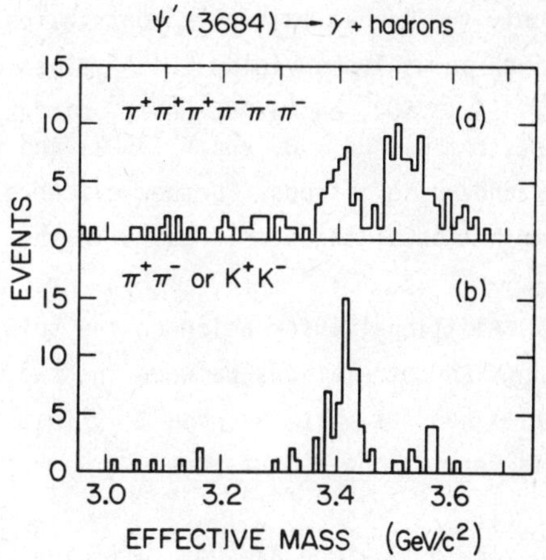

Fig. 1.10 Hadronic decays of C = +1 states between ψ(3684) and ψ(3095).

The determination of the quantum number assignment of these three states is crucial to a test of the model. Clearly, since each state is seen in conjunction with a photon from a ψ(3684) decay, we can immediately conclude that C = +1. The spin and parity assignment [13)] is rather more involved. There is an immediate simplification for the 3.414 and 3.555 GeV/c^2 states; both decay to $\pi^+\pi^-$ or K^+K^-. Since 2 pseudoscalars can only form natural spin-parity states, $J^{PC} = 0^{++}, 1^{--}, 2^{++}$ ..., and C = -1 is excluded, we are led to consider even spin and C = P = +1 for these two states.

A simple test of the spin is the angular distribution of the photon in reaction (1.14). The general form of the angular distribution must be

$$\frac{d\sigma}{d\Omega} \propto 1 + A \cos^2\theta_\gamma \qquad (1.15)$$

based on angular momentum alone, where θ_γ is the angle between the photon and the positron direction. The question is what relation A has to the spin of the χ state. For J = 0, A is unambiguously 1.0, while for other spins no unambiguous value can be predicted because of the mixture of possible multipoles. If, however, one assumes that only the lowest multipole contributes, then A = -1/3 for J = 1 and A = +1/13 for J = 2. Such an analysis vields the A values of 1.4 ± 0.4, 0.1 ± 0.4, and 0.3 ± 0.4 for the 3.414, 3.502, and 3.555 GeV/c^2 states respectively. It is clear that χ(3414) is consistent with J = 0, and P_c(3502) and χ(3555) are inconsistent with J = 0 by about 2 standard deviations. Further evidence in favor of J = 0 for χ(3414) is the angular distribution of the π or K pairs in the χ rest frame; this is consistent with isotropy.

Additional information on the spin of the P_c(3502) may be obtained by studying the angular correlations between the two photons in reaction 1.13. This is a somewhat complex subject, but the result is that J = 1 is favored, J = 2 is disfavored and J = 0 is very strongly disfavored.

The data alone are not sufficient to make an unambiguous spin assignment to these three states. The state of the knowledge is summarized in Table 1.3. There is no strong evidence for the χ(3552) having J = 2, but neither is there any contradictory evidence. Thus if we believe that these three states are indeed the P-states, then a unique assignment can be made. (Note: in no case can any of these three states be a pseudoscalar which could have confused the issue).

Finally let us summarize the knowledge of the $c\bar{c}$ states by Fig. 1.11. This shows the identified states and some relevant branching ratios. For example $\psi(3684) \rightarrow \psi(3025)\gamma$ proceeds with a branching ratio of 4 %. The γ^* indicates decay by means of an intermediate photon (vacuum polarization effect). Clearly the general picture agrees with that expected from Fig. 1.7.

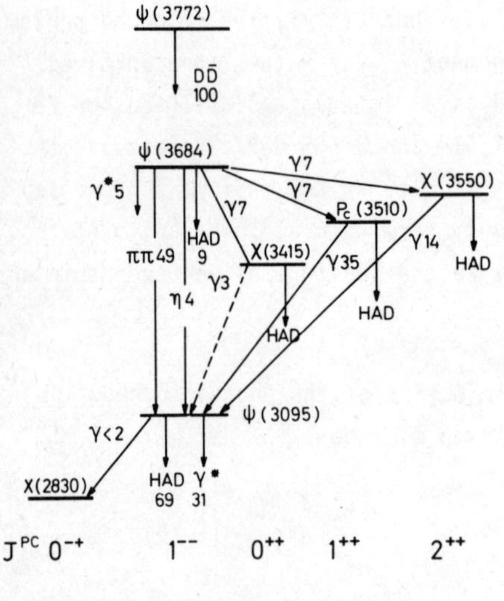

Fig. 1.11. Experimental level scheme for $c\bar{c}$ states (c.f. fig. 1.7). A character like γ 7 indicates photon emission with branching ratio of 7 % for a transition between the indicated states. γ^* indicates virtual photon (vacuum polarization) effect.

Table 1.2 Branching Ratios of P-States

Mode	χ(3415)	P_c or χ(3510)	χ(3555)
$\pi^+\pi^-$	0.010 ± 0.003	-	0.0029 ± 0.0015 (for $\pi^+\pi^-$ and K^+K^- together)
K^+K^-	0.010 ± 0.003	-	
$\pi^+\pi^-\pi^+\pi^-$	0.044 ± 0.008	0.015 ± 0.006	0.0023 ± 0.0006
$\pi^+\pi^-K^+K^-$	0.037 ± 0.010	0.009 ± 0.004	0.020 ± 0.006
$\pi^+\pi^-p\bar{p}$	0.005 ± 0.002	0.0014± 0.0011	0.0029 ± 0.0014
$3(\pi^+\pi^-)$	0.019 ± 0.007	0.024 ± 0.008	0.011 ± 0.007
γψ(3095)	0.033 ± 0.010	0.234 ± 0.08	0.16 ± 0.03

Table 1.3 Spin-Parity Assignments of P-States

State	ππ/KK	Ang.Dict.	Cascade Corr.	Suggested J^P
χ(3415)	0^+, 2^+ ...	J = 0?	-	0^+
P_c or χ(3510)	-	J ≠ 0 (2σ) J = 1?	J ≠ 0 (5σ) J = 1?	1^+
χ(3555)	0^+, 2^+ ...	J ≠ 0 (2σ)	-	2^+

II A. D-States' Characteristics

While the ψ family fits nicely into the description of charm, the only way to make a convincing case for the existence of charm is to see other $q\bar{q}$ combinations in addition to $c\bar{c}$. This means that one must see the $c\bar{u}$ and $c\bar{d}$ systems (D-states) as well as $c\bar{s}$ systems (F-states). The traditional way to identify a particle is to be able to reconstruct a well defined peak in an invariant mass plot and then be able to use characteristics of the decays to establish its spin and parity etc.

To begin looking for a peak in an invariant mass plot, one should have some idea in which modes to look. One of the characteristics of the charm model is that the major quark transition should be $c \to s$. Note two things about this:
(a) A D-state consisting of $c\bar{u}$ or $c\bar{d}$ will probably have a particle with negative strangeness in its decay products
(b) The transition $c \to s$ involves the escape of 1 unit of charge. Thus

$$D^0 \to K^-\pi^+ \qquad (2.1)$$
$$D^+ \to K^-\pi^+\pi^+ \qquad (2.2)$$

are natural decay modes. (As an exercise, make possible quark diagrams to convince yourself that this is the case; take particular care to note the charges of the particles, for this is crucial).

Evidence for the existence [14,15] of D's first came from data samples at w = 4.028 and 4.413 GeV. Fig. 2.1 shows a collection [16,17] of invariant mass plots for different choices of final state particles. There are clear peaks seen at an invariant mass of ~ 1.87 GeV/c^2 in both charged and neutral modes. We shall soon see that these signals are strongly correlated with peaks in missing mass of about 2 GeV/c^2 corresponding to D^* production. Of particular interest is the mass plot of Fig. 2.2, which displays one of the unambiguous predictions of the charm model: Fig. 2.2a shows the reaction 2.2, while Fig. 2.2b shows the reaction

$$D^+ \not\to K^+\pi^+\pi^- \quad . \qquad (2.3)$$

The reaction 2.2 is called "exotic", while 2.3 is "non exotic". The reason for this is that $K^-\pi^+\pi^+$ state can have isospin 3/2 or 5/2, but not 1/2; therefore it cannot be made up of 2 quarks. The preference for the exotic decays is a valuable tool for distinguishing the D's from "old physics" and experimentally a particularly clean way to study backgrounds to $D^\pm$ production: the non exotic combinations have the same topology but no $D^\pm$ signal; they are therefore a good estimator of backgrounds. In addition the exotic decay is a clear indication that the D is not just another kind of K^*.

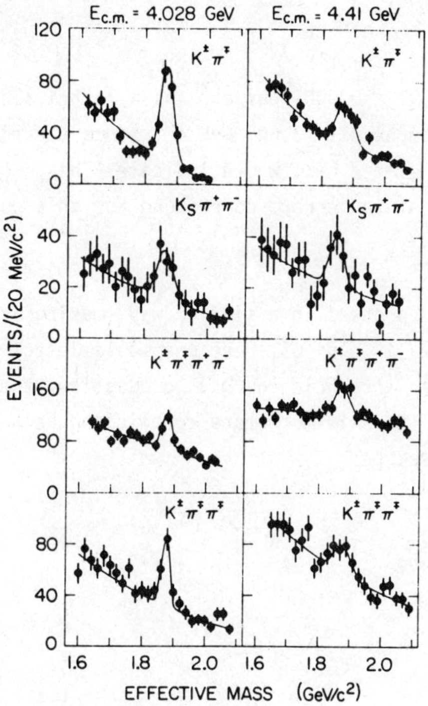

Fig. 2.1 Invariant mass spectra showing D^0 and $D^{\pm}$ decays at center of mass energies of 4.028 and 4.413 GeV.

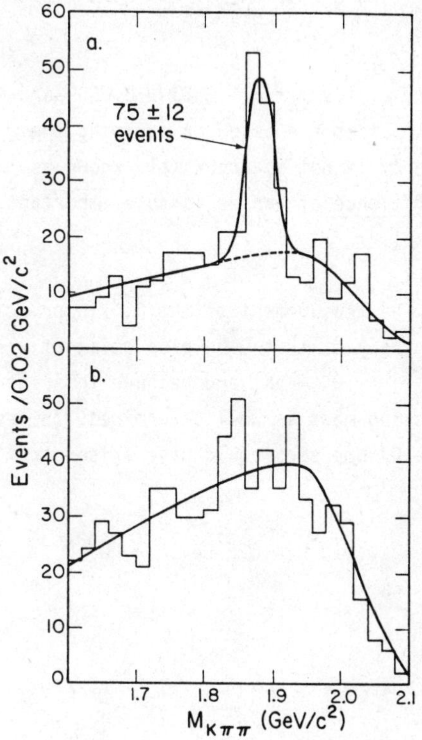

Fig. 2.2 Invariant mass spectra for $D^{\pm}$ decays for (a) the exotic channel $K^{\mp}\pi^{\pm}\pi^{\pm}$ and (b) the non exotic channel $K^{\pm}\pi^{+}\pi^{-}$

The whole field charm physics has been one of remarkable strokes of good fortune. One of these is the peak in R at w = 3.772 GeV, which we have not discussed. This resonance has the delightful property of being just barely above the threshold for $D\bar{D}$ production, and further that it appears to decay almost exclusively to $D\bar{D}$. This "D-factory" added another observed decay mode, $D^{\pm} \to K^0\pi^{\pm}$, and also allowed very precise mass determinations and absolute branching ratios.

The determination of the D^0 and D^+ masses [18] may be made very precisely at w = 3.772 GeV, because the D's are produced nearly at rest. This means that their momentum is rather low and well measured. Furthermore, the ring energy is well known. Therefore if one observes one D having a momentum P and assumes the D's to be pair produced, then the mass may be calculated simply,

$$m_D^2 = (\frac{w}{2})^2 - \vec{p}^2 \quad . \tag{2.4}$$

This yields $m_{D^+} = 1.8683 \pm 0.0009$ and $M_{D^0} = 1.8633 \pm 0.0009$ GeV/c^2. (As a technicality, all masses are based on assuming the ψ mass to be exactly 3.095 GeV/c^2, since the ring energy is not so accurately known as the errors stated here would indicate. The difference of masses is more important, and the stated errors correspond for this purpose).

The measurement of the D^{*0} in principle could proceed in a similar way, making use of the fact that $D^* \bar{D}^*$ production at $w = 4.028$ GeV is copious. Here one would detect $D^{*0} \to \pi^0 (D^0 \to \pi K)$ and because this energy is near threshold for $D^* \bar{D}^*$ production, the missing mass is well determined. In practice [19] this is much more complicated because the D^0 one sees could have arisen from several sources

$$e^+ e^- \ (4.028 \ GeV) \ \to \ D^0 \ \bar{D}^0 \tag{2.5a}$$
$$\big\lfloor_{\to \ K\pi}$$

$$\to \ D^0 \ D^{*0} \tag{2.5b}$$
$$\big\lfloor_{\to \ K\pi}$$

$$\to \ D^0 \ D^{*0} \tag{2.5c}$$
$$\big\lfloor_{\to \ (\gamma \ or \ \pi^0) \ D^0}$$
$$\big\lfloor_{\to \ K\pi}$$

$$\to \ D^- \ D^{*+} \tag{2.5d}$$
$$\big\lfloor_{\to \ \pi^+ \ D^0}$$
$$\big\lfloor_{\to \ K\pi}$$

$$\to \ D^{*0} \ \bar{D}^{*0} \tag{2.5e}$$
$$\big\lfloor_{\to \ (\gamma \ or \ \pi^0) \ D}$$
$$\big\lfloor_{\to \ K\pi}$$

$$\rightarrow D^{*+} D^{*-} \qquad (2.5f)$$
$$\quad \lfloor\rightarrow \pi^+ D^o$$
$$\qquad\quad \lfloor\rightarrow K\pi$$

The result is that the observed momentum spectrum of the D^o is very complicated. (Experimentally it is cleaner to deal directly with the observed momentum than with the missing mass which in turn is coupled to the assumed D^o mass). Fig. 2.3 shows the contributions to the momentum spectra from these various sources along with the observed spectra. Extracting the D^{*o} mass involves a detailed fit to the observed distribution. The result is $m_{D^{*o}} = 2.006 \pm 0.0015$ GeV/c^2.

The determination of the D^{*+} mass could proceed the same way as the D^{*o}, but as it happens the statistics are not very good. A different trick can be employed. One can reconstruct the D^{*+} mass directly by observing $D^{*+} \rightarrow \pi^+ D^o$. To do this one must use higher energy data, because the π will be too soft to observe at 4 GeV. The result is $m_{D^{*+}} = 2.0086 \pm 0.0010$. For reasons which will become apparently shortly, the same scheme cannot be used for the D^{*o}.

Fig. 2.4 shows a level diagram of the D system. Also shown are the Q-values (available kinetic energy) for various possible D^* decay modes. For example the Q-value for $D^{*+} \rightarrow \pi^+ D^o$ is 5.7 ± 0.5 MeV. Note that the pionic decays generally have very low Q-values, so that even though the decay is strong, the rate is heavily suppressed by phase space. It appears that $D^{*o} \rightarrow \pi^- D^{*+}$ is kinematically forbidden.

Now let us turn to the subject of cross section and branching ratios for D production. Experimentally this means simply measuring the inclusive cross section for observing D's in any particular decay channel. Table 2.1 summarizes knowledge on inclusive production cross section times branching ratios for various D^o and D^+ decay modes at several energies [21]. There are three things to note from this table. (a) Both D^o and D^+ are observed to decay into 2 and 3 pseudoscalars. This will have important consequences when we discuss the spin assignment and the weak nature of the D decay. (b) Production of D's is copious. (c) Decay modes such as $D^o \rightarrow \pi^+\pi^-$ are suppressed. This is just as it should be, because a c → d quark transition amplitude is suppressed by the sine of the Cabbibbo angle.

The measurement of absolute branching ratios for D decays is more difficult. One must know both the number of D's decaying a certain way and the number of D's produced. Getting the total number of D's produced is a little difficult. Fortunately the ψ(3772) provides this mechanism, for it appears [22] to decay almost exclusively into $D\bar{D}$. If we

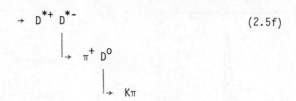

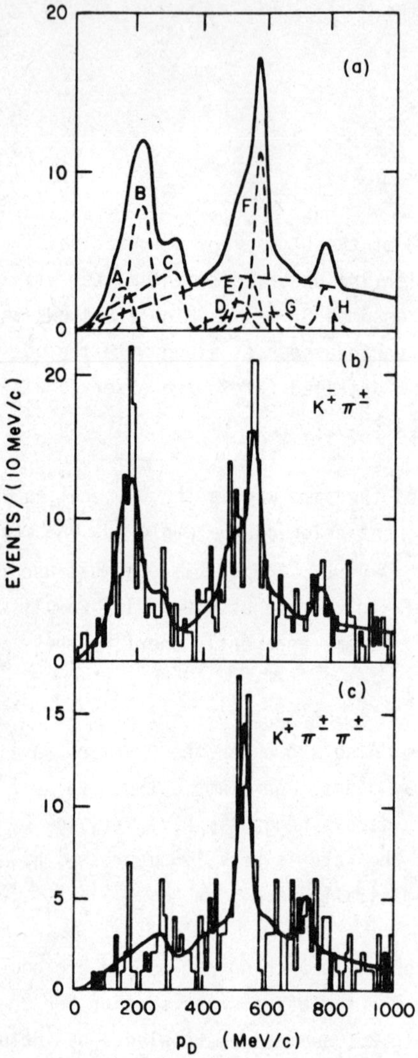

Fig. 2.3 Observed D momentum spectrum for production at 4.028 GeV center of mass energy. (a) shows the contributions to the D^0 spectrum. A, B, and C are the contributions from $D^*\bar{D}^*$ with A: $D^{*+} \to D^0\pi^+$, B: $D^{*0} \to D^0\pi^0$, C: $D^{*0} \to D^0\gamma$. D, E, F, and G are contributions from $D^*\bar{D}$ and $D\bar{D}^*$ production with D: $D^{*+} \to D^0\pi^+$, E: $D^{*0} \to D^0\pi^0$, F: direct D^0, G: $D^{*0} \to D^0\gamma$. H is the contribution from $D^0\bar{D}^0$ production. (b) shows the data for D^0 and the fit to all the contributions of (a). (c) shows the data for $D^\pm$ and the result to the fit.

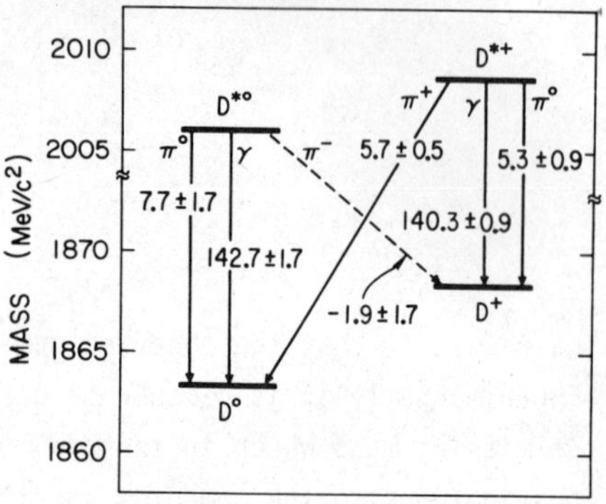

Fig. 2.4 Level scheme for the D-system showing Q-values for measured transitions. For example $D^{*+} \to \pi^+D^0$ proceeds with 5.7 MeV free kinetic energy.

further assume $\psi(3772)$ to be in a definite I-spin state, 0 or 1, then we know $\Gamma_{D^0\bar{D}^0} = \Gamma_{D^+D^-}$, except for some small phase space factors. We now have an absolute inclusive cross section and cross section times branching ratios, hence we can compute branching ratios. Table 2.2 lists the resulting branching ratios.

By measuring the inclusive electron yield at the $\psi(3772)$ one can deduce the semileptonic decay branching rates [23,24]. Fig. 2.5 shows the inclusive electron cross section as a function of center of mass energy. Unfortunately in measuring just an inclusive electron yield, one does not know what fraction comes from D^0 and what part from D^+. Thus one only obtains some average of the two. The result is a $B(D \to eX) \sim 10\,\%$.

Now that we have both $\sigma \cdot B$ and B, we are in a position to determine the inclu-

Table 2.1 Production Cross Section Times Branching Ratio for D^0 and D^+

	Mode	w = 3.772	4.028	4.414 GeV
	$K^{\mp}\pi^{\pm}$	0.25 ± 0.05	0.53 ± 0.10	0.28 ± 0.08 nb
	$K^0\pi^+\pi^-$ + cc	0.46 ± 0.12	1.01 ± 0.28	0.85 ± 0.32 nb
D^0	$K^{\mp}\pi^{\pm}\pi^+\pi^-$	0.36 ± 0.10	0.77 ± 0.25	0.85 ± 0.36 nb
	$\pi^+\pi^-$	-	<0.04 nb	-
	K^+K^-	-	<0.04 nb	-
	Sum	1.07 ± 0.21	2.3 ± 0.4	2.0 ± 0.5 nb
	$\bar{K}^0\pi^+$ + cc	0.14 ± 0.05	<0.17 nb	-
	$K^{\mp}\pi^{\pm}\pi^{\pm}$	0.36 ± 0.05	0.37 ± 0.09	0.31 ± 0.11 nb
	$\pi^{\pm}\pi^+\pi^-$	-	<0.03 nb	-
	Sum	0.50 ± 0.07	0.37 ± 0.09	0.31 ± 0.11

Table 2.2 Branching Ratios of the D's

D^0 Mode	Branching Ratio		D^+ Mode	Branching Ratio
$K^-\pi^+$	0.018 ± 0.005		$\bar{K}^0\pi^+$	0.015 ± 0.006
$K^0\pi^+\pi^-$	0.044 ± 0.011		$K^-\pi^+\pi^+$	0.039 ± 0.010
$K^-\pi^+\pi^0$	0.12 ± 0.06		$K^+\pi^+\pi^-$	<0.002
$K^-\pi^+\pi^-\pi^+$	0.035 ± 0.009			
$e^{\pm}$ + anything	0.098 ± 0.014	=	$e^{\pm}$ + anything	0.098 ± 0.014

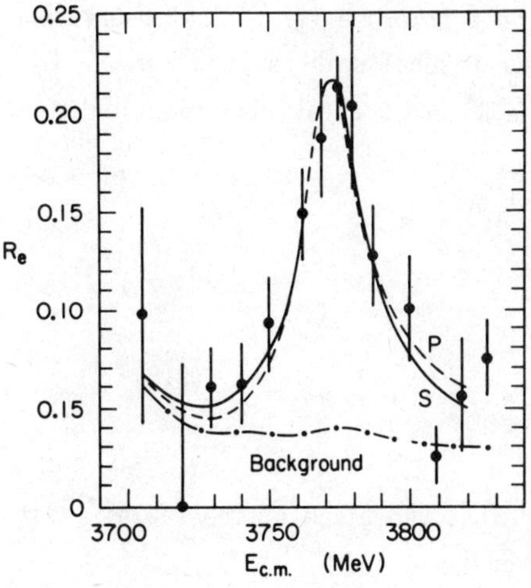

Fig. 2.5 The cross section for seeing an electron as a function of center of mass energy near the $\psi(3772)$.

sive cross section for D production. The D production is very copious indeed. Define:

$$R_D = \sigma_D/(2 \cdot \sigma_{\mu\mu}) \qquad (2.6)$$

as the contribution to the total R-value due to D production. (The factor 2 comes from the inclusive cross section accepting either D produced). There are well established peaks in R at 4.028 and 4.413 GeV; the question is how much is due to R_D? If we make the measured value of R (corrected for τ production) and substract 2.4 for old u-d-s type physics we obtain R_{new}. Table 2.3 summarizes knowledge at these two energies. It is clear that D production can essentially saturate R_{new}, however if the whole charm picture is to hold together there must be room for F production where it is kinematically allowed.

Table 2.3 D-Production Cross Section in Units of μ-Pair Cross Section

	w = 4.028 GeV	w = 4.414 GeV
R_{D^o}	2.3 ± 0.6	2.3 ± 0.8
R_{D^+}	0.9 ± 0.4	0.9 ± 0.4
Sum	3.2 ± 0.7	3.2 ± 0.9
$R_{\text{"new"}}$	3.1 ± 1.0	2.5 ± 1.0

Now let us turn to the determination of the spin of the D's and D^*'s. The predicted values of J^P for (D, D^*) are (0^-, 1^-). As with the K's there is no way to establish the absolute parity of the D's, because they are either occur in pairs or if they are single then parity violating weak interactions communicate to the world of known-parity particles. Thus we may arbitrarily define the D parity to be -. We shall assume that the $\psi(3772)$, like the $\psi(3095)$ and $\psi(3684)$ has the quantum numbers of photon, $J^P = 1^-$, even though there are no experimental data to prove it. *) The observation of $D \to 2$ pseudoscalars suggest that the D's have natural spin-parity, i.e. $J^P = 0^+$, 1^-, 2^+ ... . Assuming parity conservation in the D^* and D decay, the reaction $D^* \to \pi(D \to \pi K)$ restricts the possible choices:

a) If the D has J = 0 then the parity of D^* must be

$$P_{D^*} = P_\pi P_D (-1)^{J_{D^*}} = -(-1)^{J_{D^*}} P_D \qquad (2.7)$$

b) If the D^* has J = 0 then

$$P_{D^*} = P_\pi P_D (-1)^{J_D} = -(-1)^{J_D} P_D \qquad (2.8)$$

(The oribital angular momentum must be J_D to get $J_{D^*} = 0$).

---

*) If this assumption is false we would have a gross violation of the law of least amazement (and an even more important discovery!).

c) If both D and D* have J = 0 then we have $P_{D^*} = -P_D$, but a 1^- state ($\psi(3772)$) cannot decay into a 0^+0^- final state without violating parity.

A natural method of determining the spins [25] is to study the production angular distribution, i.e. the angle between the direction of the D and the incident positron. General considerations of angular momentum say

$$\frac{d\sigma}{d\Omega} \propto 1 + \alpha \cos^2\theta_D \tag{2.9}$$

For the case of $D\bar{D}$ production this distribution can be studied using $\psi(3772)$. If $J_D = 0$ then $\alpha = -1.0$. For other spins the value of α is not unambiguous. The measured value is $\alpha=-1.00\pm0.09$ for D^0 and -1.04 ± 0.10 for D^+. While this does not prove that $J_D = 0$, it is strongly suggestive. Supporting evidence for $J_{D^0} = 0$ comes from the decay angular distribution for the mode $D^0 \to K^-\pi^+$. This is quite consistent with isotropy (50 % confidence) and inconsistent with $J_D = 1$ (< 1 % confidence). Similar measurements at 4.028 GeV for $D^*\bar{D}^*$ production yield $\alpha = -0.30 \pm 0.33$. Thus $J_{D^*} = 0$ is disfavored by 2 standard deviations. The production angular distribution for $D\,D^*$ gives complementary information. If $(J_D, J_{D^*}) = (0, 1)$ or $(1, 0)$ then α is unambiguously + 1.00. Data at 4.028 GeV are quite consistent with this value and inconsistent with (0, 0). This test, naturally, cannot distinguish between (0, 1) and (1, 0). The joint production-decay distributions have more statistical power than the projections alone. This very strongly favors (0, 1) vs. (1, 0) (confidence level 2×10^{-3}).

Another way to study the D^+ spin is to study the Dalitz plot for $D^+ \to K^-\pi^+\pi^+$ decays [26]. If $J_D = 0$ the plot should be uniformly populated. Other natural parity states should have a zero at the boundary of the Dalitz plot. Fig. 2.6 shows the Dalitz plot for $D^\pm$ decays according to exotic reaction 2.2 and the equivalent plot for the non exotic (background) configuration. Qualitatively, both distributions are compatible with uniform populations. In order to make a quantitative test of the uniformity, the plot has been divided into two regions respectively favored and disfavored for spins other than zero (viz. J = 1 and 2). The countour is chosen in such a way that equal areas are enclosed, so that J = 0 should have relative populations of 1 : 1 for the 2 regions. For spin 1 the regions should have the relative populations 1 : 8.2; one observes 34 ± 9 and 38 ± 9 events, after correcting for background. Thus spin 1 has a confidence level nearly 2×10^{-5}. For spin 2 one expects a relative population of 1: 5.6 and one observes 31 ± 9 and 35 ± 9 events. Spin 2 has a confidence level nearly 2×10^{-3}. The same argument can be extended to higher spins, because of the general property of a vanishing density on the boundary.

Therefore all evidence seems to point to the ground state D's having J = 0. This, however, poses a problem: A spin 0 object cannot decay into both 2 and 3 pseudoscalars

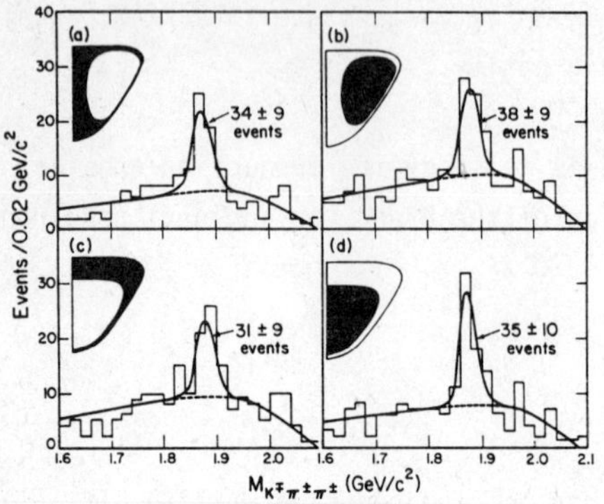

Fig. 2.6 Exotic mass contributions for $K^{\mp}\pi^{\pm}\pi^{\pm}$ (c.f. Fig. 2.2) for shaded parts of the Dalitz plot. (a) and (b) show the result of dividing the Dalitz plot into 2 equal areas along a contour of constant matrix element for J = 1. For spin 1 one expects (a) and (b) to have $D^{\pm}$ events in the ratio: 1 : 8.2. For J = 0 the ratio should be 1 : 1. (c) and (d) show same distributions using J = 2. Relative ratio should be 1 : 5.6 for J = 2 and 1 : 1 for J = 0.

without violating parity! It is easy to see that 2 pseudoscalars having a relative angular momentum L = J necessarily have a parity $P = (-1)^J$. A system of 3 pseudoscalars can be decomposed into a system of 2 having natural spin-parity and a single pseudoscalar. To make J = 0 in the end, the relative angular momentum of the un-paired pseudoscalar must be the total angular momentum of the pair. Thus the parity is (-1). Since all evidence points to J = 0 for the D we therefore have positive evidence that parity is violated in the D-decay. This in turn indicates that weak interactions must be responsible for D decays.

Let us summarize the reasons for believing that the D's are evidence for charm: (a) the widths are very small, (b) D decays involve strange particles, (c) the D decays involve an exotic channel, (d) semileptonic D decays are copious, (e) D decays involve parity violation, and (f) Cabbibbo suppressed decays are indeed suppressed. All these agree with the predictions of the charm model, and strongly disagree with interpreting the D family as just another K^*.

II B. F-States' Characteristics

Now that the ψ-family and D-family fit nicely into the charm description, it is urgent to find the F-family, i.e. those states consistenting of $c\bar{s}$ quark configuration. Where should one look for them? As discussed for the case of D's one expects mainly a $c \to s$ quark transition. Thus the final state of F decay should consist of $s\bar{s}$ plus other non strange particles. Examples of this $s\bar{s}$ system could be ϕ + pions or η + pions; similarly K^+K^- + pions is also possible. In analogy to the D^*, there must also be an F^*, which lies only slightly higher in mass than the F. The F-system, however, should be an I = 0 system, so that $F^* \to \pi F$ would be prohibited by isospin, however the reaction

$$F^* \to \gamma F \qquad (2.10)$$

should exist. The photon energy should be fairly low, $\sim$ 100 MeV.

Finding the F [27,28] has proved more difficult than finding the D. The published evidence for the existence of the F is of two types (a) circumstantial with many events and (b) direct with few events. In both cases one is relying on finding η's as part of the signature for the F.

The circumstantial evidence came from measuring the inclusive cross section for η production, in which the η is detected by its decay into two photons. If there is a substantial F production and subsequent decay into an η plus other things, one should see a significant increase in inclusive η production. Figure 2.7 shows the invariant mass spectrum of photon pairs in events at various center of mass energies. The events were selected by requiring at least 2 and not more than 6 photons and at least 2 and not more than 6 charged particles. To make a γ-γ mass combination both photons had to be in the energy range 0.14 to 1.00 GeV, and the vector sum of the photon momenta had to lie between 0.3 and 1.4 GeV/c. At all 6 energies, Fig. 2.7 shows a clear π^0 peak. In addition there is a very prominent peak at the η mass in the 4.42 GeV data, and perhaps at other energies as well. If the reaction 2.10 takes place, then one should expect to see a low energy photon accompaning the η. Figure 2.8 shows the γ-γ mass spectrum for those events with at least one photon having an energy of less than 140 MeV. There is a large peak at the η mass in the 4.36 to 4.49 GeV data, whereas at other energies no statistically significant peak exists. The crucial point is that not only is there an η signal in events having a low energy photon, but there is a strong correlation; stated differently the probability of finding an η in an event is much larger if there is also a low energy photon in the event.

If F production occurs in pairs, and semileptonic decays of F's are plentiful as

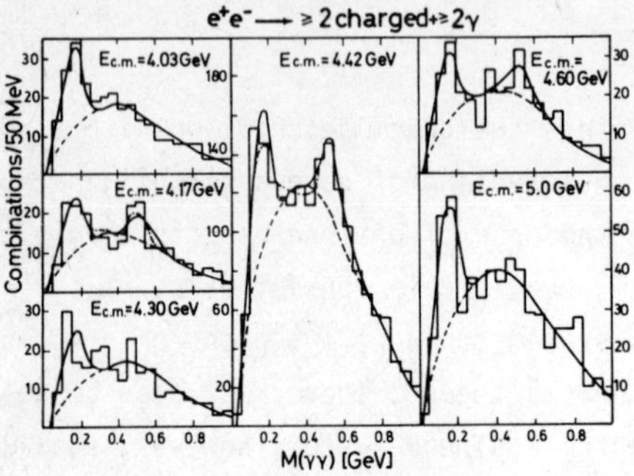

Fig. 2.7 Invariant mass spectra at various center of mass energies for events having at least 2 charged particles and 2 photons. The dashed line represents the estimated background.

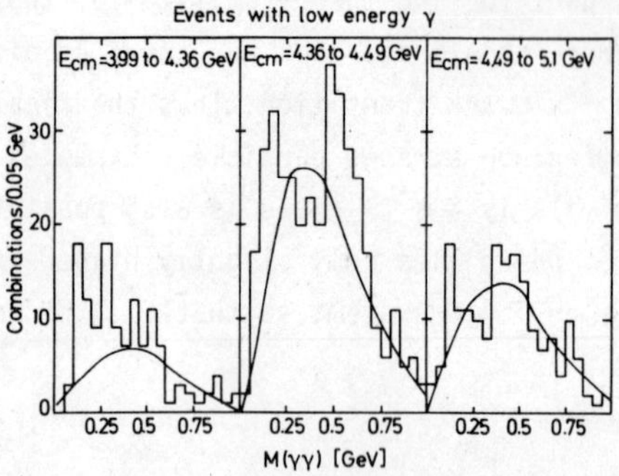

Fig. 2.8 Invariant mass spectra for the subset of Fig. 2.7 having in addition a low energy photon (< 140 MeV).

expected, then one should also expect to see events having an η from the decay of one F and an electron from the deday of the $\bar{F}$. An indication of this behavior is shown in Fig. 2.9, where η production is correlated with electron production.

Now let us look at the direct evidence of F's, in particular the production processes

$$e^+e^- \to F \quad F^* \qquad (2.11)$$

$$\quad\quad\quad |\to \gamma(F)$$

$$|\to \pi\eta$$

$$\quad |\to \gamma\gamma$$

$$e^+e^- \to F^* \quad (\bar{F}^*) \;, \qquad (2.12)$$

$$|\to \gamma \; F$$

$$\quad |\to \eta\pi$$

$$\quad\quad |\to \gamma\gamma$$

where particles in parentheses are not observed. Note that reactions 2.11 and 2.12 show only one of several possibilities to see a low energy photon, a charged pion, and

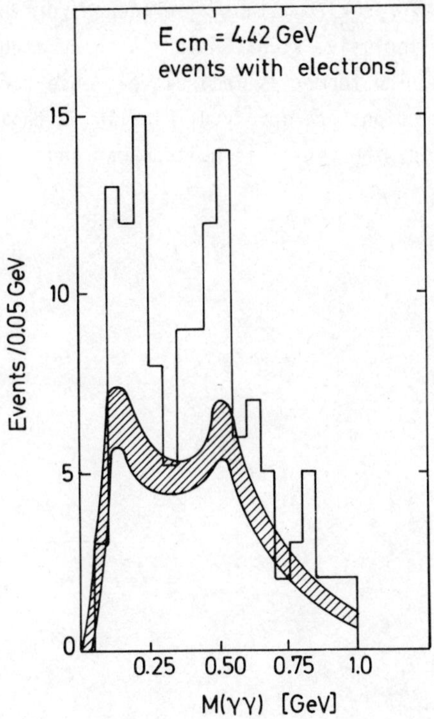

Fig. 2.9 Invariant mass spectrum for events at 4.42 GeV having at least 2 charged particles, one of which is identified as an electron, and at least 2 photons.

an η decay in the final state. In particular for reaction 2.11 one can see either the original F decay or the F resulting from the F^* decay; a similar confusion exists for reaction 2.12. Fortunately the precise origin of the F observed does not play a very important role, because of favorable kinematics. Events were fitted to reaction 2.11, assuming that the missing mass is the same as the observed F mass; the same events were also fitted to reaction 2.12 but the missing mass was required to be the same as the observed Fγ invariant mass. Figure 2.10 shows the fitted F and F^* masses from reaction 2.11, both for data near 4.42 GeV (11 events) and for all other energies combined (10 events). There is a clear cluster of 6 events near 2 GeV for the F and F^* masses in the 4.42 GeV data, but there is no clustering in the other data, which come from over 3 times the integrated luminosity. Using the latter data, the background estimated for the 6 events FF^* signal is less than 0.2 events. If one fits to reaction 2.12 instead, the fitted masses change slightly. The result is that the F mass is 2.03±0.06 GeV, and the F^*-F mass difference is 0.11±0.05 averaged over the two cases.

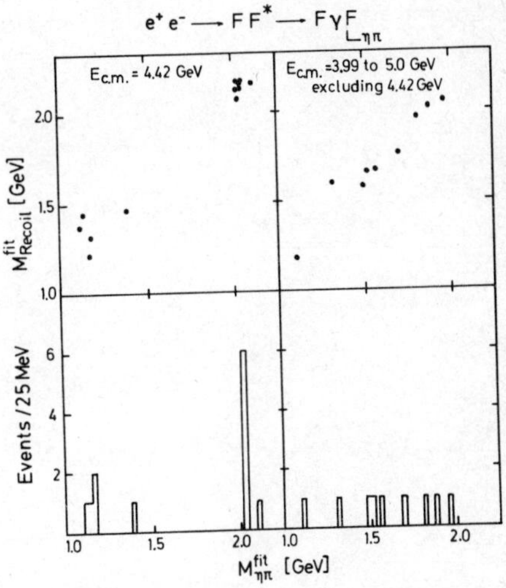

Fig.2.10 Fitted ηπ mass vs. fitted missing mass, assuming $e^+e^- \to FF^*$ where $F^* \to \gamma F$ and $F \to \eta\pi$. (a) shows data at 4.42 GeV and (c) the sum of all other energies. Projections on the ηπ mass axis of (a) and (c) appear in (b) and (d).

To summarize the situation of the F family, there is circumstantial evidence for the existence of both an F and F^* by virtue of the η inclusive signal which is correlated with a low energy photon and also with electrons. The former is indirect evidence for an F^* and the latter is evidence that weak interactions are involved. Finally, although only 6 events have been seen due to FF^* production, the signal is quite clean and free from background.

II C. τ: The Other Side of QCD, and in Indication of more to come

In the grand scheme of things we are told that there should be a quark-lepton symmetry (u,d), (c,s) : (ν_e,e), (ν_μ,μ). With u, d, s, and c quarks we just fit into the e and μ families of leptons. If one finds another lepton and its neutrino, then one should also expect another pair of quarks, (t,b) : (ν_τ,τ), and vice versa. Today we have evidence for both a heavy lepton τ, and its neutrino, and a b quark. Clearly we should expect to have a t quark. For now we shall discuss the τ. The next lecture will discuss the b and some future lecture should presumably be able to discuss the t.

The traditional method of indentifying a particle by reconstructing its mass from its decay products is doomed to failure for a heavy lepton, because there will always be a missing neutrino. Other methods depend upon the particle living long enough to see its flight path; this is also hopeless, because a high mass lepton will decay very rapidly. Therefore we shall have to content ourselves with rather more indirect evidence. In particular we shall study anomalous events which are not easily explained by other hypotheses, study the threshold behavior of these anomalous events, and finally study branching ratios of the decays of the object. In the end we shall have no excape but to conclude that a new lepton has been found.

The first evidence for the existence [21)] of the τ came from the observation of

$$e^+e^- \rightarrow \mu^\pm e^\mp + \text{"nothing"} \qquad (2.13)$$

where "nothing" means neither charged particles nor photons, but substantial energy and momentum. Experimentally such events are particularly valuable to study, because normal QED processes have great difficulty in contributing. (For example $e^\pm e^\mp$ + "nothing" is possibly due to just a radiatively degraded elastic scattering event). The first observations of these events were numerous, but troubled by rather large backgrounds due to difficulties in event identification. Soon other experiments also saw such events in smaller number, but much lower backgrounds. Thus one had to accept the existence of these peculiar events, but their interpretation remained unclear. One would expect a heavy lepton τ to have leptonic decays of the form

$$\tau^\pm \rightarrow \nu_\tau \quad e^\pm \, \nu_e \qquad , \qquad (2.14)$$

$$\tau^\pm \rightarrow \nu_\tau \quad \mu^\pm \, \nu_\mu \qquad . \qquad (2.15)$$

If τ's are produced in pairs such decay schemes would naturally lead to events like reaction (2.13) and not to final states as $\mu^\pm e^\pm$ "nothing", which are not observed above a background level. This consistency argument, however, is not sufficient to claim with confidence that a new lepton had been found. As it happens, nature played a dirty trick and set the energy threshold for τ pair production rather near the D-pair production threshold. Thus there was the nagging doubt that the anomalous events were somehow due to charm.

The study of μ-inclusive and e-inclusive spectra became the means to conclusively exclude charm as a possible source for the anomalous events. First, a heavy lepton would be expected to decay with a rather low charged multiplicity, whereas charmed mesons should have much higher charged multiplicities. Thus the majority of τ decays should have only 1 charged track, meaning that a $\tau^+\tau^-$ pair should mainly appear in the laboratory as 2 charged prong events. On the other hand $D\bar{D}$ production should only rarely appear as only 2 charged tracks. This means that making a cut on the charged multiplicity should discriminate strongly between the two cases. Figures 2.11 and 2.12 show the inclusive electron spectra for high multiplicity events [30] (charm) and low multiplicity events [31] (τ). Although the errors are fairly large the general

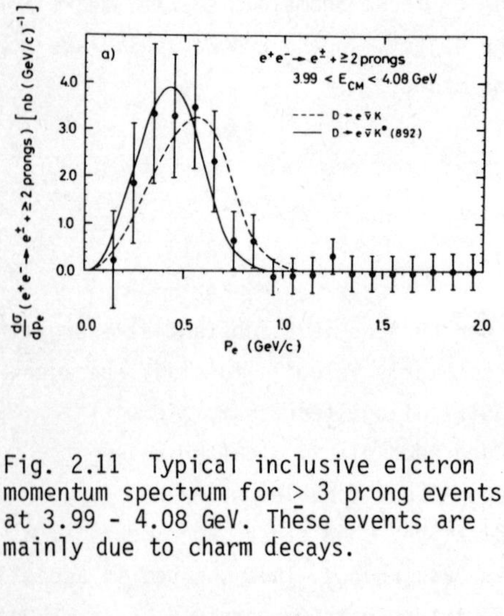

Fig. 2.11 Typical inclusive elctron momentum spectrum for $\geq$ 3 prong events at 3.99 - 4.08 GeV. These events are mainly due to charm decays.

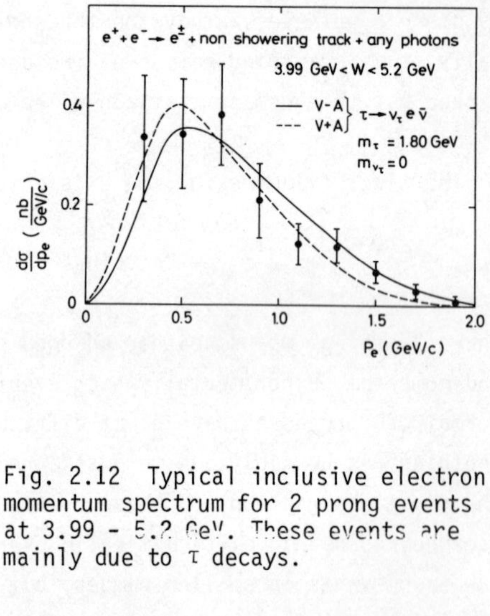

Fig. 2.12 Typical inclusive electron momentum spectrum for 2 prong events at 3.99 - 5.2 GeV. These events are mainly due to τ decays.

shapes of the two spectra are unmistakably different: the low multiplicity spectrum is much richer in high energy electrons. In fact the data of Fig. 2.12 are well fit by the expected spectrum from decays of a heavy lepton having a mass $\sim$ 1.8 GeV.

Compelling evidence for heavy lepton production came from measuring the center of mass energy dependence of the inclusive electron cross section. If one defines σ_e as the integral of the inclusive cross section (e.g. Fig. 2.12) and plots $\sigma_e / (\sigma_\mu = 4\pi\alpha^2 / (3s))$ vs w as shown in Fig. 2.13, one sees a very clear threshold behavior[32]. The rise is extremely rapid, and the value is rather large. A detailed fit of these

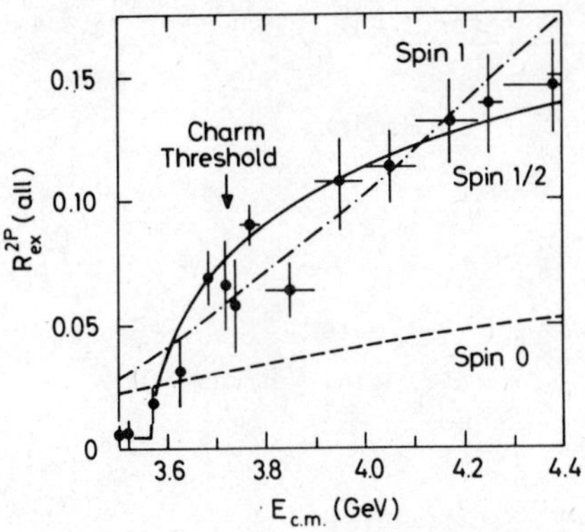

Fig. 2.13 The ratio of the cross section for seeing an electron to the QED μ-pair cross section as a function of center of mass energy. The dashed line shows the best fit to the data assuming J = 0; the dash-dot line corresponds to J = 1 and the solid line corresponds to J = 1/2.

data to point-like pair production of various types of objects leads to some very clear conclusions: (a) Spin 0 is excluded because the observed rate is too large. This is because one expects

$$R_e = 2\ B_e \frac{1}{4} \beta^3 \quad \text{(spin 0)} \qquad (2.16)$$

where B_e is the branching ratio to e + "nothing" and β is the velocity of the τ. The figure shows the best fit possible with B_e = 1.0. (b) Spin 1 point-like can be made to at least vaguely resemble the data shown, but the R_e value will continue to grow with energy like w^2, which is quite inconsistent with higher energy data. (c) Spin 1/2 should have

$$R_e = 2\ B_e\ \beta \frac{3 - \beta^2}{2} \quad \text{(spin 1/2)} \quad (2.17)$$

which is a very good fit to the data. (d) Higher spins will be at least as divergent at spin 1, and thus can be excluded. We therefore have good evidence for τ having spin 1/2 and that these e-events are not related to charm production. The really important thing, however, is that from the fit to the excitation curve one finds a mass of $m_\tau = 1.782 \begin{array}{l} +\,0.002 \\ -\,0.007 \end{array}$ GeV/c^2. This is unambiguously below the lowest D mass (1.8633 ± 0.0009 GeV/c^2), and the last possible doubt that charm mysteriously caused the anomalous events is removed.

Let us turn now to measurements of branching ratios [38)] of the τ. Table 2.4 shows a selection of measured branching ratios along with that expected for a heavy lepton. In all cases the agreement of measurements and predictions is good. It is important to note that πν mode comes from an axial vector current, while the ρν mode involves a vector current. This is additional confirmation of the expected weak interaction.

In summary we know quite a bit about the τ. The mass is rather accurately determined, and most significantly it lies below the lowest lying charmed meson. The spin is quite clearly 1/2. The branching ratios agree well with predictions, so that the expected weak coupling seems to be correct.

Table 2.4 Measured and Predicted Branching Ratios for τ

Mode	Measured	Predicted	Remark
$e\,\bar{\nu}_e\,\nu_\tau$	0.167 ± 0.010	0.166	
$\pi\,\nu_\tau$	0.082 ± 0.020	0.095	(Axial Vector)
$\rho\,\nu_\tau$	0.24 ± 0.09	0.025	(Vector)
$A_1\nu_\tau$	0.104 ± 0.024	0.081	(Axial Vector)
$\nu_\tau + \geq$ 3 Prongs	0.32 ± 0.04	0.26	

III A. The Υ Family Characteristics

As mentioned previously, once one has a heavy lepton τ and its neutrino, one also expects another pair of quarks, called t and b. From analogy to the ψ system one would therefore expect to find first bound $q\bar{q}$ states and then a release of mesons carrying a new quantum number. This process should happen once each for the t and the b. The big uncertainty is the mass scale on which all this should happen.

The first evidence supporting such a picture came from a Fermilab experiment [34)] which observed μ-pairs with large invariant mass coming from hadron-hadron collisions. Somewhat later the ψ-equivalent of a storage ring experiment at DESY confirmed the FNAL finding [35-37)] that there are at least 2 very narrow states. Figure 3.1 shows the μ-pair mass spectrum from the FNAL experiment (after background subtraction) along with some typical Υ formation results from $e^+e^- \to \Upsilon \to$ hadrons from DESY. Note in particular the widths of the peaks are all limited by the resolution of the experiments. The storage ring experiments allow rather precise mass determinations and width determinations, using the same method used for the ψ's. There is, however, a major difference in the level of knowledge of the Υ's compared to the ψ's: There is no direct experimental proof that the Υ has the same quantum numbers as the photon. This is because there are orders of magnitude fewer events available for the Υ than for the ψ's. This has several reasons: (a) The energy resolution of the ring is a factor of 10 worse at Υ compared to ψ, due to increased synchrotron radiation. This in turn reduces the apparent peak height of the cross section by a factor of 10. (b) The production cross section is reduced by at least another factor of 10 for fixed Γ_e due to the higher mass; (c.f. Eq 1.4). (c) As we shall soon see the value of Γ_e is also a factor of 3 smaller than the ψ. (d) It took a heroic effort to make the DORIS storage ring work at the required energies, which are above the design energies.

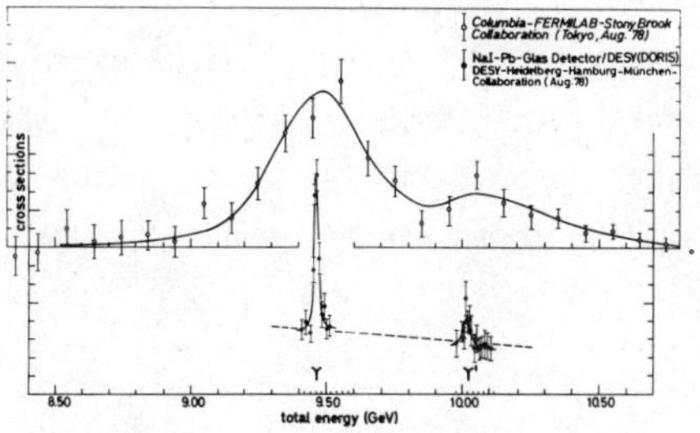

Fig. 3.1 Evidence for Υ and Υ' production (a) the background subtracted μ-pair spectrum for hadron-hadron collisions (b) the cross section $e^+e^- \to$ hadrons.

If one assumes that both the Υ's have $J^{PC} = 1^{--}$, and that its decays are nearly all hadronic, the value of Γ_e can be obtained rather easily. Determination of the total width, however, is very difficult, because the observed number of $\Upsilon(9460) \to \mu^+\mu^-$ decays [38)] is so small; for the $\Upsilon(10020)$ no such information is available at all. The state of the knowledge of the Υ family is summarized in Table 3.1 and compared with ψ parameters. Note, the limits given for Γ of Υ(9460) are on one hand derived from Γ_e and $B_e = B_\mu$ and on the other by the observed width of the excitation curve, which is given by the ring energy resolution.

Table 3.1 Parameters of Heavy $q\bar{q}$ States

State	Γ_e	Γ	Remark
ψ(3095)	4.8 ± 0.6	69 ± 12 keV	$Q_q = 2/3$
ψ(3684)	2.2 ± 0.3	228 ± 56 keV	
Υ(9460)	1.33 ± 0.14	>23, <18000 keV	$Q_q = 1/3$
Υ(10020)	0.32 ± 0.13	<1800 keV	

The fact that Γ_e for Υ is a factor of $\sim$3 smaller than Γ_e for the ψ indicates that the Υ is probably a bound state of charge 1/3 quarks rather than 2/3 quarks like the ψ. The reason for this is that experimentally the ratio of Γ_e to the quark charge is about the same for the lowest lying members of all the known vector mesons, if one choses 1/3 quark charges in the Υ.

III B. Recent PETRA Results

1. R

The first results from PETRA became available early this year. A whole new energy domain has been opened and many questions need to be answered. Data are now available [39,40,41] at center of mass energies of 13.0, 17.0, and 27.4 GeV *). One of the most basic questions is the total cross section for e^+e^- annihilation into hadrons, or equivalently the value of R (Eq. 1.1). All three running PETRA experiments have published R values at 13 and 17 GeV. The results at 13 GeV are R = 5.0 ± 0.5 ± 1.0, 4.6 ± 0.5 ± 0.7, and 5.6 ± 0.7 ± 1.1 for the PLUTO, MARK J, and TASSO collaborations respectively. The first quoted error is pure statistical, and the second is systematic. At 17 GeV the values are 4.3 ± 0.5 ± 0.8, 4.8 ± 0.6 ± 1.0, and 4.0 ± 0.7 ± 0.8. The prelimimary value at 27.4 GeV from TASSO is 4.3 ± 0.7 ± 0.7. Clearly all the experiments are in agreement, and the R-values are about the same as at much lower energies (c.f. Fig. 1.2). This is not a surprise.

2. Jet Structure and Relation to New Physics

The topology of the events themselves is an interestic topic, because within any kind of quark model one expects to have jets, and the jets should become more pronounced with increasing center of mass energy. Figure 3.2 shows one event from the 27.4 GeV TASSO data. This is a view along the e^+ beam line, and one sees the tracks of charged particles bent by the magnetic field. This event was selected from the 43 events available because it clearly shows the kind of 2-jet structure one expects. (In all honesty, this event is more jet-like than most, but it is not atypical). One of the interesting things is that not all events are very jet-like, and Fig. 3.3 shows another of the 43 events at 27.4 GeV. This is indeed a spectacular event showing no tendency to form jets at all. This event is not typical, but neither is it the only one of its kind. Clear one needs some parameter to quantify how jet-like individual events are. Several variables have been discussed in the literature for the purpose,

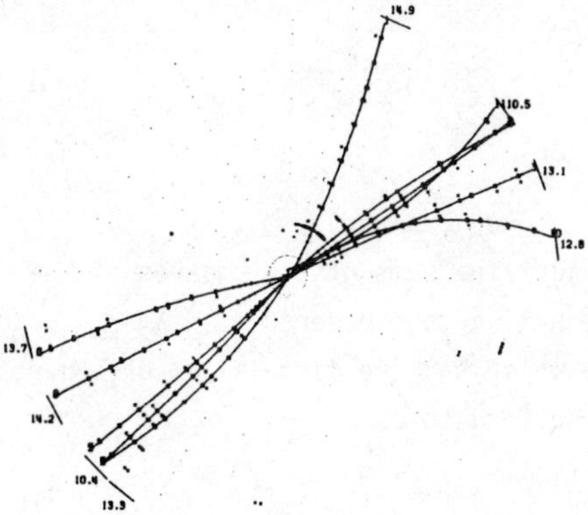

Fig. 3.2 View along magnetic field or a typical jet-like event at 27.4 GeV.

---

*) The 27.4 GeV data were not available at the time the lectures were given, but where available at the time of writing; they have been included for completeness. These data should be considered preliminary.

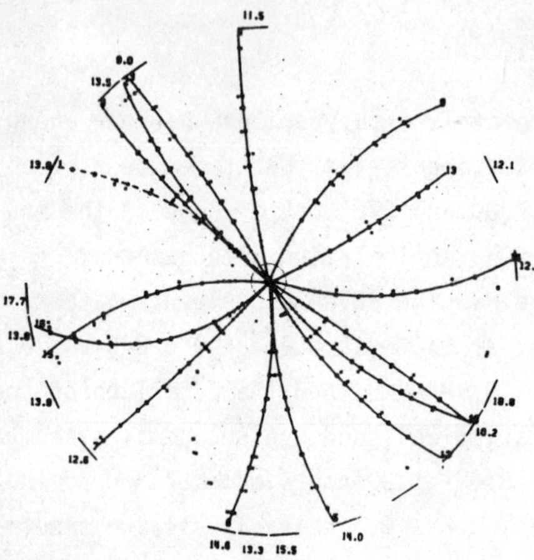

Fig. 3.3 View along magnetic field for a non-jet-like event at 27.4 GeV.

sphericity, sperocity, and thrust. For reasons which will soon be clear this discussion will concentrate on sphericity. Consider the 3 dimensional second rank tensor formed by summing over all N particles α in an event

$$T^{ij} = \sum_{\alpha=1}^{N} p_\alpha^i \, p_\alpha^j \quad , \tag{3.1}$$

where p_α^i represents the cartesian coordinates of the momentum of particle α.(Historically this is not the tensor which was considered, but the others differ only by a multiple of the unit tensor. The choice made here is pedagogically preferrable). Suppose that by some technique one has found the 3 mutually orthogonal unit eigenvectors of T^{ij}, called u_β^i, where β = 1, 2, 3 denotes the particular eigenvector. To these eigenvectors belong 3 eigenvectors λ_β. From the definition of an eigenvector and using Eq. 3.1 one immediately sees that

$$\sum_{i,j} u_\beta^i \; T^{ij} \; u_\gamma^j = \lambda_\beta \, \delta_{\beta\gamma} \tag{3.2}$$

and therefore

$$\lambda_\beta = \sum_{\alpha=1}^{N} (\vec{p}_\alpha \cdot \vec{u}_\beta)^2 \quad . \tag{3.3}$$

Physically this says that the eigenvectors are just the sums of the squares of the momenta parallel to the eigenvector. If the eigenvalues are ordered $\lambda_1 \leq \lambda_2 \leq \lambda_3$, then the eigenvector u_3 corresponds to the direction which has the largest sum of squares of longitudinal momenta; the sphericity can be defined to be

$$S = \frac{3}{2} \left(1 - \frac{\lambda_3}{\lambda_1 + \lambda_2 + \lambda_3}\right) \tag{3.4}$$

$$= \frac{3}{2} \, \frac{\text{Min} \sum_\alpha p_{T_\alpha}^{\;2}}{\sum_\alpha p_\alpha^2} \tag{3.5}$$

The variable S ranges from 0 for the completely jet-like event ($\lambda_1 \leq \lambda_2 << \lambda_3$) to 1 for a completely isotropic event ($\lambda_1 = \lambda_2 = \lambda_3$). Figure 3.4 shows observed sphericity distributions with simple Monte Carlo calculations for the expected shapes based on

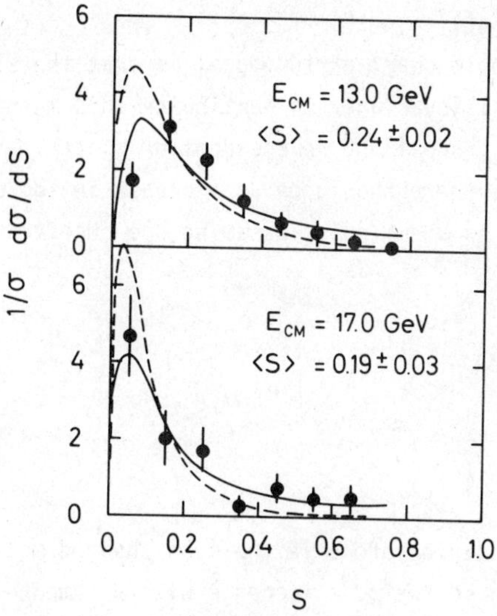

Fig. 3.4 Sphericity distribution. The dashed line shows the result of a Monte Carlo calculation having only u, d, and s quarks. The solid line shows the same calculation with u, d, s, c, and b quarks.

the model of Field and Feynman, which has only u, d, and s quarks. Clearly this shape is too narrow to well represent the data. If, however, one introduces c and b quarks into the model in the same way, then the result looks very much like the data. The lesson is that the heavier quarks are required to make the distribution wide enough. This is, however, not a proof that b quarks are required; most of the effect comes from the c quark. Figure 3.4 shows the mean sphericity vs. center of mass energy. One sees that <S> decreases with energy just as it should indicating that the jet character of the events becomes increasingly pronounced.

The variables spherocity and thrust were invented at least in part because there have problems theoretically with calculating sphericity within the framework of QED. The reason for this has to do with infrared problems, where very soft but poorly understood gluons give a significant contribution to S. The variable T, thrust

$$T = \mathrm{Max} \; \frac{\sum\limits_{\alpha} |p_{\alpha \parallel}|}{\sum\limits_{\alpha} |p_{\alpha}|} \tag{3.6}$$

conveniently sums over these soft gluon effects, but it is experimentally not so well defined as sphericity. The reason for this is that T is not an analytical function of the momenta, because of the absolute value which is used. Therefore during a mimimization process one must allow for the possibility that $p_{\parallel}$ is counted in the direction of the jet or against the direction of the jet. On the other hand if all particles of a jet lie in one hemisphere then the jet direction so determined coincides with the vector sum of the momenta of the particles in that jet.

3. Inclusive Distributions

One of the clear properties expected of a simple quark-parton model is that the single particle inclusive cross sections should be scale invariant. In particular this means that the quantity $sd\sigma/dx$, where $x = E_{hadron}/E_{beam}$ should be independent of s, at least for x above some minimum value. Below this value there should be an increase in $sd\sigma/dx$ with s due to an increasing multiplicity. There is a sum rule, assuming one observes only charged particles,

$$\int_0^1 s \frac{d\sigma}{dx} dx = \langle N_{ch} \rangle \; s\sigma_{tot} \quad . \qquad (3.7)$$

If R is a constant then $s \cdot \sigma_{tot}$ is also a constant. Therefore if the mean charged multiplicity $\langle N_{ch} \rangle$ increases with s as it should, then $sd\sigma/dx$ must increase with s somewhere. Figure 3.5 shows an artists conception of data at several energies. (The representation is accurate; but the real data are hard to display well within the confines of the printed paper). For practical reasons the variable x_p is now chosen to be p_{hadron}/E_{beam}, because the momentum is directly measured but E_{hadron} depends having the particle mass, which is not always known. The data are indeed more or less scale invariant for $x \gtrsim 0.3$, but below $x = 0.3$ the cross sections for higher s are above the lower s cross sections. An exception to this rule are the data near 4.0 GeV, where the scaling works above $x = 0.3$, but the expected behavior at lower x is not observed; there is an excess of events at low x compared to higher s data. We now know that this came about because of charm production, which resulted in copious production of low x particles. For comparison look at Fig. 3.6 which shows $sd\sigma/dx_p$ at the PETRA energies from TASSO along with a couple of lower energy measurements. The general trend we saw before is correct with scaling above $x \simeq 0.3$ and the general increase with s at lower x. An exception to this, however, are the 13 GeV data, which lie above the 17 GeV data. The first question one must ask is how much does it really exceed the 17 GeV data? The answer is 2 standard deviations, including a 10 % systematic uncertainty between the two energies. Secondly, since the R value at 13 GeV is higher than at 17 GeV, then at least a part of the effect is due to this fact. It appears on close examination, however, that the shape really is different. If the shape is different, then we must may be seeing a phenomenon similar to 4 GeV, only now we are seeing the effect of b production. Whether or not this is really the case will have to be settled with more data.

Another quantity of interest is the distribution of momenta transverse to the jet axis. Figure 3.7 shows the quantity $1/\sigma_{tot}\ d\sigma/dp_T^2$, vs p_T^2, where the sphericity method has been used to define the jet axis. The data at all 3 energies look similar and fall like $\exp(-p_T^2/0.08)$ for small p_T. At larger p_T the data lie more and more above the

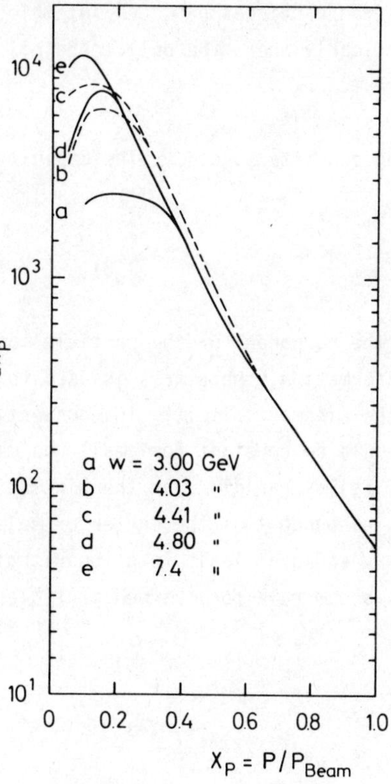

Fig. 3.5 Artist's conception of the inclusive cross section s dσ /dx at various energies. Note different character of data at 4.03 and 4.4 GeV.

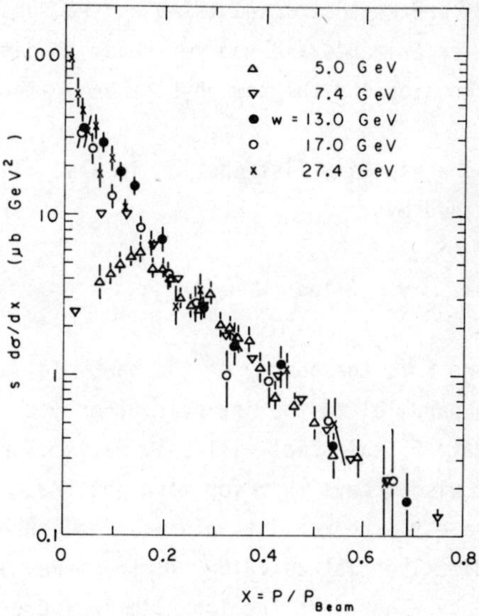

Fig. 3.6 Inclusive cross section s dσ/dx for various energies.

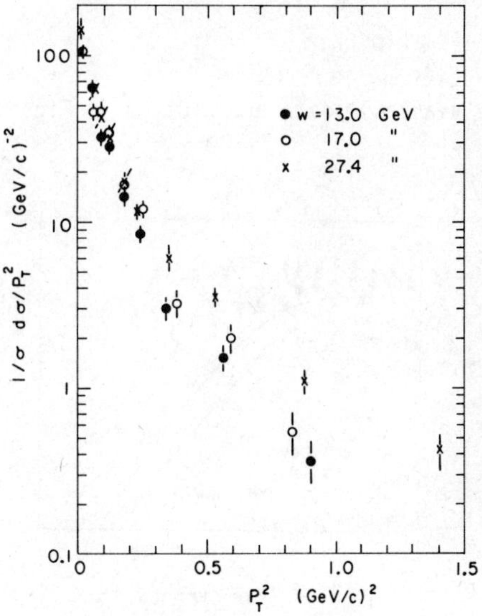

Fig. 3.7 Inclusive cross section (1/σ) $d\sigma/dp_T^2$ for various energies.

simple exponential (Gaussian) form. One must ask whether this departure is due to interesting physics or dull technical details. For example one eventually expects to see evidence for gluon bremsstrahlung, which will cause the p_T^2 distribution to become wider than the simple quark jets. Unfortunately a Monte Carlo calculation which begins with a Gaussing distribution of p_T for its quarks ends up looking rather much like the real data after real particles are formed and decay and the jet axes are reconstructed using only a subset of all the particles produced. At this time the experiments are not yet ready to distinguish clearly where the dull technical details stop and the new physics begin to be important.

The rapidity distribution is also a interesting quantity to study. The rapidity is defined by

$$y = \frac{1}{2} \log \left(\frac{E + p_{\parallel}}{E - p_{\parallel}}\right) \tag{3.8}$$

where E is the energy of the particle, and $p_{\parallel}$ is the component of the particle momentum parallel to the jet axis. (For lack of any information, the π mass is used to calculate E; this fact will only matter for y near the kinematic limit). The conventional wisdom says that for each jet $1/\sigma_{tot}\ d\sigma/dy$ should be constant for small and moderate y (the rapidity plateau) and then fall off fairly rapidly near the kinematic limit. The plateau value, furthermore, should be independent of the center of mass energy. Therefore the logarithmic increase of the y endpoint leads to a logarithmic growth of the mean multiplicity, because there is a sum rule for charged particles

$$\int_0^{y_{max}} \frac{1}{\sigma} \frac{d\sigma}{dy}\, dy = \langle N_{ch} \rangle \quad . \tag{3.9}$$

The available data are plotted in Fig. 3.8 using both TASSO data and low energy data for comparison purpose. (For historical reasons an extra factor of 1/2 is included so that the distribution is for each of 2 jets). Clearly there is a plateau region which remains fairly constant until the kinematic limit is approached, and then the rates decrease just as expected. There is, however, a substantial s dependence of the plateau value. The reason for this is quite simple, and it is related to the sum rule (3.9). The mean charged multiplicity at PETRA energies is roughly 10 whereas a simple logarithmic increase would have predicted about 6.5. Since the plateau cannot extend enough to accomodate this increased multiplicity, the only place to go is up.

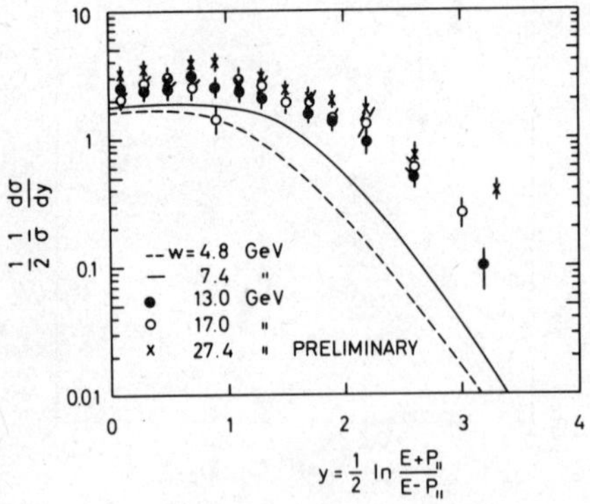

Fig. 3.8 Rapidity distribution per jet (1/2) $(1/\sigma)\ d\sigma/dy$.

4. What is to come?

We have just begun to explore a whole new energy range of e^+e^- physics and a great deal needs to be done. To put this into perspective look at Fig. 3.9. This figure is

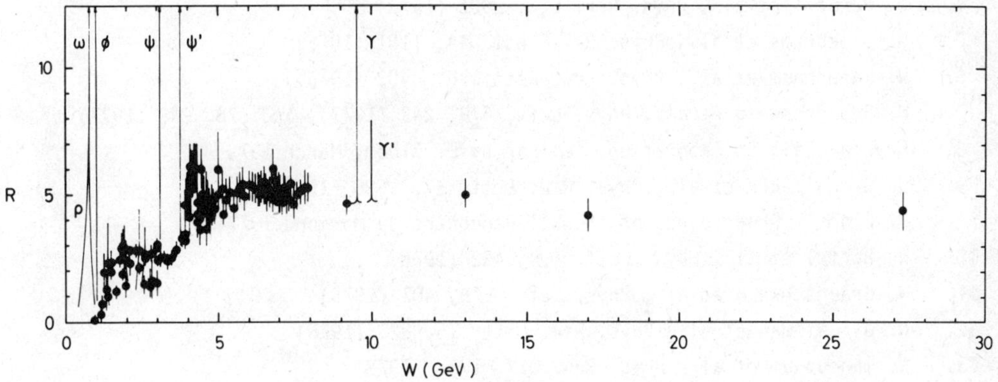

Fig. 3.9 The ratio R vs center of mass energy (c.f. Fig. 1.2)

essentially Fig. 1.2 discussed in the first lecture but replotted with a larger energy scale to accomodate PETRA measurements. One is immediately struck by how much of the plot is empty! Recall that the region just above ψ production near 4 GeV is very complex with many peaks and the whole field of D and F physics. One sees clearly Υ and Υ' production near 10 GeV and there is every reason to believe that there should be a complicated region somewhat above Υ production replete with "bottom" physics. This is a region which will probably be best studied by the CESR machine which has just been built at Corenll. We anxiously await exciting results there. Somewhere one must find the "top" quark effects. Estimates of where this should begin range as low as 20 GeV to much higher values. Present data are not adequate to say whether or not we have passed the t threshold. This is the main topic to be pursued at PETRA in the immediate future. When one thinks (as an experimenter) how much work was required to understand the region from 4 - 5 GeV, it is rather humbling to see how much remains to be done just to look for what we expect to see! On top of that, experience tells us that there are usually things to be discovered which were not expected!

References

1. A. Boyarski et al., Phys. Rev. Lett. 34, 1357 (1975)
2. V. Lüth et al., Phys. Rev. Lett. 35, 1124 (1975)
3. B. Jean-Marie et al., Phys. Rev. Lett. 36, 291 (1975)
4. I. Peruzzi et al., Phys. Rev. D17, 2901 (1978)
5. G. S. Abrams et al., Phys. Rev. Lett. 34, 1181 (1975)
6. W. Tanenbaum et al., Phys. Rev. Lett. 36, 402 (1976)
7. W. Braunschweig et al., Phys. Lett. 67B, 243 (1977), and 67B, 249 (1977)
8. Crystal Ball Collaboration, Seminar by E. Bloom, March 1979
9. J. S. Whitaker et al., Phys. Rev. Lett. 37, 1596 (1976)
   V. Blobel, Proceedings of the XII Rencontre de Moriond 1977
10. W. Bartel et al., Phys. Lett. 79B, 492 (1978)
11. W. Braunschweig et al., Phys. Lett. 57B, 407 (1975)
12. G. J. Feldman et al., Phys. Rev. Lett. 35, 821 (1975)
13. W. Tanenbaum et al., Phys. Rev. D17, 1731 (1978)
14. G. Goldhaber et al., Phys. Ref. Lett. 37, 255 (1976)
15. I. Peruzzi et al., Phys. Rev. Lett. 37, 569 (1976)
16. M. Piccolo et al., Phys. Lett. 70B, 260 (1977)
17. D. L. Scharre et al., Phys. Rev. Lett. 40, 74 (1978)
18. I. Peruzzi et al., Phys. Rev. Lett. 39, 1301 (1977)
19. G. Goldhaber et al., Phys. Lett. 69B, 503 (1977)
20. G. Feldman et al., Phys. Rev. Lett. 38, 1313 (1977)
21. M. Piccolo et al., Phys. Lett. 70B, 260 (1977)
22. V. Vuillimin et al., Phys. Rev. Lett. 41, 1149 (1978)
23. J. M. Feller et al., Phys. Rev. Lett. 40, 274 (1978)
24. W. Bachino et al., Phys. Rev. Lett. 40, 671 (1978)
25. H. K. Nguyen et al., Phys. Rev. Lett. 39, 262 (1977)
26. J. E. Wiss et al., Phys. Rev. Lett. 37, 1531 (1976)
27. R. Brandelik et al., Phys. Lett. 70B, 132 (1977)
28. R. Brandelik et al., Phys. Lett. 80B, 412 (1977)
29. M. L. Perl et al., Phys. Rev. Lett. 35, 1489 (1975)
30. W. Braunschweig et al., Phys. Lett. 63B, 47 (1976), and
    R. Brandelik et al., Phys. Lett. 70B, 125 (1977), and Phys. Lett 70B, 387 (1977)
31. R. Brandelik et al., Phys. Lett. 73B, 109 (1978)
32. W. Bachino et al., Phys. Rev. Lett. 41, 13 (1978)
33. G. Feldman, Proc. of XIX International Conference on High Energy Physics, Tokyo 1978, and SLAC-PUB-2224, for a comprehensive summary.
34. S. W. Herb et al., Phys. Rev. Lett. 39, 252 (1977)
    W. R. Innes et al., Phys. Rev. Lett. 39, 1240 (1977)
35. Ch. Berger et al., Phys. Lett. 76B, 243 (1978)

36. C. W. Darden et al., Phys. Lett. 76B, 246 (1978)

37. J. K. Bienlein et al., 78B, 360 (1978)
C. W. Darden et al., Phys. Lett. 78B, 364 (1978)

38. C. W. Darden et al., Phys. Lett. 80B, 420 (1979)
Ch. Berger et al., Z. Phys. C1, 343 (1979)

39. Ch. Berger et al., Phys. Lett. 81B, 410 (1979)

40. D. Barber et al., Phys. Rev. Lett. 42, 1113 (1979)

41. R. Brandelik et al., Phys. Lett. 83B, 261 (1979)

MASSES AND MASS GENERATION IN CHROMO AND FLAVOUR DYNAMICS +*

Harald Fritzsch

Institute of Theoretical Physics, University of Bern

and

CERN, Geneva

Abstract:

m_{ν_e} m_e m_{ν_μ}

g_3

g_2 m_μ

g_1 m_{ν_τ}

m_τ

m_u

m_d

m_s

m_c

m_b

m_t m_W

θ_1 δ^1

θ_2 δ

θ_3 θ_1^1 θ_2^1 θ_3^1

m_H

+ Lectures given at the 10th GIFT Seminar on Theoretical Physics, Jaca, Spain (June 1979). To be published in the Proceedings by Springer Verlag

** Supported in part by the Schweizerische Nationalfonds

1. INTRODUCTION

During the last fifteen years we have seen a remarkable progresss in the understanding of high energy physics phenomena. The rôle of the quarks has become relatively clear; although nobody has (yet) seen a free quark, it has become obvious that the quarks are the elementary building blocks of the hadronic matter. Non-Abelian gauge theories have turned out to be of crucial importance for the understanding of all interactions. At present most of the particle theorists have great confidence that the electromagnetic and weak interactions can be described by a system of non-Abelian gauge fields, and that the masses of the W and Z bosons are generated by a spontaneous breakdown of the gauge symmetry. Furthermore the prospects have become very good that the strong interactions are a manifestation of a pure, unbroken non-Abelian gauge theory, the theory of colored quarks and gluons (QCD). This is all very satisfying, in particular in view of the fact that the theory which is most simple and beautiful in its structure, Einstein's theory of General Relativity, is nothing other than a non-Abelian gauge theory, employing the symmetry of space and time as the underlying gauge symmetry.

The theories of the electromagnetic, weak and strong interactions which have been developed during the last ten or fifteen years, have done something similar to the understanding of particle physics, as the theory of quantum mechanics developed during the years 1924-1927 has done to the understanding of atomic physics. Before the development of quantum mechanics as a full theory in 1924-1927 there was essentially no real understanding of the physics of atoms. Of course, a lot was known about the atomic spectra, and many regularities had been observed, sometimes "explained" by phenomenological rules. The situation changed suddenly with the development of quantum mechanics. The rules of quantum mechanics allow (at least in principle) to calculate the spectra of all atoms. Furthermore all the regularities and rules, observed previously by studying the atomic spectra in great detail, were finally understood within the theory. However, the theory of quantum mechanics had its limitation. The atomic physics phenomena could not be calculated from first principles, like causality or unitarity, but only in terms of a few parameters which had to be taken from outside, in particular the electric charges of the electron and proton, the fine structure constant α, and the masses of the electron and the nuclei. It was generally accepted by almost everybody who worked on the development of quantum mechanics that the fundamental parameters like α or the masses of the particles cannot be understood within the theory, i.e. that quantum mechanics itself is a theory describing only a limited range of phenomena.

In particle physics the development was quite similar. At first one accumulated a lot of experimental material. Many regularities and symmetries (e.g. isospin, SU(3) ...) were found. Very often in the development of particle physics during the last thirty years it was difficult to decide which phenomena were fundamental and useful to study, and which phenomena were not. For example ten years ago only very few people worried about the existence or non-existence of neutral currents. Yet it has turned out that it was extremely useful to worry about this question.

Since a few years ago the situation in particle physics has changed entirely. Now one has confidence that the SU(2)xU(1) theory of the electromagnetic and weak interactions is indeed the correct theory. QCD is believed to be the correct theory of the strong interactions. Of course, it is not yet clear if it is justified to think that the correct theories of the strong, weak and electromagnetic interactions are found. For the purpose of this lecture I shall simply assume that this is the case. The pattern which then arises is quite different from the one which people thought of years ago. Fifteen or twenty years ago most theorists were willing to believe that most of the questions left open at that time will be answered once the correct theories of the strong, electromagnetic and weak interactions are found, questions like: Why is the proton about 2000 times heavier than the electron? What determines the value of α ? Today one knows the structure of the theory, but not yet the answers to the questions mentioned above, and many others. It looks as if the development in particle physics is similar to the one in atomic physics. We can express the particle physics phenomena in terms of a few fundamental parameters (e.g. α, Λ[QCD scale], m_e, m_μ, ...). There is no way to determine these parameters within the theory, but once they are given everything else is calculable, at least in principle (of course, we realize that in reality such a calculation may be extremely difficult or even impossible, e.g. the calculation of the mass ratio M_Δ/M_P up to a precision of less than 1 o/oo). The similarity between the developments in quantum mechanics and particle physics is expressed below.

	Atoms	Particles
1924:	atomic spectra many regularities (Hund's rule, Laudé's rule, ...)	particle spectra, many regularities (SU(3), scaling, neutral currents and their absence in K_{oL}-decay, ...)
	↓	↓
1924-1927:	quantum mechanics	quark model, QFD[SU(2)xU(1)], QCD[$SU(3)^c$]
	↓	↓
	atomic physics - explained in principle in terms of a few parameters (m_e, M_p,α).	particle phenomena - explained in principle in terms of a few parameters (α, Λ[QCD], $m_{leptons}$, m_{quarks} ...).

A difference between quantum mechanics and the modern theories of particle physics consists in the number of free parameters. While the number of free parameters in quantum mechanics is relatively small, the number of parameters in particle physics is sizeable. First of all the three coupling constants of QFD and QCD are essentially free parameters ($\alpha_1, \alpha_2, \alpha_3 = \alpha_s$). Furthermore the lepton and quark masses are free parameters, and moreover all elements of the lepton - and quark mass matrix (e.g. weak mixing angles). In case of the conventional six-quark scheme the free parameters (more than twenty) are shown in the abstract.

The question which arises is: What determines the yet free parameters ? Of course, it may be that the theories of QFD and QCD mentioned above are not quite correct, and there exists a better theory of the weak, electromagnetic and strong interactions in which all parameters are fixed. I find this possibility rather unlikely. Today it looks as if the situation is similar to the one in atomic physics after the establishment of quantum mechanics. The free parameters entering in the description of the weak and strong interactions are indeed free parameters within QCD and QFD, and have to be determined by the next, deeper layer of interactions (or whatever structure there may be) which exists, i.e. interactions or structures which have not yet been seen. Thus I think that flavor dynamics and chromodynamics will not be the final theories explaining everything in particle physics. There will exist further interactions, which are essential for understanding the mass spectrum of leptons and quarks and all other, yet unresolved questions in particle physics. At present it is unknown at which level these interactions will show up. If we are lucky, the new interactions come in at a scale of $\sim 10^{-16}$ ... 10^{-17} cm. It may also be that they come in at a very small scale of $\sim 10^{-30}$ cm ($\sim 10^{16}$ GeV), in which case only theorists can play with these interactions. (To my astonishment I found out recently that some theorists believe accelerators reaching 10^{16} GeV will be constructed in the future.)

In this lecture I shall not speak about a definite theory, but rather about the various aspects of the mass problem in chromo and flavor dynamics. After a general discussion of the mass problem in physics I review shortly the conventional mechanism which is used in QFD to generate the masses of various particles. It follows a discussion of the mass generation via radiative corrections, which allows (under certain assumptions) to predict the mass of the scalar boson in terms of the W boson mass. Subsequently the mass problem in QCD will be discussed, especially the distinction between the nonperturbative scale Λ in QCD and the quark masses. The possible rôle of the quark masses for the problem of weak interaction mixing will be discussed. Finally we study some of the aspects of the mass problem within the unified theories of the strong, electromagnetic, and weak interactions.

2. MASS SCALES IN PHYSICS

Without any exaggeration one can say: the history of physics is the history of mass scales. For some reason (which is not at all clear) the various mass scales come in on hierarchic order. From a practical point of view this is very helpful, since one is able to study the phenomena related to each mass scale in an isolated way. For example the muon is about 200 times heavier than the electron. For this reason atomic physics is the physics determined in its scale essentially by the electron mass (of course, apart from the physics dealing with muonic atoms). Imagine a fictitious world in which the muon is much lighter, for example only just a little heavier than the electron. In this case the muon can be rather long lived or can be absolutely stable (if $m_{\nu_\mu} \neq 0$). As a result the conventional atomic physics would be much more complicated, dealing with the two scales m_e and m_μ. (For example there would be two types of water, an electron type water, and a muon type water). Probably the chemical industry would enjoy such a world, since the number of different products one can manufacture would be tremendously increased.

In particle physics we can distinguish at present four different mass scales which seem to be different in a qualitative way.

1. The mass scale of the fermions, starting with the very light neutrino masses (in case $m_\nu \neq 0$) and the electron mass, and reaching all the way up to the b-quark mass

of the order of 5 GeV. Very likely there exist even heavier fermions.

2. The QCD mass scale of the order of 1 GeV, which determines the masses of hadrons in the absence of the quark masses.

3. The mass scale of the order of 100 GeV given by the masses of the weak intermediate bosons.

4. The scale given by gravity (Planck length) given in energy units by $\sim 10^{19}$ GeV.

Besides these mass scales which are observed either directly or indirectly there may exist further scales, for example a mass scale of the order of 10^{15} ... 10^{16} GeV, which plays a rôle in some of the unified theories, which will be discussed later. Furthermore the smallness of CP violation may be attributed to a mass scale which is about 10 ... 100 times larger than the W mass scale. The various mass scales are shown in Fig. (1).

3. MASSES IN CONVENTIONAL QFD

The process of mass generation in the conventional gauge theory framework (spontaneous symmetry breaking) is well-known: the gauge fields and the fermion fields are coupled in a gauge-invariant manner to scalar fields. The coupling of the scalars to the gauge fields is given by the gauge coupling constant g, i.e. the same coupling constant which describes the coupling of the gauge fields among themselves and to the fermions. On the other hand the coupling of the scalar fields to the fermions is parametrized by the Yukawa coupling constant G, which is independent of g and can be chosen differently for the different fermion representations.

In the minimal SU(2)xU(1) theory [1] one introduces one doublet of scalar fields:

$$\phi = \begin{pmatrix} \phi_1 \\ \phi_2 \end{pmatrix}.$$

The potential $U(\phi)$ of the scalar field is assumed to have the form

$$V(\phi) = \mu^2 \phi^+ \phi + \lambda(\phi^+ \phi)^2. \tag{3.1}$$

For $\mu^2 < 0$ one obtains a spontaneous breaking of the SU(2)xU(1) symmetry. The scalar field acquires a vacuum expectation value $v = \sqrt{-\mu^2/\lambda}$, which has the dimension of a mass. By a suitable choice of the SU(2) frame the vacuum expectation value of ϕ can be brought into the form:

$$<0|\phi|0> = \frac{1}{\sqrt{2}} \begin{pmatrix} 0 \\ v \end{pmatrix}. \tag{3.2}$$

This defines the direction of the electric charge, since only the neutral component of ϕ can acquire a vacuum expectation value, i.e. ϕ can be written as

$$\phi = \begin{pmatrix} \phi^+ \\ \phi^0 \end{pmatrix}$$

($<0|\phi^+|0> = 0$, $<0|\phi^0|0> = v/\sqrt{2}$).The doublet of complex scalars (3.3) contains four hermitean fields, namely the two hermitean fields composing ϕ^+ and $\phi^- = (\phi^+)^*$, and the fields

$$\chi = \frac{1}{\sqrt{2}i} (\phi^0 - \bar{\phi}^0)$$

$$H = \frac{\phi^0 + \bar{\phi}^0}{\sqrt{2}} \tag{3.4}$$

By a suitable arrangement of the phases we can make v real, i.e. $<0|\chi|0>= 0$, $<0|H|0> = v$. The fields ϕ^+, ϕ^-, χ disappear as actual particles, and their degrees of freedom reappear as the longitudinal components of the massive vector bosons W^+, W^-, and Z. What is left is the field H, which can be redefined such that its vacuum expectation value is zero:

$$H = H' + v. \tag{3.5}$$

The field H' corresponds to a massive scalar field with the $(\text{mass})^2 = -2\,\mu^2 > 0$, the neutral Higgs particle.

Since the mass of the W boson is directly related to the Fermi constant G, the actual value of v is given by

$$\frac{G}{\sqrt{2}} = \frac{g^2}{8M_W^2} = \frac{1}{2v^2} \tag{3.6}$$

$$v = 246 \text{ GeV}.$$

The mass of the Higgs particle is given by:

$$M_H = \sqrt{-2\mu^2} = \sqrt{2\lambda v^2} = \sqrt{\lambda} \cdot 348 \text{ GeV}. \tag{3.7}$$

Thus e.g. for $M_H = 11$ GeV one has to have $\lambda = 1/1{,}000$, i.e. λ is a rather small coupling constant.

It is impossible to make λ arbitrarily small, which according to eq. (3.) means lowering m_H to arbitrarity small values. Since the coupling of the scalar fields to the gauge bosons is required by gauge invariance to be of the order of e, radiative corrections involving loops of virtual gauge bosons lead to induced self-couplings of the scalars which are of the order of e^4. Thus λ cannot be smaller than $\sim \alpha^2$, and one finds a lower limit for m_H, which is of the order of 5 GeV [2)].

Since we are dealing with a doublet of scalar fields, one has the special situation that the spontaneous breaking of all the three SU(2) generators happen with equal strength. Thus the gauge boson mass matrix takes the special form

$$(W^1\ W^2\ W^3;\ B) \begin{pmatrix} g^2 & 0 & 0 & 0 \\ 0 & g^2 & 0 & 0 \\ 0 & 0 & g^2 & g'g \\ 0 & 0 & g'g & g'^2 \end{pmatrix} \begin{pmatrix} W^1 \\ W^2 \\ W^3 \\ B \end{pmatrix} \tag{3.8}$$

(B: U(1) gauge boson);

it has a global SU(2) symmetry. This implies that in the limit $g' \to 0$, i.e. $\sin^2\theta_W \to 0$, the photon coincides with the U(1) boson B, and the mass of the Z boson is equal to M_W. In the general case the deviation of the Z mass from M_W is given by the W^3 - B mixing, described by $\theta_W = \text{arc sin} \frac{g'}{\sqrt{g^2 + g'^2}}$:

$$M_Z = \frac{M_W}{\cos\theta_W} = \frac{37.3 \text{ GeV}}{\sin\theta_W \cos\theta_W}. \tag{3.9}$$

We emphasize the importance of the "isospin" relation (3.9), which is the first relation between the masses of elementary particles based on an actual theory and which is well in agreement with the experimental data on neutral current processes. The latter can be parametrized by a special strength parameter ρ which in terms of M_Z and M_W is given by

$$\rho = \frac{M_Z \cos\theta_W}{M_W}. \tag{3.10}$$

One expects $\rho = 1$ if the spontaneous breaking of the SU(2)xU(1) theory is generated by one or several doublets of scalar fields. As soon as other SU(2) representations are present, ρ differs from one. The present experiments give $\rho = 0.98$ with an error of $\sim 5\%$ (see e.g. ref. (3)), i.e. the spontaneous breaking of the SU(2)xU(1) gauge symmetry must indeed be dominated rather strongly by scalar fields transforming as doublets of the weak isospin.

The masses of the fermions are generated spontaneously by this Yukawa-type coupling to the scalar fields. Unlike the gauge couplings the Yukawa couplings are not related to each other, i.e. different fermions are coupled differently to the scalars. For example the electron mass is given by

$$m_e = G_e \left(\frac{v}{\sqrt{2}}\right) \tag{3.11}$$

where G_e is the Yukawa coupling constant relevant for the electrons. It is found to be extremely small:

$$G_e = \sqrt[4]{8} \cdot m_e \cdot \sqrt{G} \approx 2.9\ 10^{-6}. \tag{3.12}$$

We should like to remark that the Yukawa coupling constant G can be arbitrarily small, in contrast to the scalar selfcoupling λ. The radiative corrections of G do not receive contributions of order e^2, but only contributions of order G^2. Thus it is consistent to have $G \ll e$, i.e. to have the fermion masses much smaller than the gauge boson masses.

Analogously the masses of all fermions (quark masses, lepton masses) are generated. Furthermore the Yukawa couplings of the scalars to the fermions do in general mix the various weak eigenstates, i.e. the mass eigenstates obtained after the diagonalization of the fermion mass matrix are not weak interaction eigenstates, and one obtains the weak interaction mixing described by the Cabibbo angle and related angles.

4. MASS GENERATION BY RADIATIVE CORRECTIONS

We have remarked earlier that it is not possible to set the self-interaction coupling constant λ of the scalars in QFD to zero. Radiative corrections will take over in case $\lambda \to 0$, and the smallest possible coupling constant λ, obtained as a result of radiative corrections, is of the order of e^4. The effective potential is modified by the radiative corrections due to gauge boson loops, scalar loops, etc. Thus one may ask (as done in ref. (4)) what happens if we set the scale μ^2 of QFD to zero.

The most simple situation to study in this respect is scalar electrodynamics, where the mass of the scalar particles ist set to zero [4].

The Lagrangian of massless scalar electrodynamics is given by

$$\mathcal{L} = -\frac{1}{4}(F_{\mu\nu})^2 + \frac{1}{2}(\partial_\mu\phi_1 - e A_\mu\phi_2)^2 + \frac{1}{2}(\partial_\mu\phi_2 + e A_\mu\phi_1)^2 - \frac{\lambda}{4!}(\phi_1^2 + \phi_2^2)^2 \tag{4.1}$$

(the complex scalar field ϕ is given by $\phi = (\phi_1 + i\,\phi_2)/\sqrt{2}$.

By taking into account the renormalization effects both due to the electromagnetic interaction and the quartic self-interaction one obtains [4]

$$V(\phi) = \frac{\lambda}{4!}\phi^4 + \left(\frac{5\lambda^2}{1152\pi^2} + \frac{3e^4}{64\pi^2}\right)\phi^4\left(\ln\frac{\phi^2}{M^2} - \frac{25}{6}\right)$$

where M is an arbitrary mass parameter, used to define λ:

$$\left.\frac{d^4V}{d\phi^4}\right|_{\phi = M} = \lambda. \tag{4.3}$$

A change of M can simply be reabsorbed by a change of λ such that the potential (4.2) takes the same form replacing λ by λ' and M by M' (λ' is related to λ by a logarithmic change).

Since M is arbitrary we can choose M to be the position of the minimum of $\langle\phi\rangle$, which is then the parameter of the theory fixing all mass scales. In this case we obtain

$$V = \frac{\lambda}{4!}\phi^4 + \frac{3e^4}{64\pi^2}\phi^4\left(\ln\left(\frac{\phi^2}{\langle\phi^2\rangle}\right) - \frac{25}{6}\right), \tag{4.4}$$

and the λ coupling constant is given by e:

$$0 = V'(\phi)\Big|_{\phi = \langle\phi\rangle} = \left(\frac{\lambda}{6} - \frac{11 e^4}{16\pi^2}\right) \phi^3\Big|_{\phi = \langle\phi\rangle}, \tag{4.5}$$

$$\lambda = \frac{33}{8\pi^2}\, e^4 .$$

$$V(\phi) = \frac{3\, e^4}{64\pi^2}\, \phi^4 \left[\ln \left(\frac{\phi^2}{\langle\phi^2\rangle}\right) - \frac{1}{2}\right]. \tag{4.6}$$

One obtains a spontaneous symmetry breaking, i.e. both the scalar boson and the vector boson acquire masses. It follows

$$m^2(\text{scalar}) = V''(\phi)\Big|_{\phi = \langle\phi\rangle} = \frac{3e^4}{8\pi^2} \langle\phi\rangle^2$$

$$m^2(\text{vector}) = e^2 \langle\phi\rangle^2$$

$$\frac{m^2(\text{scalar})}{m^2(\text{vector})} = \frac{3}{2\pi} \cdot \frac{e^2}{4\pi} .$$

The result is surprising, since the scalar boson mass is some sort of radiative correction to the vector mass. Furthermore both the vector boson and the scalar boson masses are generated spontaneously by radiative effects.

The same phenomenon occurs in nonabelean gauge theories. The radiative corrections due to the gauge boson loops and the fermion loops cause a spontaneous breakdown of the theory even in the absence of the negative scalar mass μ^2. For example in the minimal SU(2)xU(1) theory one finds neglecting the fermion loops:

$$m^2(H) = \frac{3}{32\pi^2} \left[2\, g^2\, m^2(W) + (g^2 + g'^2)\, m^2(Z)\right] \tag{4.8}$$

(H: Higgs meson).

Of course, in deriving this relation we have set $\mu^2 = 0$. Within the SU(2)xU(1) theory there is no reason to suppose that this is the case, apart from the economical point of view that one parameter less is better than not. We shall see later that the assumption $\mu^2 = 0$ is quite natural in order to understand the hierarchy of interactions in the unified theories including the strong interactions.

Using the relations

$$g = \frac{e}{\sin\theta_W},\ g' = \frac{e}{\cos\theta_W}$$

$$m_W = \sin^{-1}\theta_W \cdot M_o \qquad (4.9)$$

$$m_Z = \sin^{-1}\theta_W \cdot \cos^{-1}\theta_W \cdot M_o$$

$$(M_o = (\frac{\pi\alpha}{\sqrt{2}\cdot G})^{1/2} = 37.29\ \text{GeV})$$

one finds [4)]

$$m(H) = (\frac{3\alpha}{8\pi})^{1/2} \frac{\sqrt{2 + \cos^{-4}\theta}}{\sin^2\theta_W} \cdot M_o . \qquad (4.10)$$

This gives m(H) = 9.08 GeV for $\sin^2\theta_W = 0.23$. The mass of the Higgs boson is rather sensitive to $\sin^2\theta_W$ (see table below):

$\sin^2\theta_W$	0.19	0.20	0.21	0.22	0.23	0.24	0.25
m(H) [GeV]	10.8	10.3	9.85	9.45	9.08	8.74	8.43.

(4.11)

If the Higgs boson mass is indeed of the order of 9 - 10 GeV, as predicted by the minimal SU(2)xU(1) theory if the mass generation is due to radiative corrections, there exists the possibility to observe it in the Υ-decay. One has [5)]

$$B = \frac{\Gamma(V(\bar{b}b) \to H + \gamma)}{\Gamma(V(\bar{b}b) \to \mu^+ \mu^-)} = \frac{G\, M_W^2}{4\sqrt{2}\pi\alpha}\,(1 - \frac{m_H^2}{m_V^2}) \qquad (4.12)$$

where $V(\bar{b}b)$ is a $\bar{b}b$ vector state (γ, γ', γ''). The emission of the Higgs boson proceeds via the process described by the diagram Fig. (2).

The Yukawa coupling of the H meson to the b quark is given by m_b. One expects $B \sim 10^{-3} \ldots 10^{-4}$ in case of the γ decays. The branching ratio (4.12) is independent of the quark charge, i.e. formula (4.12) is valid also for a possibly existing heavy t quark. The branching ratio (4.12) is quite sizeable for m(T) $\gtrsim$ 30 GeV, namely $\sim$ 7% (T : ($\bar{t}t$) - vector meson).

5. MASS SCALE IN QCD

Chromodynamics, the gauge theory of colored quarks and gluons [6], is supposed to be the theory of hadrons and of the strong interactions. The formal Lagrangian is given by

$$\mathcal{L} = -\frac{1}{4} G^{A}_{\mu\nu} G^{\mu\nu}_{A} + \bar{q} \left[i \gamma^{\mu} (\partial_{\mu} - i g B^{A}_{\mu} \frac{\lambda_A}{2}) - \mathcal{M} \right] \tag{5.1}$$

$$q = \begin{pmatrix} u \\ d \\ s \\ \cdot \\ \cdot \\ \cdot \end{pmatrix}$$

where $\mathcal{M}$ denotes the quark mass matrix. As far as the strong interactions are concerned, the quark mass matrix is an arbitrary matrix in the flavor space (e.g. containing γ_5 terms). By suitable unitary transformations of the righthanded and lefthanded quark fields we can always achieve that $\mathcal{M}$ takes the form

$$\mathcal{M} = \begin{pmatrix} m_u & & & \\ & m_d & & \\ & & \cdot & \\ & & & \cdot \\ & & & & \cdot \end{pmatrix} \tag{5.2}$$

where m_u, m_d, ... are positive real numbers with the dimension of a mass, i.e. the quark masses. This can be seen as follows. In general the quark mass term is given by

$$q^{+}_{R} \gamma_{o} M q_{L} + \text{h.c.} \tag{5.3}$$

(M: mass matrix).

An arbitrary matrix M can be transformed into a diagonal matrix with positive elements M_d by a biunitary transformation

$$V^{-1} M U = M_d . \tag{5.4}$$

(A simple way to see this is to consider the hermitean matrixes M^+M and MM^+. Both matrices can be diagonalized by unitary matrices: $U^{-1}(M^+M)U^{-1} = M_d^2 = V^{-1}(MM^+)V$ $= U^{-1}M^+VV^{-1}MU = V^{-1}MUU^{-1}M^+V$. It follows $V^{-1}MU = U^{-1}M^+V = (M_d^2)^{1/2}$.)

In the modern gauge theories of flavor dynamics the quark and lepton masses are generated spontaneously, as discussed in the previous section. Furthermore the QFD charges commute with the color generators, and the QFD coupling constants are much smaller than $\alpha_s = \frac{g_3^2}{4\pi}$. Thus it seems reasonable to consider the limit g(flavor) $\to 0$, $\mathfrak{m} \to 0$, i.e. to turn off the electromagnetic and weak interactions as well as the quark masses. We do not know exactly whether this limit can actually be taken in a consistent way, but it seems very reasonable to assume that it can. Furthermore nothing seems wrong in assuming that the various quark masses can be varied independently, i.e. one may make some of the quarks very heavy, in which case they essentially drop out of the world of strong interaction dynamics at energies below the energy given by the heavy quark masses.

First we study the limit $\mathfrak{m} = 0$ (all quark masses vanish), and g(flavour) = 0. Here no scale parameter is left in the QCD Lagrangian. If we suppose that the observed strong interactions are indeed described by QCD, we may ask: What happens to the strong interactions (e.g. π - N coupling constant, proton mass, slope of particle trajectories etc.) in the limit $\mathfrak{m} \to 0$?

In the limit $\mathfrak{m} \to 0$ the QCD Lagrangian is formally invariant under the group $U_n^L x U_n^R$ (n: number of flavors). Both the vector charges and the axial vector charges in the flavor space are conserved, i.e.

$$\partial^\mu (\bar{q}_i \gamma_\mu q_j) = 0 \tag{5.5a}$$

$$\partial^\mu (\bar{q}_i \gamma_\mu \gamma_5 q_j) = 0 \tag{5.5b}$$

(i, j: indices, denoting the various quark flavors).

However the conservation laws (5.5b) are broken by anomalous contributions [7] in case of i = j. In this case the triangle anomaly generates a new term:

$$\partial^\mu (\bar{q}_i \gamma_\mu \gamma_5 q_i) \sim g_3^2 \Sigma \tilde{G}^A_{\mu\nu} G^{\mu\nu}_A \tag{5.6}$$

(G^A : gluon field strength, $\tilde{G}^A_{\mu\nu} = \frac{1}{2} \varepsilon^{\mu\nu\rho\sigma} G^A_{\mu\nu}$).

For this reason we shall consider only the subgroup $SU_n^L x SU_n^R$ (the vector U(1) generator is, of course, identified with the baryonic charge and not considered any further).

Suppose we act with one of the generators of $SU_n^L x SU_n^R$, e.g. with the isospin generator $F_{d,u} = \int d^+ u \, d^3x$, and its axial counterpart $F_{d,u}^5 = \int d^+ \gamma_5 u \, d^3x$, on a proton state. In this case we find

$$F_{d,u} |p> = |n> \tag{5.7a}$$

$$F_{d,u}^5 |p> = |\hat{n}> \tag{5.7b}$$

(p proton state, n neutron state)

where $|\hat{n}>$ denotes a neutron state with the "wrong" parity, i.e. in the $SU_n^L x SU_n^R$ symmetry limit one would have parity doubling for baryons. On the other hand there is no sign of a parity doubling in nature. There are two ways to deal with this problem:

a) The limit $\mathcal{M} \to 0$ is such that the world of the strong interactions in this limit resembles in no way the real world.

b) The real world is not very far from the limit where at least the "light" quarks u, d, and s are massless. The chiral symmetry $SU(3)_L x SU(3)_R$ is realized in the Nambu-Goldstone manner. There exist eight massless Goldstone bosons, (π, K, η), while all other hadrons remain massive in the limit $m_u = m_d = m_s = 0$. Especially we have

$$F_{d,u}^5 |p> = |\pi^-, p> \tag{5.8}$$

i.e. the proton is accompanied by zero-momentum pion.

It is generally believed that the real world resembles case b, due to the success of the chiral symmetry (current algebra, soft pion physics). If we furthermore assume that the world of the strong interactions is described by QCD, we are led to the following interesting conjecture:

In the limit $\mathcal{M} \to 0$ the chiral symmetry of the QCD Lagrangian is realized in the Nambu-Goldstone manner. The masses of all hadrons, except the lowest lying pseudoscalar mesons, stay finite in the limit $\mathcal{M} \to 0$.

Thus far nobody has shown that the conjecture described above is correct. However it is not difficult to believe that there exists an intrinsic mass scale in QCD even in the limit $\mathcal{M} \to 0$, since a mass scale is introduced in any case by the renormalization procedure. For example, we need to introduce a mass scale in QCD in order to define the renormalized coupling constant g_s. Using the lowest loop approximation one finds

$$\alpha_s = \frac{g_s^2}{4\pi} \cong \frac{1}{B \ln \left(\frac{q^2}{\Lambda^2}\right)} , \quad B = \frac{11 - \frac{2}{3} n}{4\pi} \tag{5.9}$$

(n: number of flavors)

where Λ is the mass parameter mentioned above. It is a quantity which has to be determined by experiment. The latter gives: $\Lambda \sim 400$ MeV. Using eq. (5.9) one finds that α_s becomes large ($\alpha_s >> 1$) for $q^2 \to \Lambda^2$ ("Landau pole"), which simply means that the perturbative expansion breaks down near $q^2 = \Lambda^2$. It is remarkable that Λ^{-1} comes out to be about 1/2 Fermi, i.e. Λ^{-1} is of the order of the extension of the hadrons. Thus we are led to another conjecture: The distance Λ^{-1} at which α_s becomes large is nothing else than the confinement radius describing the extension of the quark wave function of the hadrons. Furthermore all hadron masses (nucleon mass, ρ meson mass, etc.) are in the limit $\mathcal{m} = 0$ directly proportional to the scale parameter Λ.

We should like to remark that the scale Λ in QCD should not be regarded as a free parameter in itself. In fact, the QCD scale is something which can be chosen freely. All physical quantities of the dimension of a mass can then be expressed in terms of Λ. There is only one QCD theory, i. e. if we change $\Lambda \to 2\Lambda$ we obtain exactly the same theory as before, describing the same physics, except all quantities of the dimension of a mass are scaled up by a factor of two. In this sense QCD (in the limit $\mathcal{m} = 0$) can be regarded as a theory without any free parameter. Within QCD it has no a priori physical meaning to say: Λ is 400 MeV. However it is extremely important to remark: the proton mass is $c \cdot \Lambda$, where $c \approx 2.3$. This number 2.3 is one of the numbers which eventually have to be calculated within QCD.

It is a different matter to compare Λ with parameters outside QCD, e.g. with the electron or muon mass. It is certainly impossible to explain within QCD why Λ is about four times as large as m_μ. On the other hand it is hard to believe that the relation $\Lambda \sim 4 \cdot m_\mu$ is purely accidental. Thus we are led to the conjecture that there must exist eventually a mechanism which ties together the QCD parameter Λ and e.g. the lepton masses. Needless to say that until now there exists no theory which can do that. Of course, we do not know how direct the relationship between the lepton masses and Λ is; perhaps it is very indirect, and it is foolish trying to make a connection. Nevertheless it is remarkable that the scale of the lepton masses and the QCD scale Λ are of similar orders of magnitude; there would be nothing wrong, it these two scales would be totally different, e.g. $\Lambda = m_\mu \cdot 10^{10}$.

6. QUARK MASSES IN QCD

If the quarks are permanently confined, the question arises, how one can define the quark masses (elements of the quark mass matrix $\mathcal{m}$) such that one can assign actual values (in MeV) to them. In the field equations of QCD the quark masses appear always as multiplied by quark fields. For example the divergence of an axial vector current (color singlet) is given by

$$\partial^{\mu}(\bar{q}_i \gamma_{\mu} \gamma_5 q_j) = i(m_i + m_j)\, \bar{q}_i \gamma_5 q_j \tag{6.1}$$

where q_i and q_j are two different quark flavors (we choose two different flavors in order to avoid the anomalous gluon contribution, discussed before). The l.h.s. of eq. (6.1) is a quantity which is observable, i.e. it must be independent of the renormalization point μ in QCD. On the other hand a scalar or pseudoscalar density like $\bar{q}q$ or $\bar{q}\gamma_5 q$ is μ-dependent. Thus the quark masses m_i must also be μ-dependent. The change of m_i under a change of μ is described by

$$\frac{dm}{m} = \frac{d\mu}{\mu} \cdot \gamma_m (g, \frac{m}{\mu}) \tag{6.2}$$

where γ_m is the associated anomalous dimension. For $\mu >> 1$ GeV, where the lowest loop expansion of the QCD renormalization group equation should be valid, one has:

$$\gamma_m = - \frac{g^2(\mu)}{2\pi^2} = - \frac{2\alpha_s}{\pi}$$

$$\frac{m(\mu_1^2)}{m(\mu_2^2)} = \frac{1}{1 + \frac{\alpha_s}{\pi} \ln (\frac{\mu_1^2}{\mu_2^2})} . \tag{6.3}$$

Thus the quark masses shrink as μ increases; this effect is caused by the emission and reabsorption of virtual gluons. At decreasing distances ($\mu^{-1} \rightarrow \infty$) less self-energy corrections are seen. Therefore m slides to zero logarithmically.

In the region where eq. (6.3) can be applied, the ratio of two masses of different flavours is independent of μ^2, i.e. the ratios of two quark masses approaches rapidly a finite number:

$$\lim_{\mu \rightarrow \infty} \frac{m_i(\mu)}{m_j(\mu)} =: \frac{m_i^o}{m_j^o} . \tag{6.4}$$

The ratio (m_i^o / m_j^o) is defined to be the bare quark mass ratio. The latter is a number which can be interpreted as the QCD analog of a lepton mass ratio, e.g. m_μ/m_e. It is a number which cannot be calculated within QCD, but only within an extended theory including the flavor interaction and, perhaps, further ones.

We can calculate the bare quark mass ratios using PCAC [8]. The $(\text{mass})^2$ of a pseudoscalar meson is given by

$$M^2(ps) = - F^{-2} \int d^4x\, d^4y \langle 0| \left[\mathcal{F}_0^5(x) \left[\mathcal{F}_0^5(y), \bar{q}\, \mathcal{M}\, q\right]\right] |0\rangle \qquad (6.5)$$

where $\mathcal{F}_\mu^5$ is the axial vector current corresponding to the quantum number of the corresponding meson, and F is the associated meson decay constant. Using eq. (6.5) one obtains

$$M^2(\pi^+) \cong M^2(\pi^o) = 4\, F_\pi^{-2}\, (m_u \langle\bar{u}u\rangle + m_d \langle\bar{d}d\rangle) \qquad \text{etc.} \qquad (6.6)$$

where $\langle\bar{q}q\rangle$ is the vacuum expectation value of $\bar{q}q$. The latter has to be nonzero, since otherwise the Goldstone picture of the pseudoscalar mesons in the limit $\mathcal{M} = 0$ would not hold. Like the confinement phenomenon the fact that $\bar{q}q$ has a nonvanishing v.e.v. has not been demonstrated to be true in QCD. It is <u>assumed</u> to be the case here.

If we assume further that the v.e.v.'s of $\bar{q}q$ obey an SU(3) symmetry as well as the meson decay constants (this would be true exactly in the limit $m_u = m_d = m_s = 0$), we find:

$$\begin{aligned} M_\pi^2 &= (m_u + m_d)\, M_o + \gamma(\pi) \\ M_{K^+}^2 &= (m_u + m_s)\, M_o + \gamma(K^+) \\ M_{K^o}^2 &= (m_d + m_s)\, M_o + \gamma(K^o) \\ (M_o &= 4\, F^{-2} \langle\bar{q}q\rangle). \end{aligned} \qquad (6.7)$$

Here we have included the electromagnetic self-energies of the corresponding particles denoted by γ (those were neglected in eq. (6.6)). In the limit of chiral symmetry one has

$$\begin{aligned} \gamma(\pi^o) &= \gamma(K^o) = 0 \\ \gamma(\pi^+) &= \gamma(K^+). \end{aligned} \qquad (6.8)$$

Using these relations as well as eq. (6.7) one arrives at the following quark mass ratios:

$$\frac{m_d}{m_s} \approx \frac{1}{20}, \frac{m_d}{m_u} \approx 1.8. \qquad (6.9)$$

There are some uncertainties in these relations, due to the uncertainty in the quality of the SU(3)xSU(3) symmetry, and of the relations (6.8). However it is expected that the errors induced by those uncertainties are not large. In general we can say that m_d/m_u should be between 1.5 and 2, and m_s/m_d between 15 and 25. The pattern for the light quark masses which arises leads also to a consistent picture for the baryon mass differences (see e.g. ref. (9)).

We emphasize the surprisingly large violation of isospin exhibited by the quark mass matrix. The mass difference m_d-m_u is of the same order as m_d or m_u. Consequently the observed isospin symmetry has nothing to do with a possible degeneracy $m_d = m_u$. It is rather a consequence of the smallness of m_d and m_u with respect to the basic scale Λ of QCD: m_d, $m_u << \Lambda$.

We conclude our discussion of the quark masses in QCD with a few comments about the values of the quark masses. In the limit $m_q \to \infty$ (i.e. $m_q >> \Lambda$) we can define the absolute value of a quark mass by identifying it with the quark mass value which enters in the naive nonrelativistic bound state description of the $\bar{q}q$-systems. This would be the quark mass value m_q at $\mu^2 \sim m_q^2$. This procedure can probably be applied to the b-quark, in which case the b-quark mass is about 4.8 GeV. With some caution we may apply the same idea to the c-quark and find $m_c \simeq 1.5$ GeV.

It is more difficult to introduce absolute values for the light quarks u, d, and s. One possibility is the following one [8)]. In the nonrelativistic limit where the "small" components of the quark fields dominate, one has $\bar{q}q = q^+q$. Let us suppose that this relation is true for the s-quarks if we set $\mu \approx 1$ GeV (mass of $\bar{s}s$ bound states like the ϕ-mesons). In this case m_s is equal to the typical mass difference between the strange and nonstrange hadrons, which is about 180 GeV. If we normalize m_s to be 180 MeV, we find

$$\begin{matrix} m_d = 9 \text{ MeV} \\ m_u = 5 \text{ MeV} \end{matrix} \quad (m_s =: 180 \text{ MeV}). \qquad (6.10)$$

These are the quark mass values which we shall use for our subsequent considerations. We emphasize that m_s is of the same order as the QCD scale Λ, while m_u and m_d are much smaller than Λ. The u and d quarks are essentially massless. The d-quark is about 4 MeV heavier than the u-quark . This is the reason for the fact that e.g. the neutron is heavier than the proton.

We emphasize that the quark masses quoted above should not be confused with the constituent u and d quark masses, which are used e.g. for the calculation of the magnetic moments and which are essentially $M/3 \approx 300$ MeV. The latter are effective masses which are acquired by the quarks since they move inside a confining potential. They are not present once we "look" at a quark inside a hadron at distances smaller than Λ^{-1} (e.g. in deep inelastic scattering); in this case the quark mass values given in eq. (6.10) are relevant.

The connection between the constituent quark masses and the quark masses given in eq. (6.10) can be described as follows. The constituent quark masses are generated since the v.e.v. of $\bar{q}q$ in nonzero. Since the matrix element $\langle 0|\bar{q}q|0\rangle$ has dimension three, we can write, taking into account the effect of spontaneous symmetry breaking[10]:

$$m_q(\mu^2) = m_q^o(\mu^2) + (v(\mu^2)/\mu^2) \cdot c(\mu^2) \tag{6.11}$$

where $v(\mu^2)$ is given by

$$v(\mu^2) = \langle 0|\bar{q}q|0\rangle,$$

$m_q^o(\mu^2)$ is the quark mass in the absence of spontaneous symmetry breaking (it follows eq. (6.3)), and $c(\mu^2)$ is a logarithmically varying function of μ^2, depending of $g(\mu^2)$ (see e.g. ref. (10)). One observes that according to eq. (6.11) the nonperturbative effects disappear like $(1/\mu^2)$ for $\mu^2 \to \infty$, i.e. they disappear very quickly, and only the "current" quark masses $m^o(\mu)$ are left over.

7. QUARK MASSES AND FLAVOR MIXING

The leptons and quark flavors observed thus for seem to come in three generations:

$$\text{I.}\quad \begin{pmatrix} \nu_e & u \\ e^- & d \end{pmatrix}$$

$$\text{II.}\quad \begin{pmatrix} \nu_\mu & c \\ \mu^- & s \end{pmatrix} \qquad (7.1)$$

$$\text{III.}\quad \begin{pmatrix} \nu_\tau & t \\ \tau^- & b \end{pmatrix}$$

The neutrinos are observed to be very light and perhaps massless. Only the effects due to the five quarks u, d, c, s, and b have been observed. The t-quark (if it exists) must be heavier than 8 GeV. The mass ratios of the leptons and quarks belonging to different generations are observed to be large, e.g.

$$\frac{m_s}{m_d} \approx 20 \qquad \frac{m_\mu}{m_e} \approx 210 \qquad \frac{m_c}{m_u} \approx 300 \ . \qquad (7.2)$$

The three generations denoted above are classified according to the masses of the leptons and quarks. It is remarkable that the weak interactions cause dominantly transitions between the quarks in the same family, e.g. $u \leftrightarrow d$, $c \leftrightarrow s$, ..., while all other transitions are observed to be small, e.g. $u \leftrightarrow s$, $c \leftrightarrow d$, ... (suppressed by the weak mixing angles like the Cabibbo angle). This shows that there must exist a connection between the weak mixing angles and the quark masses. Of course, no such connection exists if the quark mass matrix is completely arbitrary. Thus we must look for a special form of the quark mass matrix.

One particular form of the quark mass matrix is as follows (we discuss first the case of four quarks)

$$\overline{(u_o, c_o)}_L \begin{pmatrix} 0 & a \\ a & b \end{pmatrix} \begin{pmatrix} u_o \\ c_o \end{pmatrix}_R + \text{h.c.} \qquad (7.3)$$

An analogous form is supposed to be valid for the d,s-system (q_o: weak interaction eigenstate). The mass matrix given in eq. (7.3) describes a situation where the u and d quark masses are massless in the absence of the weak interaction mixing, i.e. in the limit $\theta_c \to 0$ (θ_c: Cabibbo angle). The u and d quark masses are generated solely by the weak interaction mixing. It is not possible to obtain a matrix like the one given in eq. (7.3) in the minimal SU(2)xU(1) theory in a natural way, but it arises in gauge theories involving the group SU(2)xSU(2)xU(1)[11].

Diagonalizing the matrix given in eq. (7.3) one can express the Cabibbo angle in terms of the quark mass ratios:

$$\theta_c \cong \left| \sqrt{\frac{m_d}{m_s}} - e^{i\alpha} \sqrt{\frac{m_u}{m_c}} \right| \qquad (7.4)$$

(α: undetermined phase parameter). Relation (7.4) gives $9.5^o < \theta_c < 15^o$, which agrees well with experiment .

A simple generalisation of this idea to the six quark case is to assume the following structure of the quark mass matrix in terms of the weak interaction eigenstates

$$\overline{(u_o, c_o, t_o)}_L \begin{pmatrix} 0 & a & 0 \\ a & 0 & b \\ 0 & b & c \end{pmatrix} \begin{pmatrix} u_o \\ c_o \\ t_o \end{pmatrix}_R + h.c. \qquad (7.5)$$

(analogously for d, s, b). In this case the c quark mass is generated by the weak mixing, and subsequently the u quark mass. This would explain the hierarchy of the quark masses: $m_u \ll m_c \ll m_t$; $m_d \ll m_s \ll m_b$.

Diagonalizing the matrix given in eq. (7.5) we can calculate the weak mixing angles in terms of the quark mass ratios (for details we refer the reader to ref. (13)). The results are in good agreement with the experimental results.

If we request that a matrix of the type given in eq. (7.5) is generated by the minimal scheme of scalar mesons using a flavor group SU(2)xSU(2)xU(1) one finds the relation 13), 14)

$$\frac{m_t\, m_c\, m_u}{m_b\, m_s\, m_d} = \left(\frac{m_t - m_c + m_u}{m_b - m_s + m_d}\right) . \qquad (7.6.)$$

This relation predicts $m_t \approx 11$ GeV, although the uncertainty in the prediction is quite large, due to the uncertainties in the light quark mass ratios. Taking into account those uncertainties, we expect the t quark mass to have a mass in the range $9\ \text{GeV} < m_t < 14\ \text{GeV}$. The experiments now under way at DESY will soon find the t quark if its mass is indeed given by eq. (7.6).

8. UNIFICATION OF THE STRONG AND ELECTROWEAK INTERACTIONS

There are various reasons to believe that the gauge theory based on the group U(1)xSU(2)xSU(3) could not be the final answer. If this were the case, it would be impossible to obtain information about the various coupling constants g_1, g_2, g_3, or to understand why the electric charges are quantized. Furthermore all the free parameters of QCDxQFD would remain free parameters. There are two possible ways to proceed.

I. The quarks, leptons, and gauge bosons are not elementary, but composed of subunits. In this case one may interpret the various flavor families as different excitations of one basic configuration, e.g. the muon would be an excitation of the electron. Thus far not much progress has been made in this direction.

II. The gauge group U(1)xSU(2)xSU(3) is viewed as a subgroup of a larger simple group, denoted by G [15]. The corresponding gauge theory would have to exhibit at least two stages of symmetry breaking as shown below:

$$\begin{array}{c} G \\ \downarrow \text{I.} \\ U(1)\times SU(2)\times SU(3) \\ \downarrow \text{II.} \\ U(1)\times SU(3)\ . \end{array}$$

At the first stage I all gauge bosons which are not gluons, W, Z or γ, acquire a large mass, say of the order of M, while at the second stage II the masses of the W and Z (of the order of m) are generated. One has to have $M \gg m$, in order to understand why the subgroup U(1)xSU(2)xSU(3) plays such an important rôle at relatively low energies. At energies much below M the unified interactions can be neglected for many purposes, and we are left with an effective gauge theory, based on the group U(1)xSU(2)xSU(3). Within such an approach the fermions (leptons, quarks) form irreducible representations of the large group G, and one may have the hope to learn something about their mass spectrum.

Since the group G is assumed to be simple (or perhaps semisimple such that the various factors are related by discrete symmetry, e.g. parity), only one coupling constant is allowed. As a consequence the coupling constants g_1, g_2, and g_3 of the theory based on U(1)xSU(2)xSU(3) are related to each other by group factors. Furthermore the electric charges of the leptons and quarks are quantized, since the electric charge operator is one of the generators of G (in the U(1)xSU(2) theory of QFD the electric charges are not quantized due to the presence of the U(1) factor).

There are various schemes of unified theories discussed in the literature. Before I will discuss some of the details, I would like to mention first some general aspects of unified theories. Usually the discussion of gauge theories starts with a discussion of the fermions. If we have an idea how many fundamental fermions there are, or we want to consider, there exists an obvious limit on the extension of the group G. It cannot be larger than the maximal group G^{max}, which is just the symmetry group of the kinetic energy term in the fermion Lagrangian. In case of N Weyl spinors this group is SU(N). As an example we consider the fermions

$$\begin{pmatrix} \nu_e & \vdots & u \\ e^- & \vdots & d \end{pmatrix}_L \qquad (e^+, \bar{d}, \bar{u})_R$$

which are 15 fermions (including the color quantum number). Thus the maximal group is SU(15). This group cannot be used as a gauge group, since it is not anomaly free, and only a subgroup of SU(15) can be gauged, e.g. SU(5). If we are interested in the relation between the coupling constants g_1, g_2, and g_3, it is sufficient to look at the relations imposed by the maximal group G^{max}. Obviously the relations obtained by using any subgroup will be the same.

Let us look at the fermions mentioned above, and use G^{max} = SU(5). We normalize the SU(15) generators such that tr F^2 = 1 (F: SU(15) generator). On the other hand the U(1)xSU(2)xSU(3) generators are defined in the usual way. Thus the W_3, γ and Z bosons are related by

$$\frac{T_3}{\sqrt{\mathrm{tr}\, T_3^2}} = \cos\theta_W \cdot \frac{Z}{\sqrt{\mathrm{tr}\, Z^2}} + \sin\theta_W \cdot \frac{Q}{\sqrt{\mathrm{tr}\, Q^2}} \tag{8.1}$$

(tr T_3^2 etc. means the trace in the (15)-plet of fermions, θ_W: weak mixing angle). The electric charge, T_3 and the weak hypercharge Y are related by

$$Q = T_3 + \frac{1}{2} Y . \tag{8.2}$$

From eq. (8.1) we find:

$$\sin\theta_W = \frac{\operatorname{tr}(T_3 \cdot Q)}{\sqrt{\operatorname{tr} T_3^2}\sqrt{\operatorname{tr} Q^2}} . \tag{8.3}$$

On the other hand eq. (8.2) implies

$$\operatorname{tr} T_3^2 = \operatorname{tr}(Q T_3) , \tag{8.4}$$

and we obtain

$$\sin^2\theta_W = \frac{\operatorname{tr} T_3^2}{\operatorname{tr} Q^2} . \tag{8.5}$$

In the case of the 15 fermions given above one has

$$\operatorname{tr} T_3^2 = \frac{1}{2} \cdot 4 \qquad \operatorname{tr} Q^2 = 2\,(1 + \frac{5}{3})$$
$$\sin^2\theta_W = \frac{3}{8} . \tag{8.6}$$

Analogously we can calculate the strong coupling constant $\alpha_s = \frac{g_3^2}{4\pi}$ in terms of $\alpha = \frac{e^2}{4\pi}$. With our choice of normalization we have

$$\operatorname{tr} Q^2 \cdot \frac{e^2}{4\pi} = \frac{g^2}{4\pi} . \tag{8.7}$$

The strong coupling constant g_3 is defined such that the coupling of the gluons to the quarks is given by $g \cdot \frac{\lambda_i^c}{2}$ where the color matrices are normalized by $\operatorname{tr}(\lambda_i \lambda_j) = 2\,\delta_{ij}$. Taking into account that the fermions include both quarks and antiquarks we find

$$\operatorname{tr} (\frac{\lambda_i}{2})^2 \cdot 2\,f \cdot \frac{g_3^2}{4\pi} = \frac{g^2}{4\pi} \tag{8.8}$$

(f: number of quark flavors).

Comparing eq. (8.7) and (8.8) we find

$$\frac{g_3^2}{4\pi} = \alpha \cdot \frac{\operatorname{tr} Q^2}{f} . \tag{8.9}$$

In case of the fifteen elementary fermions given above one finds

$$\alpha_s = \frac{8}{3}\,\alpha . \tag{8.10}$$

If we are dealing with several families of leptons and quarks, i.e. repetitions of the basic pattern studied before the relations (8.6) and (8.10) remain unchanged.

Both the relations (8.6) and (8.10) are bad as far as the phenomenological consequences are concerned, if we take those relations at face value. Eq. (8.6) gives $\sin^2\theta_W = 0.37$, while the experimental value is $\sin^2\theta_W = 0.23 \pm 0.02$. Eq. (8.10) predicts $\alpha_s = 8/3\,\alpha = 1/51$ while the experimental value of α_s at distances of the order of ~ 1/10 Fermi is ~ 0.4.

One possible way out of this situation is to assume that the unification mass M is a very large mass, and the renormalization effects are large enough to renormalize both $\sin^2\theta_W$ and α_s such that there is no conflict with experiment [16]. At energies much below M the renormalization of the coupling constants is given by

$$\mu \frac{d}{d\mu} g_i(\mu) = \beta_i\left[g_i(\mu)\right] \approx b_i\; g_i^3(\mu). \tag{8.11}$$

It follows

$$\frac{1}{g_i^2(\mu)} - \frac{1}{g_i^2(M)} = 2\, b_i \ln\left(\frac{M}{\mu}\right) \tag{8.12}$$

$$\left(b_3 = -\frac{1}{(4\pi)^2}\cdot\frac{33}{3} + b_1\;,\; b_2 = \frac{1}{(4\pi)^2}\cdot\frac{22}{3} + b_1\;,\; b_1 = -\frac{1}{(4\pi)^2}\cdot\left(-\frac{2}{3}f\right)\right)$$

(f: number of flavors in the sequential QFD).

At $\mu \approx M$ the electromagnetic and strong coupling constants are related in case of the conventional set of leptons and quarks by relation (8.10). At $\mu \ll M$ one finds using the relations (8.11, 8.12)

$$\left(\frac{1}{\alpha(\mu)} - \frac{8/3}{\alpha_3(\mu)}\right) \cong \frac{11}{\pi} \ln\left(\frac{M}{\mu}\right). \tag{8.13}$$

The relations (8.11, 8.12) are valid only in the region $M_W < \mu < M$ (for $\mu < M_W$ one has to take into account the threshold effects for the W and Z bosons). For this reason we choose $\mu = 100$ GeV, and use $\alpha_s(100) \approx 0.13$, $\alpha(100) \approx 1/127$ (for a calculation of the renormalization of α see e.g. ref. (17)). This gives, following eq. (8.13)

$$M \approx 1.6 \cdot 10^{15}\ \text{GeV}.$$

Analogously we can calculate the renormalization of $\sin^2\theta_W$. Since $\tan\theta_W = g_1/g_2$, g_2 increases and g_1 decreases as μ descends from M down to ~ 100 GeV, it is clear that

$\sin^2\theta_W$ as a function of μ will decrease. One finds:

$$\sin^2\theta_W = \frac{1}{6}\left(1 + \frac{10}{3}\cdot\frac{\alpha(\mu)}{\alpha_s(\mu)}\right). \tag{8.14}$$

In the symmetry limit α/α_s is equal to 3/8, and it follows $\sin^2\theta_W = 3/8$. In the extreme case $\alpha/\alpha_s = 0$ $\sin^2\theta_W$ approaches its lower limit 1/6 = 0.167. Choosing $\alpha_s(100) = 0.13$ and $\alpha(100) = 1/127$, one obtains

$$\sin^2\theta_W = 0.20\,. \tag{8.15}$$

The value of $\sin^2\theta_W$ depends on α_s. For example for $\alpha_s(100) = 0.10$ one finds $\sin^2\theta_W = 0.21$, and for $\alpha_s(100) = 0.20$ one obtains 0.19. Due to the uncertainties in the QCD scale parameter Λ and in the extrapolation of $\alpha_s(\mu)$ from the energies investigated thus far ($\mu \sim 10$ GeV) to the energy $\mu \approx 100$ GeV (number of quark flavors, etc.) it is not possible to calculate $\sin^2\theta_W$ to better than 5%. All one can say is: If the effective gauge theory at $\mu < M$ is U(1)xSU(2)xSU(3), one expects $M \approx 10^{15} \ldots 10^{16}$ GeV and $\sin^2\theta_W = 0.19 \ldots 0.21$. If it should turn out that $\sin^2\theta_W$ is much above 0.20, e.g. 0.23 ... 0.25, it would mean that either the ideas about a unification of the strong, electromagnetic, and weak interactions are not correct or the unification of the interactions proceeds in several steps, involving several mass scales, instead of the one step discussed above.

9. UNIFICATION BASED ON SU(5)

The group U(1)xSU(2)xSU(3) has rank 4. For this reason the unifying group G must have a rank larger or equal 4. The fermions transform under SU(2)xSU(3) as follows:

$$f = \begin{pmatrix} \nu_e \\ e^- \end{pmatrix}_L, \begin{pmatrix} u \\ d \end{pmatrix}_L, e^+_L, \bar{u}_L, \bar{d}_L + \text{other generations} \tag{9.1}$$

$$= (2,1) + (2,3) + (1,1) + (1,\bar{3}) + (1,\bar{3}) + \text{other generations}.$$

This representation is complex, and each generation is composed of 15 fermions.

The smallest group which can serve as a unifying group is SU(5), which has rank 4. The (5)-representation of SU(5) decomposes under SU(2)xSU(3) as

$$(5) = (2,1) + (1,3). \tag{9.2}$$

The 10-representation of SU(5) is obtained as the antisymmetrized product $(5x5)_a$. It is easy to work out its SU(2)xSU(3) content:

$$10 = (5x5)_a = \{[(2,1) + (1,3)]\ [(2,1) + (1,3)]\} = (1,1) + (2,3) + (1,\bar{3}). \quad (9.3)$$

Thus the representation (9.1) is obtained as the reducible representation $\bar{5}$ + 10. We make the following comments about this scheme:

a) The basic set of fermions including the electron and its neutrino as well as the light quarks u and d appear in a reducible representation:

$$\bar{5} = \left\{ \begin{matrix} \nu_e \\ e^- \end{matrix} \;\vdots\; \bar{d} \right\}$$

$$10 = \left\{ \begin{matrix} u \\ d \end{matrix} \;\vdots\; \bar{u}, e^+ \right\} \quad (9.4)$$

b) The other generations of fermions (μ, τ, ...) are simply interpreted as replications of the first one.

c) In the 10-representation of SU(5) appear both quarks and antiquarks. Therefore baryon number could not be a conserved quantity.

d) In SU(5) there is no room for a righthanded counterpart of the lefthanded neutrino.

e) The gauge bosons (24-representation of SU(5)) transform under SU(2)xSU(3) as:

$$24 = (1,8^c) + (3,1^c) + (1,1^c) + (2,3^c) + (2,\bar{3}^c)\ . \quad (9.5)$$

Besides the gauge bosons belonging to the $U(1)xSU(2)xSU(3)^c$ subtheory (γ, Z, W; gluons) one has 12 additional gauge bosons which are color (anti)triplets as well as weak doublets. We shall denote them by the doublet:

$$\begin{pmatrix} X^c \\ Y^c \end{pmatrix} + \text{antiparticles.} \quad (9.6)$$

The adjoint representation of SU(5) can be obtained by multiplying the 5-representation with its complex conjugate: $\bar{5}$x5 = 24 + 1. Using the decomposition of the 5-representation in terms of quarks and leptons as given in eq. (9.2) it is easy to work out the coupling of X and Y. They couple to a di-antiquark configuration ($\bar{q}\bar{q}$) and to a leptoquark configuration ($q\bar{l}$). The X and Y interactions lead to the decay of the proton into leptons and mesons in the second order of the gauge coupling. As a consequence the masses of the X and Y bosons must be very large (not less than ~ 10^{15} GeV). We emphasize that the X and Y bosons are colored and supposedly confined. Thus the limit on the X and Y masses quoted above should be interpreted as a limit for the effective mass.

Symmetry breaking in SU(5):

In SU(5) the symmetry breaking must proceed in two stages. At the first stage the SU(5) gauge symmetry is broken down to $U(1)\times SU(2)\times SU(3)^c$. The X and Y bosons acquire large masses. At the second stage the W and Z masses are generated:

$$\begin{array}{c} SU(5) \\ \downarrow \;\; \text{I} \; (M_X, M_Y) \\ U(1)\times SU(2)\times SU(3)^c \\ \downarrow \;\; \text{II} \; (M_W, M_Z) \\ U(1)\times SU(3)^c . \end{array}$$

The minimal scheme to arrange the symmetry breaking as described above is to introduce both a 24-plet and a 5-plet of scalars. The 24-representation of SU(5) is used to accomplish the first stage of the symmetry breaking. The potential is given by

$$U(\phi) = -\frac{1}{2}\mu^2 \operatorname{tr}(\phi^2) + \frac{a}{4}(\operatorname{tr}\phi^2)^2 + \frac{b}{2}\operatorname{tr}\phi^4. \tag{9.7}$$

Using this potential the SU(5) symmetry can be broken down in two possible ways:

a) $SU(5) \to U(1)\times SU(2)\times SU(3)^c$
(desired breaking)

b) $SU(5) \to U(1)\times SU(4)$
(undesired breaking).

It depends on the values of the coupling constants a and b, which way the SU(5) breaking proceeds. Obviously we have to arrange them in such a way that the symmetry breaking of type a results. Unfortunately nobody has yet found a good argument why it is the type a of symmetry breaking, which is realized (in case the group SU(5) is relevant for the real world at all). It is hard to believe that the symmetry breaking proceeds via the a-scheme simply by accident. Perhaps the radiative correction in the SU(5) scheme are such that one is always driven towards the a-type of breaking. Thus far the radiative corrections to the potential (9.7) have not yet been studied in detail.

The second stage of symmetry breaking is accomplished in the most simple way be using a 5-representation H of SU(5). The latter consists of three color-triplet scalars which must not acquire any v.e.v. (color conservation), and a doublet of color singlet scalars.

It is always possible to perform a rotation in the SU(2) space such that only one of the SU(2) components acquires a v.e.v.:

$$H = \begin{pmatrix} 0 \\ v \\ \hline 0 \\ 0 \\ 0 \end{pmatrix}. \qquad (9.8)$$

This choice of the SU(2) axes defines the direction of the electric charge. One obtains automatically:

$$\begin{aligned} Q(\nu_e) &= 0 \\ 3\,Q(\bar{d}) &= Q(e^+), \end{aligned} \qquad (9.9)$$

i.e. the electric charges are quantized.

We add the following comments:

A. The masses of the X and Y bosons are generated in the primary stage of the symmetry breaking. Since at this stage the weak SU(2) group remains unbroken, one has:

$$M(X) = M(Y)\ . \qquad (9.10)$$

This degeneracy is broken by the secondary symmetry breaking, however only very slightly. The relation (9.10) is violated by a very tiny correction of order $M_W/M_X \sim 10^{-13}$. For any practical purpose we can neglect it.

B. The masses of the X and Y bosons are determined by the v.e.v. of the ϕ-field (24-plet), while the W and Z masses are given by the v.e.v. of the H-field (5-plet). For phenomenological reasons we have to assume that there exists a huge gap between the W/Z-masses and the X/Y masses: $M_{W,Z} \approx 10^2$ GeV, $M_{X,Y} \approx 10^{15}$ GeV (the "great desert" in particle physics). As a consequence we have to assume: $v << V$ ($v/V \sim 10^{-13}$). It is hard to understand this large gap in the boson mass spectrum, since in general one would expect that there exist interactions between the ϕ and H bosons (e.g. $\lambda(\bar{H}H)\,\mathrm{tr}\,\phi^2$). Such interactions are even induced by radiative corrections. Thus it seems not natural to have $v/V << \alpha$ [18]. We shall come back later to the problem of gauge hierarchies.

C. The decay of the proton is induced by the interactions due to X/Y exchange. The leading decay modes are: $p \to e^+ \pi^o, \rho^o, \omega, \ldots, p \to \bar{\nu}_e \pi^+, \rho^+, \ldots$. The life time of the proton can be calculated in terms of M_X and $|\psi(0)|^2$ ($\psi(0)$: wave function of the proton; the decay amplitude is proportional to $\psi(0)$ since two quarks in the proton have to come close to each other in order to activate the X/Y interactions). One obtains :

$$\tau(\text{proton}) \approx 10^3 \ldots 10^4 \cdot M_X^4/M_p^5. \tag{9.11}$$

The mass of the X-bosons is estimated to be about 10^{15} GeV (see eq. (8.13)), and the proton life time is expected to be of the order of 10^{30} yrs. However the calculated life time is rather sensitive to M_X, and it is not possible to improve our knowledge about M_X such as to determine it to better than a factor of ten. Thus we can say that within the SU(5) scheme the proton life time is expected to be ~ $10^{28} \ldots 10^{32}$ yrs.

On the other hand the present experimental limit on the proton life time is ~ 10^{29} yrs [20], i.e. very close to the theoretical value. For this reason it is important to improve the present experimental limits on the proton stability by 2 - 3 orders of magnitude. If the decay of the proton is found at a level of e.g. $\tau(\text{proton}) \approx 10^{31}$ yrs [20], it would, of course, not establish that the SU(5) scheme is correct, but it would support the ideas about a unified theory of all interactions.

D. Mass relations in SU(5)

The fermion masses are generated by the coupling of the H-bosons to the fermions. The coupling terms have the following algebraic structure.

$$\begin{aligned} &\text{I.: } \bar{5}_f \times 10_f \times \bar{5}_H \\ &\text{II.: } 10_f \times 10_f \times 5_H \end{aligned} \tag{9.12}$$

Using the subgroup $SU(2)xSU(3)^c$ we find

$$A: \left[(1,\bar{3}^c) + (2,1^c)\right]_f \times \left[(2,3^c) + (1,\bar{3}^c) + (1,1^c)\right]_f \left[(1,\bar{3}^c) + (2,1^c)\right]_H \quad (9.13)$$

$$B: \left[(2,3^c) + (1,\bar{3}^c) + (1,1^c)\right]_f \left[(2,3^c) + (1,\bar{3}^c) + (1,1^c)\right]_f \left[(1,3^c) + (2,1^c)\right]_H$$

Only the upper component of the $(2,1^c)$-term in the H-representation acquires a v.e.v. Thus only those terms in the fxf-products which can form a color and SU(2) singlet when multiplied by $(2,1^c)_H$ contribute to the fermion mass matrix. It is easy to work out what they are in terms of the fermion fields:

$$\begin{aligned} &\text{I: const} \cdot v \cdot (\bar{d}d + \bar{e}e) \\ &\quad \text{(d-quark and electron mass)} \\ &\text{II: const} \cdot v \cdot \bar{u}u \\ &\quad \text{(u-quark mass)} . \end{aligned} \quad (9.14)$$

There are two mass parameters, namely m_u and m_e; the d quark and the electron are degenerate in mass.

If we have several generations of fermions one can have mixing between the various generations. However it is easy to see that the mass terms can always be brought into the following form:

$$\begin{aligned} \bar{f}\, \mathcal{M}\, f &= m_e\, (\bar{e}e + \bar{d}d) + m_\mu\, (\bar{\mu}\mu + \bar{s}s) + \ldots \\ &+ \bar{U}MU \end{aligned} \quad (9.15)$$

where U stands for the quarks of charge 2/3:

$$U = (u,\ c,\ t,\ \ldots)$$

and M is an arbitrary hermitean matrix. The latter has off-diagonal matrix elements which lead to the weak interaction mixing (Cabibbo angle, ...). Thus the important prediction of the minimal SU(5) scheme is:

Mass of charged lepton = Mass of quark of charge (- 1/3).

This prediction is valid at $\mu = M$, and renormalization effects have to be taken into account in order to compare this prediction with experiment [19]. The main correction is the QCD correction for the quark mass, which causes the quark mass to increase as the renormalization point is lowered, and we have:

$$m_q\ (\mu = M) = m_\ell$$
$$m_q(\mu) > m_\ell\ (\mu < M). \tag{9.16}$$

For example one has $m_b(M) = m_\tau$. If we set μ = few GeV, we find: $m_b \approx 5$ GeV in case of six quark flavors [19], which agrees well with the experimental value for m_b.

However there are problems concerning the SU(5) mass relations. Although the radiative corrections change the absolute values of the quark masses, they do not change the mass ratios. As a consequence we have:

$$\frac{m_d}{m_s} \cong \frac{m_e}{m_\mu}\ ,\ \frac{m_s}{m_b} \cong \frac{m_\mu}{m_\tau}\ . \tag{9.17}$$

Both relations are not true if we take the quark mass ratios derived previously on the basis of PCAC. Thus far nobody has found a satisfactory solution to this problem, and it seems that only one of the following two possibilites are open:

a) The SU(5) mass relations are not valid, e.g. due to the presence of the 45-representation of scalars besides the 5-representation.

b) The estimates of the quark mass ratios on the basis of PCAC are wrong.

E. B-L conservation in SU(5).

It is interesting to see that in the SU(5) scheme with the minimal set of scalars (24, 5) there exists an exactly conserved quantum number, namely B-L (baryon number minus lepton number) [21]. This can be seen as follows. The mass terms in the minimal SU(5) scheme transform like $10_f \times 10_f \times 5_H + \bar{5}_f \times 10_f \times \bar{5}_H$. Let us define a global quantum number X as follows:

$$X(10_f) = +1$$
$$X(\bar{5}_f) = -3 \tag{9.18}$$
$$X(5_H) = -2\ .$$

Note that these quantum numbers are such that the mass term has X = 0.

After the superstrong breaking the following SU(5) generator is still conserved:

$$Y = \begin{pmatrix} -\frac{3}{2} & & & & 0 \\ & -\frac{3}{2} & & & \\ & & 1 & & \\ & & & 1 & \\ 0 & & & & 1 \end{pmatrix}.$$

Both X and Y are broken by the secondary stage of symmetry breaking, but the combination $X - \frac{4}{3}Y$ remains conserved, since $(X - \frac{4}{3}Y)$ of the component of H which acquires a v.e.v. vanishes. It is easy to verify that $(X - \frac{4}{3}Y)$ is proportional to B-L for both the $\bar{5}_f$ and 10_f representation. Thus B-L is exactly conserved.

The B-L conservation law implies that the neutrinos in the minimal SU(5) scheme must be exactly massless. Since the neutrinos are two-component fields, they can acquire only a Majorana mass term. The latter which violates the conservation of lepton number is forbidden to exist due to the B-L conservation.

F. We should like to mention the following problem which arises in the SU(5) scheme, but also in all other unification schemes. The 24-representation of scalars causes the symmetry breaking

$$SU(5) \rightarrow U(1) \times SU(2) \times SU(3)^c .$$

The QCD gauge group $SU(3)^c$ is left as an unbroken subgroup. The further breaking involving the 5-representation causes the breaking of SU(2) such that the W and Z bosons acquire masses and the photon remains massless. It is possible to arrange that the group $SU(3)^c$ remains unbroken (only the colorless components of 5 acquire a nonzero v.e.v.). However we see no reason why this should be so in a natural way. In general the v.ev. of the 5-representation is expected to point towards some direction in the SU(5)-space, and this need not be a direction which is orthogonal to all eight $SU(3)^c$-generators. In other words: there is no reason why only the flavor group SU(2), but not the QCD gauge group $SU(3)^c$ is broken by the 5-representation. Thus far nobody has given a convincing reason why the color group remains unbroken. This problem remains one of the unsolved outstanding problems of unified theories.

G. Other unified theories

The SU(5) scheme is the minimal scheme which unifies the strong and electroweak interactions. No other group of rank 4 is able to do that. However many other schemes become possible, if we increase the rank of the unifying group. For example an interesting scheme is the one based on the rank 5-group SO(10)[22]. Here the fermions are described by the 16-dimensional spinor representation. Other schemes include the rank 12-group $[SU(4)]^4$, discussed by Pati and Salam [23], or the exceptional groups E(6), E(7), and E(8) [24]. For details of these more extended schemes we refer the reader to the original literature.

10. DYNAMICAL MASS GENERATION AND GAUGE HIERARCHIES

One of the puzzles of unified theories is the appearance of the "great desert" between 10^2 GeV and 10^{15} GeV. Of course, it may be that there is no such desert, and the space in energy above M_W is filled with many new energy scales, related to new interactions. Let us suppose that this is not the case. (We note that the new experimental results on $\sin^2\theta_W$ support this idea). Then the question arises: are there ways to understand the large gap between 10^2 GeV and 10^{15} GeV ? Let us recall that in the unified theories there exists another large mass gap, namely the gap between the unification mass M and the QCD mass Λ. The latter is the mass scale at which the QCD coupling constant α_s becomes of order unity, while M is the mass scale, where α_s is equal to (8/3) , i.e. M and Λ are related by the relation

$$\Lambda \approx M \exp\left(- \frac{16\ \alpha^2}{g^2(M)}\right) \cdot C \qquad (10.1)$$

where C is a constant of order unity. The large gap between Λ and M is naturally explained by the large exponential factor entering in eq. (10.1) On the other hand $M_{W,Z}$ and Λ are not very much different (viewed from the height of the unifying mass scale M), and it would be desirable to have a relation like (10.1) relating M and M_W. There are two different ways to obtain such a relation, which have been discussed recently, and below we shall describe them both.

I. The case $m^2 = 0$

Let us look again at the potential of the H-fields in the SU(5)-theory. For $\mu \ll M$ only the interaction of the H-fields among themselves and with the gauge bosons (except X, Y) is relevant. In general we have

$$V(H) = - m^2 H^2 + \frac{\lambda}{4!} H^4 + \text{radiative corrections} \qquad (10.2.)$$

Since we must have $v \sim \langle H \rangle \ll M$, it is necessary that $m^2 \ll M^2$. There is only one natural choice for m^2 in this case, namely $m^2 = 0$. Let us assume that the H-potential does not contain an explicit mass term (like in the case of scalar electrodynamics discussed in chapter 4). The potential is of the form

$$V(H) = \frac{1}{4!}\lambda(\mu)\, H^4 + (C_1 g^4 + C_2 \lambda^2)\, H^4 \left(\ln \frac{H^2}{\mu^2} - \frac{25}{6}\right) \tag{10.3}$$

(C_1, C_2 are constants, g is the gauge coupling constant, see eq. (4.2)).

The radiative correction due to the scalar loops and the gauge boson loops are of the same order if $\lambda \sim g^2$. Thus it would be natural to suppose that at $\mu = M$ λ is of order g^2 (this is, for example, the case in supersymmetric theories). One way to calculate the minimum of the potential (10.3) is to choose a particular renormalization point μ_o, namely the one where $\lambda(\mu)$ vanishes. Using the renormalization group equations for λ and g, one finds [19]

$$\mu_o = M \exp\left(\frac{-16\pi^2}{g^2(M)}\right) F \tag{10.4}$$

where F is a function of λ and g^2, which for $\lambda \sim g^2$ is of order unity. The minimum of V(H), i.e. $\langle H \rangle$, is given by

$$\langle H \rangle = \mu_o \exp \frac{11}{6} \, . \tag{10.5}$$

Thus we arrive at an equation which is similar to eq. (10.1). A quantitative estimate has been made by Weinberg [19] in case of the minimal SU(2)xU(1) theory (6 flavors, one scalar doublet) setting $\mu_o = \exp\left(\frac{-11}{6}\right) \cdot v \approx 40$ GeV. One finds

$$\frac{\lambda(M)}{e^2} \approx 1.75$$

for $M = 10^{15}$ GeV. (One obtains $(\lambda/e^2) \approx 1$ for $M \approx 10^7$ GeV, and $(\lambda/e^2) \approx 2$ for $M \approx 10^{18}$ GeV). As one can see, the small values of v/V follow as the consequence of our previous assumption $m^2 = 0$, and $\lambda(M) \approx e^2$. The puzzle of gauge hierarchies is reduced to the question why m^2 is zero to start with, and why the scalar self-coupling λ is of the order of e^2 at $\mu = M$.

We should like to emphasize that the case $m^2 = 0$ is precisely the one needed to derive the relation (4.8) which fixes the mass of the physical H-boson in terms of the W-mass.

II. Technicolor

Another possibility is to interpret the W, Z -mass scale as the consequence of a new type of strong interactions, which uses a new gauge degree of freedom (technicolor)[26]. In order to describe such a situation, let us consider the conventional gauge group U(1)xSU(2)xSU(3) in the absence of elementary scalars. If we assume in addition that there is only one quark doublet and that in the limit of vanishing quark masses the chiral symmetry of the strong interactions is realized in the Nambu-Goldstone way (i.e. the pions are massless Goldstone bosons, while all other particles acquire a mass), the W and Z bosons acquire masses by dynamical symmetry breaking which are of

$$M_{W,Z} \approx e \cdot F_{\pi} \approx 30 \text{ MeV}. \tag{10.6}$$

At the same time the pseudoscalar mesons disappear from the spectrum of physical states; their degrees of freedom constitute the longitudinal modes of the W and Z bosons. The pattern which arises is quite satisfactory, except the masses of the gauge bosons are of the order of $e \cdot F_{\pi}$ i.e. $e \cdot \Lambda$ (since F_{π} is of the order of the QCD parameter Λ). It was proposed by Susskind [26] to introduce another type of strong interactions (technicolor) with the following properties:

a) The technicolor group G^t is larger than SU(3), e.g. SU(4). Thus its confinement scale Λ^T is larger than Λ if all coupling constants are equal at $\mu = M$. The logarithm or the ratio Λ/Λ^t is given by the ratio of two group factors.

b) The quarks of the new technicolor interaction transform nontrivially under G^t and are weak doublets, but are color singlets. The techniquarks are massless, which implies in the absence of the weak interaction the existence of massless technipions.

In the presence of the weak interactions the Z and W particles acquire a mass. The technipions disappear from the spectrum of physical states. (In fact, only a linear combination of the technipion and the ordinary pion is "eaten" up by the W-Z mass generation, however the technipion configuration dominates, and the mixing between the technipion and the ordinary pion is very small.)

The masses of the W and Z bosons are of order

$$M_{W,Z} \approx e \cdot F_{\pi}^{t}, \tag{10.7}$$

(F_{π}^{t}: decay constant of technipion), and for $M_W \approx 90$ GeV one finds:

$$F^t \approx 250 \text{ GeV}.$$

This implies that the typical mass scale for the technihadrons (particles, composed of the techniquarks) is of the order $M_p\ (F^t_\pi/F_\pi) \approx 2 \ldots 3$ TeV.

It is interesting to note that the lightest hadron composed of the techniquarks and carrying the technicolor analog of the baryon number will be absolutely stable. For example in case of G^t = SU(4) the lightest stable particle has the quark composition QQQQ which is a meson. No technibaryons (i.e. technihadrons of non-integer spin) exist.

In case of G^t = SU(n) the lightest stable technihadrons are of the type $(Q^{(1)}\ Q^{(2)} \ldots Q^{(n)})$, i.e. for n = integer they are bosons and for n = non-integer they are fermions.

The electric charges of the U and D quarks depend on the assignments of the weak hypercharge, due to the relation

$$Q = T_3 + \frac{1}{2} Y . \tag{10.8}$$

Furthermore we have

$$Q(U) = +1 + Q(D).$$

There is no reason that the electric charges of U and D are 2/3 or - 1/3 respectively (if they would, the stable technihadrons would have, in general, nonintegral electric charges). One possible assignment of the electric charges is

$$Q(D) = -\frac{1}{n}$$
$$Q(U) = \frac{n-1}{n} . \tag{10.9}$$

In this case the charge of the (D D ... D)-particle is - 1, and the charge of the (U U ... U)-particle is (n - 1). We emphasize that each doublet of technicolor quarks causes the appearance of anomalies within the SU(2)xU(1) theory. If those are cancelled as usual in the sequential QFD by the lepton contributions, one needs in case of the charge assignment (10.9) n leptons of charge - 1.

10. CONCLUSION

After going through the various aspects of the mass problem in particle physics, we come to the following conclusions. Despite the impressive progress which has been made during the last ten years in particle physics the problem of mass and mass generation is still rather mysterious. At the moment there are two independent mechanisms to produce masses:

a) The dynamical mass generation due to the breakdown of perturbation theory, e.g. the mass generation in QCD in the absence of the quark masses.

b) The "kinematical" mass generation employing the mechanism of spontaneous symmetry breaking.

The idea of the technicolor quantum number or the mechanism to produce masses by radiative corrections allowed us to build a bridge between the two mechanisms discussed above. However in both cases only boson masses are related to each other; no predictions are made concerning the fermion masses. It seems not impossible to generalize the methods of dynamical symmetry breaking in order to calculate the fermion masses, although it is not clear at the moment how to proceed in this matter. Nevertheless we may speculate about the picture of particle physics which may emerge in this case. There would be a superheavy mass scale M (of the order of $\sim 10^{16}$ GeV), which sets the scale of all mass parameters in physics. The only parameter of the theory is the gauge coupling constant g(M), which describes the strength of the gauge coupling in the unified theory of all interactions except gravity. Besides the unification mass scale M there exist other mass scales m related to M by relations like

$$m = e^{-\frac{const}{g^2}} \quad .$$

Those are the mass scales of the weak boson masses, the fermion masses etc. In principle all mass ratios, e.g. m_l/m_q (l: lepton, q: quark), or m_l/Λ are calculable.

But even if it should be possible to carry out a program as described above it would still be unsatisfactory since it depends on one yet undetermined parameter, namely g(M). This parameter determines all other coupling constants in the theory, e.g. the finestructure constant α.

Besides the unification mass scale M there exists another mass scale in physics, namely the gravity mass scale given by the Plank mass $M^g = 1.22 \cdot 10^{19}$ GeV. As we have remarked earlier, there is some tendency that the unification mass M does not coincide with M^g (e.g. within the SU(5) approach one has $M \approx 10^{15}$ GeV $\approx 10^{-4} \cdot M^g$). However the estimates of M are model-dependent, and it cannot be excluded that M coincides with M^g.

One possible scheme which may emerge is the following one. One starts out with a unified theory of all interactions (including gravity). This theory has one intrinsic scale given by the Planck mass. By some yet unknown mechanism the gauge coupling constant renormalized at the Planck mass is fixed to be a specific number. Radiative corrections cause a spontaneous breaking of the gauge symmetry and the generation of gauge boson masses of the order of $M \sim 10^{15}$ GeV. Further radiative corrections cause the appearance of the other mass scales (M_W, Λ , ...). Such a theory would have only one scale parameter, namely the Planck mass. Since the scale is arbitrary, the theory has no free parameter. All dimensionless quantities are calculable, and all scale-dependent quantities can be calculated in terms of the Planck mass.

It is obvious that we are still far away from constructing a theory like the one sketched above, whose completion can be considered as the ultimate goal in physics and would imply the end of the development in fundamental theoretical physics. Nevertheless it is interesting to note that at the present time we have arrived at a stage, at which the construction of an ultimate theory of all interactions has become thinkable.

REFERENCES

1) For reviews and references see:
S. Weinberg, Proceedings of the Int. Conference on High Energy Physics, Tokyo 1978. (Physical Society of Japan, Tokyo 1979).

2) A.D. Linde, JETP Letters 23, 73 (1976);
S. Weinberg, Phys.Rev.Lett. 36, 294 (1976).

3) G. Altarelli, Proceedings of the Int. Conference on High Energy Physics, Tokyo 1978. (Physical Society of Japan, Tokyo (1979).

4. S. Coleman and E. Weinberg, Phys.Rev. 7D, 7 (1973).

5) For a review and references see:
J. Ellis, Proceedings of the Int. Universitätswochen Schladming (February 1979), CERN-preprint (June 1979).

6) For a review see e.g.:
H. Fritzsch, Acta Physica Austriaca, Suppl. XIX, 249 (1978).

7) For a review on the U(1) problem see e.g.:
R.J. Crewther, Acta Physica Austriaca, Suppl XIX (1978).

8) See e.g.:
S. Weinberg, Transactions of the New York Academy of Sciences, Series II, Vol. 38, p. 185 (1977);
H. Fritzsch, Acta Physica Austriaca, Suppl. XIX, 249 (1978).

9) S. Weinberg, unpublished;
P. Minkowski and A. Zepeda, Bern preprint (April 1979).

10) H.D. Politzer, Nucl. Phys. B117, 397 (1977).

11) See e.g.:
S. Weinberg, Transactions of the New York Academy of Sciences, Series II, Vol. 38, p. 185 (1977);
H. Fritzsch, Phys.Lett. 70B, 436 (1977);
F. Wilczek and A. Zee, Phys.Lett. 70B, 418 (1977).

12) H. Fritzsch, Phys.Lett. 73B, 317 (1978).

13) H. Georgi and D.V. Nanopoulos, Harvard University preprint HUTP-79/A001;
H. Fritzsch, CERN-preprint (March 1979)(to be published in Nucl.Phys.B).

14) H. Hagiwara, T. Kitazoe, G.B. Mainland and K. Tanaka, Phys.Lett. 76B, 602 (1978).

15) H. Georgi and S.L. Glashow, Phys.Rev.Lett. 32, 438 (1974);
J.C. Pati and A. Salam, Phys.Rev. D10, 275 (1974);
H. Fritzsch and P. Minkowski, Ann.Phys. N.Y. 93, 193 (1975).

16) H. Georgi, H.R. Quinn, and S. Weinberg, Phys.Rev.Lett. 33, 451 (1974).

17) D.A. Ross, Nucl.Phys. B140, 1 (1978);
T.J. Goldman and D.A. Ross, Caltech preprint CALT 68-704.

18) See e.g.:
E. Gildener, Phys.Rev. 14, 1667 (1976).

19) S. Weinberg, Harvard University preprint HUTP-78 A060;
See e.g. ref. 5) and:
A. Buras et al., Nucl.Phys. B135, 66 (1978).

20) See e.g.:
F. Reines and M.F. Crouch, Phys.Rev.Lett. 32, 493 (1974).

21) P. Ramond, private communication.

22) H. Fritzsch and P. Minkowski, Ann.Phys. N.Y. 93, 193 (1975);
H. Georgi, Particles and Fields, AIP, New York 1975, p. 575 (C.E. Carlson ed.).

23) See e.g.: J.C. Pati, Proceedings of the 19th Int. Conference on High Energy Physics, Tokyo 1978, p. 624.

24) F. Gürsey, P. Ramond, and P. Sikivie, Phys.Lett. 60B, 177 (1975);
Y. Achiman and B. Stech, Phys.Lett. 77B, 389 (1978);
P. Ramond, Caltech preprint CALT-68-709 (1979);
S. Weinberg, Harvard preprint

25) L. Susskind, SLAC-PUB-2142 (June 1978).

Figure captions

Fig. (1). The "observed" mass scales in physics (m_f denotes the fermion mass scale as observed thus far reaching from essentially zero (m_ν) to $m_b \sim 5$ GeV).

Fig. (2). The emission of a Higgs meson in the decay of a heavy $\bar{q}q$-resonance.

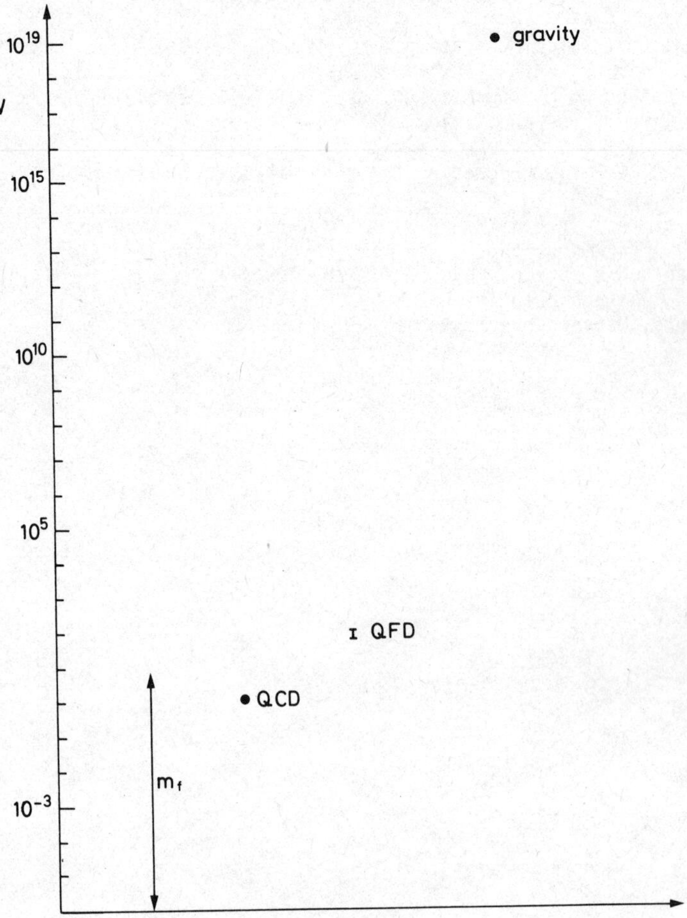

FIG.1

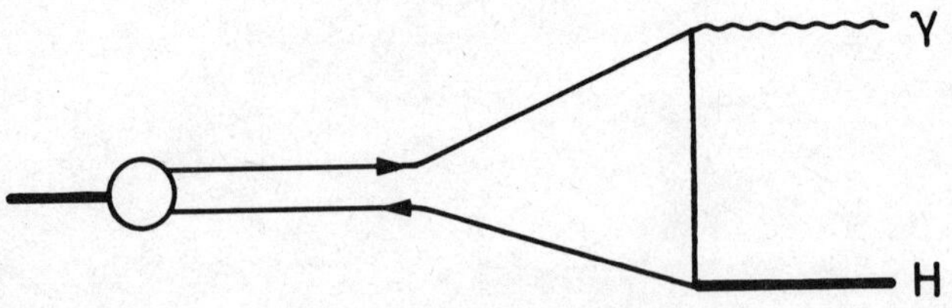

FIG.2

HADRONS

H. Leutwyler

Institute for theoretical Physics
University of Berne
Sidlerstrasse 5, CH-3012 Berne
Switzerland

Lectures given at the GIFT Seminar on Quantum Chromodynamics, Jaca (Spain), June 1979

CONTENTS

1. EQUIVALENCE PRINCIPLE OF GRAVITY [1)]

The archetype of a nonabelian gauge field is the gravitational field. The properties of gravity are well understood on the classical level. In this and in the following section I intend to review some of those features of gravity that are common to non-abelian gauge theories in general. In particular I would like to show that one of the crucial properties of QCD, viz. asymptotic freedom has its counter part in classical gravity even if perturbative quantum theory on which this notion is founded does not apply to gravity.

We know that it all emerged from the dust. The Lagrangian of dust reads

$$S = \int \frac{m}{2} \vec{v}^2 \, dt$$

for velocities small compared to the velocity of light (which in the following I put equal to 1 together with $\hbar$). For velocities comparable to c the correct action for a dust particle is

$$S = - m \int \sqrt{1 - \vec{v}^2} \, dt = - m \int ds \tag{1.1}$$

where

$$ds^2 = dt^2 - d\vec{x}^2 \tag{1.2}$$

is the Minkowski line element: the action is proportional to the Minkowski length of the world line (minimal action means maximal proper time).

The action is invariant under a group of symmetry transformations, viz. under Lorentz transformations

$$x' = \Lambda x + a \tag{1.3}$$

(Λ is a 4 x 4 Lorentz matrix and a is a translation). Let us apply the gauge principle to this symmetry group. To promote the symmetry to a gauge symmetry we have to allow the group element $\{\Lambda, a\}$ to depend on x, i.e. we have to consider general coordinate transformations

$$x' = f(x) \tag{1.4}$$

Under these "gauge" transformations the action (1.1) is not invariant as it stands.

To make it invariant we introduce a gauge field, the metric $g_{\mu\nu}(x)$. In an arbitrary curvi-linear coordinate system the Minkowski distance ds^2 is not simply given by the difference of squares (1.2) but by the general quadratic form

$$ds^2 = g_{\mu\nu}(x)dx^\mu dx^\nu = g_{oo}(dx^o)^2 + 2\, g_{o1}dx^o dx^1 + \ldots + g_{33}(dx^3)^2 \tag{1.5}$$

(Take e.g. polar coordinates for which $ds^2 = dt^2 - dr^2 - r^2 d\theta^2 - r^2 \sin^2\theta\, d\phi^2$). In an arbitrary coordinate frame the action may now be written as

$$S = - m \int ds = - m \int \sqrt{g_{\mu\nu}\, \dot{x}^\mu \dot{x}^\nu}\; dt; \; \dot{x}^\mu = \frac{dx^\mu}{dt} \tag{1.6}$$

This expression is invariant under the gauge transformation (1.4) provided we change the gauge field $g_{\mu\nu}(x)$ according to

$$g_{\mu\nu}(x) = \frac{\partial f^\rho}{\partial x^\mu} \frac{\partial f^\sigma}{\partial x^\nu} g'_{\rho\sigma}(x') \tag{1.7}$$

Up to this point we have only rewritten the action in a form that is valid in arbitrary coordinate frames - it still describes freely moving dust. To determine the motion in the coordinate frame at hand we have to specify the gauge field $g_{\mu\nu}(x)$ - by doing so we manifestly violate the gauge invariance we are looking for: the field $g_{\mu\nu}(x)$ is not gauge invariant. The only way to avoid violation of gauge invariance is not to specify the field $g_{\mu\nu}(x)$. Einstein has shown that this field indeed has a life of its own - it is the gravitational potential. Its values are not to be prescribed in advance, but are to be determined dynamically from the energy distribution of the dust. The metric obeys a second order differential equation roughly of the type

$$- \Box g_{\mu\nu} = 8\pi G\, T_{\mu\nu} \tag{1.8}$$

where $T_{\mu\nu}(x)$ is the energy-momentum density and G is Newton's constant. In contrast to an equation of the type $g_{\mu\nu}(x) = \ldots$ which necessarily destroys gauge invariance there do exist differential equations of the type (1.8) that are gauge invariant. In particular Einstein's equations

$$R_{\mu\nu} - \frac{1}{2} g_{\mu\nu} R = 8\pi G\, T_{\mu\nu} \tag{1.9}$$

are gauge invariant. The left hand side involves, in addition to the second derivatives $\sim \Box g_{\mu\nu}$ also terms proportional to the square of the first derivative - these non-linear terms are necessary to insure gauge invariance. Physically, these nonlinear terms arise as follows. The source of the gauge field $g_{\mu\nu}(x)$ is the energy density. Whoever carries energy contributes to this source, i.e. influences the time evolution of the metric. Like any other field the gravitational field itself carries energy; the corresponding energy density is proportional to the square of the field strength. The terms involving $(\partial g/\partial x)^2$ on the left hand side of (1.9) may be viewed as minus the gravitational part of the energy density. The equation is nonlinear because the gravitational field contributes to its own source - gravity couples to itself. This is a characteristic feature of nonabelian gauge fields (among which gravity plays a very special role). In general the source of a gauge field is referred to as a charge density (electric charge for the photon field, weak isospin for the gauge fields of the weak interaction, colour for the gluon fields of the strong interaction, energy for the gauge field of gravity). Whoever carries the corresponding charge influences the time evolution of the gauge field. Nonabelian gauge fields (weak gauge fields, gluons, gravity) carry their own charge, i.e. couple to themselves. Abelian gauge fields (photon) are not charged - they do not contribute to their own source. Their equations of motion (Maxwell's equations) are linear.

I return to the gauge theory of gravity. The equations of motion (1.9) may be embodied in the action principle, if the action (1.6) is modified as

$$S = - m \int ds - \frac{1}{16\pi G} \int R \sqrt{g}\, d^4x \qquad (1.10)$$

(I only book the contribution of a single dust particle). The quantity R is the curvature invariant of the gauge field $g_{\mu\nu}$. Variation of this action with respect to $g_{\mu\nu}$ leads to Einstein's equations. The equations of motion for the dust world line are

$$\delta \int ds = 0 \rightarrow \frac{d^2x^\mu}{ds^2} + \Gamma^\mu_{\rho\sigma} \frac{dx^\rho}{ds} \frac{dx^\sigma}{ds} = 0 \qquad (1.11)$$

where $\Gamma^\mu_{\rho\sigma}$ is the Christoffel symbol involving the first derivatives of $g_{\mu\nu}$. For slowly moving particles and weak gravitational fields

$$g_{oo}(x) = 1 + h_{oo}(x); \; |h_{oo}| << 1$$

(and similarly for the other components of $g_{\mu\nu}$) this becomes

$$\ddot{\vec{x}} + \frac{1}{2} \vec{\nabla} h_{oo} = 0$$

This equation of motion coincides with Newtons law if we identify the deviation h_{oo} of g_{oo} from unity with Newton's potential

$$g_{oo}(x) = 1 + 2\,V_{Newton}(x) \tag{1.12}$$

(This identification is confirmed by the corresponding approximation to Einstein's equations). The orbits of the planets are as straight as it is possible in the curved space-time around the sun: they follow lines of maximal length. The gauge field $g_{\mu\nu}(x)$ plays the role of the gravitational potential and is physically on the same level as the vector potential $A_\mu(x)$ of electrodynamics, or, more generally, on the same level as the gauge fields of QCD or QFD. The Christoffel symbol $\Gamma^\lambda_{\mu\nu}$ which collects first derivatives of $g_{\mu\nu}$ measures the acceleration (see (1.11)) - it is a force term to be compared with the field strength $F_{\mu\nu}$ of electrodynamics. Finally, the curvature $R^\kappa_{\lambda\mu\nu}$ is the derivative of the field strength (tidal force) and plays a role similar to the derivatives of the electric and magnetic fields $\vec{\nabla} \cdot \vec{E}$, $\vec{\nabla} \times \vec{B}$ that occur in Maxwell's equations.

2. WEAKENING OF GRAVITY AT SHORT DISTANCES

As a consequence of the fact that the gravitational field carries energy the energy distribution in the vicinity of the sun e.g. looks as follows

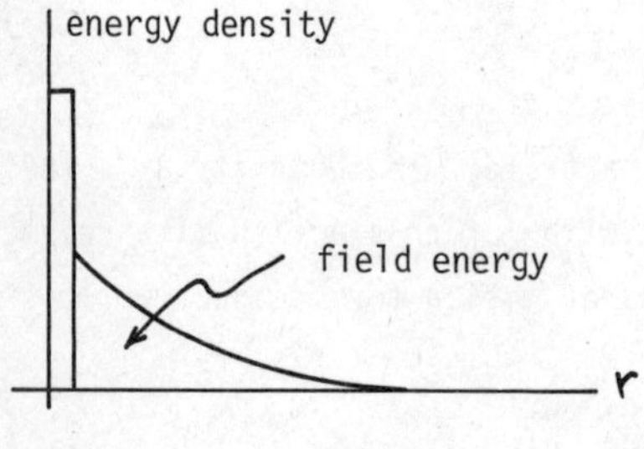

This implies that a test particle that probes the surroundings of the sun penetrates into a region where only part of the sun's total energy attracts it toward the center - part of the outer layers act in the opposite direction. The potential

$$V_{Newton} = -\frac{GMm}{r} \tag{2.1}$$

represents the interaction accurately only at large distances. In the vicinity of the sun Newton's law is modified; the effective value of the quantity GM becomes smaller than the asymptotic value known from Kepler's third law, GM = 1.475 km.

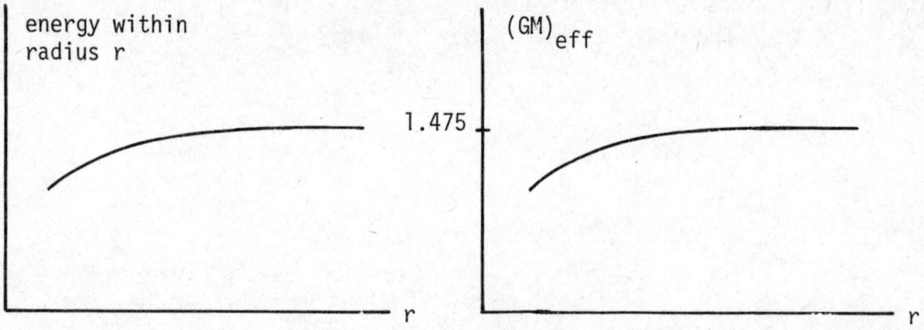

To evaluate the effect [2] of this weakening of the gravitational attraction at short distances on the orbits of the planets we make use of the fact that for a static, spherically symmetric object the space-time metric may by a suitable choice of the reference frame be brought to the form

$$ds^2 = G(r)^2\, dt^2 - H(r)^2(dx^2 + dy^2 + dz^2) \tag{2.2}$$

where $r^2 = x^2 + y^2 + z^2$; two gravitational potentials G, H suffice to describe the geometry in the vicinity of the sun (instead of 10 potentials $g_{\mu\nu}(x)$ needed in the general case). For distances large compared to the Schwarzschild radius r_o=2GM=2,95km these functions may be expanded in a power series

$$\begin{aligned} G^2(r) &= 1 - \alpha \frac{r_o}{r} + \frac{\beta}{2}\left(\frac{r_o}{r}\right)^2 + \dots \\ H^2(r) &= 1 + \gamma \frac{r_o}{r} + \dots \end{aligned} \tag{2.3}$$

(Einstein's field equations imply $\alpha = \beta = \gamma = 1$).

The planets move along geodesics

$$\delta \int ds = \delta \int dt\, G \{1 - \frac{H^2}{G^2} \dot{\vec{x}}^2\}^{1/2} = 0 \tag{2.4}$$

The velocity of a planet is small compared to c. For a circular orbit it is given by (recall that in our units c = 1):

$$\dot{\vec{x}}^2 = \frac{GM}{r} = \frac{r_o}{2r} \tag{2.5}$$

Even for elliptic orbits such as mercury's the quantity $\dot{\vec{x}}^2$ is of order r_o/r.

We expand the integrand in (2.4) up to terms of order $(\frac{r_o}{r})^2$:

$$\delta \int dt \{1 - L_1 - L_2 - L_3\} = 0$$

$$L_1 = \frac{\vec{x}^2}{2} + \frac{\alpha r_o}{2r} = O(\frac{r_o}{r})$$

$$L_2 = \frac{1}{8}\vec{x}^4 + \frac{1}{2}(\gamma + \frac{1}{2})\vec{x}^2 \frac{r_o}{r} - \frac{1}{4}(\beta - \frac{1}{2})(\frac{r_o}{r})^2 = O((\frac{r_o}{r})^2) \tag{2.6}$$

$$L_3 = O((\frac{r_o}{r})^3)$$

The Newtonian approximation retains only terms of order r_o/r. In this approximation the Lagrangian indeed reduces to the Lagrangian L_1 of Newtonian celestial mechanics, provided $\alpha = 1$ (this value of α has been used in the expression for L_2). The term L_2 which produces a distortion of the elliptic Newtonian orbits contains four different effects:

(1) Relativistic kinematics. The term proportional to $\vec{x}^4$ arises even in the relativistic treatment of classical electron orbits. It stems from the fact that the kinetic energy at velocity $\vec{x}$ is slightly smaller than $\frac{m}{2}\vec{x}^2$. Sommerfeld has evaluated the effect of this contribution on the orbits of the electron in the hydrogen atom.

(2) Space curvature. The constant γ measures the curvature of the three-dimensional space t = const. This curvature manifests itself in a tiny correction to the nonrelativistic kinetic energy term $\frac{1}{2}\vec{x}^2 \rightarrow \frac{1}{2}\vec{x}^2 (1 + \gamma\frac{r_o}{r})$.

(3) Red shift. The contribution of the classical kinetic energy to the action is further modified by the fact that a proper measurement of the velocity (and of the time interval in the action) involves a gravitational red shift factor

$$\int dt \frac{1}{2}\vec{x}^2 \rightarrow \int dt \frac{1}{2}\vec{x}^2 (1 + \frac{r_o}{2r})$$

(4) Weakening of the gravitational attraction at short distances. The effect we are looking for is the velocity independent piece of L_2 which indicates that the correct potential is not given by the Newtonian approximation $-\frac{GM}{r}$ but by $-\frac{GM}{r}\{1-(\beta-\frac{1}{2})\frac{GM}{r}\}$, to the accuracy at which we are working.

It is not difficult to calculate the distortion of the Newtonian orbits arising from each one of these contributions. Many textbooks on general relativity contain a derivation of the following formula for the shift of the perihelion due to the sum of the four terms contained in L_2:

$$\Delta\phi = \frac{3\pi r_o}{a(1-\varepsilon^2)} \frac{2-\beta+2\gamma}{3} \tag{2.7}$$

($\Delta\phi$ measures the angle by which the perihelion is displaced in one revolution of the planet, a is the semi-major axis, ε is the excentricity). According to Einstein's theory we have

$$\Delta\phi_E = \frac{3\pi r_o}{a(1-\varepsilon^2)}$$

For mercury $\Delta\phi_E$ amounts to 43 seconds of arc per century. Since the space curvature γ only affects the contribution (2) we may read off from (2.7) that

$$\Delta\phi_2 = \frac{2\gamma}{3} \Delta\phi_E$$

The red shift effect is of the same form ($\gamma \to \frac{1}{2}$):

$$\Delta\phi_3 = \frac{1}{3} \Delta\phi_E$$

The constant β directly measures the weakening of the gravitational attraction at short distances. We may read of its effect from (2.7):

$$\Delta\phi_4 = -\frac{1}{3}(\beta - \frac{1}{2}) \Delta\phi_E$$

The rest must come from Sommerfeld's term; hence

$$\Delta\phi_1 = \frac{1}{6} \Delta\phi_E$$

Inserting the values $\beta = \gamma = 1$ predicted by general relativity we thus get

$$\Delta\phi_1 = \frac{1}{6} \Delta\phi_E, \ \Delta\phi_2 = \frac{2}{3} \Delta\phi_E, \ \Delta\phi_3 = \frac{1}{3} \Delta\phi_E, \ \Delta\phi_4 = -\frac{1}{6} \Delta\phi_E \tag{2.8}$$

If there was no weakening of the gravitational attraction at short distances the shift of the perihelion would be given by $\Delta\phi_1 + \Delta\phi_2 + \Delta\phi_3 = \frac{7}{6} \Delta\phi_E$, i.e. by 50" per century. The weakening of the gravitational attraction at short distances reduces this value by $\frac{1}{6} \Delta\phi_E$, i.e. by about 7" per century. We close this discussion with two remarks:

(1) To relate the weakening of the gravitational attraction quantitatively to the energy of the field surrounding the sun one needs an expression for the local energy density of a gravitational field. Unfortunately the equivalence principle makes it impossible to localize this energy in an unambiguous manner.

(2) There is an important difference between spin 1 gauge theories such as QCD or QFD and the spin 2 gauge theory of gravity: renormalizability. For spin 1 gauge field theories the weakening of the interaction at short distances can be described in a general, quantitative manner within perturbation theory. For gravity this is not possible since perturbation theory does not work [3)]. We have to be content with the classical reasoning given above.

3. SYMMETRY OF THE DUST

We now ignore the gravitational interaction and start again with free dust. The dust of today's theories consists of quarks and leptons. We do not know why this dust occurs in so many different forms: e, ν_μ, μ, μ_μ, τ,ν_τ(?), u, d, c, s, t (?), b. The old puzzle of why there is a muon in addition to the electron remains puzzling; the standard theory of the weak, electromagnetic and strong interactions would work equally well and would even look more satisfactory if there were only one pair of leptons (e, ν_e) and one pair of quarks (u, d). What has however been achieved is a remarkable understanding of the glue that makes the dust stick together. Apparently, the only glue that nature makes use of is the gauge glue: all interactions appear to be mediated by gauge fields (perhaps with the exception of the interactions responsible for spontaneous breakdown of gauge symmetries which are as yet poorly understood). The properties of the gauge fields and of the interactions they mediate reflect symmetries of the dust - in fact these symmetries determine the interactions up to a few coupling constants. To discuss the symmetries of free dust I denote by $\psi^e, \psi^{\bar{e}}, \psi^{\nu_e}$, ψ^{u_R}, $\psi^{\bar{u}_R}$, ψ^{u_W}... the left-handed two-component fields associated with leptons and quarks (the two components of ψ^e describe left-handed electrons and right-handed positrons) and collect all these fields in the multicomponent spinor Ψ. In the absence of interactions these fields obey the equation of motion of a massless spin $\frac{1}{2}$ field:

$$\sigma^\mu \partial_\mu \psi(x) = 0 \tag{3.1}$$

The action of free dust therefore reads

$$S = i \int d^4x \, \psi^+ \sigma^\mu \partial_\mu \psi \tag{3.2}$$

It is important for the structure of the electroweak interactions that the Lagrangian does not contain any mass terms. (The fact that say the electron moves at less than the speed of light is not due to a mass term in the Lagrangian, but to a self-energy effect: the electron has to move through an asymmetric vacuum. The energy the electron needs to move at momentum $\vec{p}$ turns out to be larger than the bare energy $E = |\vec{p}|\, c$, see section 8).

The Lagrangian (3.2) has a high degree of symmetry: in the absence of any interaction the various quarks and leptons are indistinguishable. The tranformation

$$\psi'(x) = U\psi(x) \tag{3.3}$$

where U is a unitary matrix that mixes the various components of ψ leaves S invariant. (Note that U has to commute with σ^μ, i.e. must transform the two components of each one of the fermions in the same manner). If the Lagrangian contains N two component fields then it is symmetric under the group U(N). For the standard set of fermions we have 6 quarks, each in 3 colours (i.e. 36 left-handed fields) plus 3 charged leptons (6 ℓ.h. fields) and 3 neutrinos (3 ℓ.h. fields): N = 45. The global symmetry group of free dust is U(45), a huge group with N^2 = 2025 parameters !

Gauge field geometry states that any continuous global symmetry of the free dust Lagangian may be promoted to a local symmetry by introducing a suitable set of gauge fields. A symmetry group with p parameters calls for p gauge fields. In the standard model only the subgroup SU(3)xSU(2)xU(1) of U(N) is a local symmetry: there is evidence only for 12 gauge fields (8 gluons associated with $SU(3)_{colour}$, plus 3 vector mesons $W^{\pm}$, Z and the photon which gauge the group $SU(2)xU(1)_{flavour}$). It is not clear at this time whether nature makes use of any of the 2025-12 other gauge fields that the geometry of the Lagrangian describing free dust offers. (Gauge fields that happen to acquire a lot of self-energy after symmetry-breakdown may be difficult to observe. Whether they are there and are very heavy or are not there at all may make little difference in the low energy region explorable today). In the following I will only discuss the standard model and first describe some qualitative features of QCD, the gauge field theory of $SU(3)_{colour}$.

4. EQUIVALENCE PRINCIPLE OF THE STRONG INTERACTIONS

In the standard model the strong interactions are associated with the colour symmetry of the dust. The fermions that are unaffected by this symmetry are called leptons, those that transform according to the representation 3 (3*) are called quarks (antiquarks) - recall that we are using separate two component fields for particles and antiparticles. There is no evidence for dust that would transform according to a higher dimensional representation of SU(3), i.e. dust seems to be either neutral, red, green, blue, antired, antigreen or antiblue. The equivalence principle of the strong interactions states that colour is not only a global symmetry of the Lagrangian, but is a local symmetry: it is impossible to distinguish a red quark from a blue one even if one was able to deposit a red quark for reference in Paris.

x	x		x	x
		equivalent to		
u_B	u_R		u_R	u_R

The free Lagrangian does not have this property. The equivalence principle can only hold if the fermions interact (compare equivalence principle of gravity). In fact the interaction has to have very specific properties. If the transformation $u_R \underset{\bar{B}R}{\rightarrow} u_B$ is to be a local equivalence transformation then there must exist a gauge field $B^{\bar{B}R}$ that couples to this transition

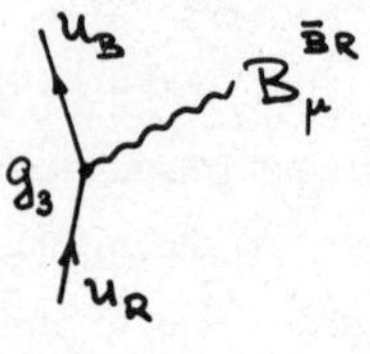

SU(3) requires 8 gauge fields interacting in this manner. The coupling of these gauge fields (gluons) is specified by the equivalence principle: every field that carries colour [i.e. transforms in a nontrivial manner under SU(3)] emits and absorbs gluons with a universal strength. Like gravity the theory is characterized by a single coupling constant which I denote by g_3.

As a consequence, quarks interact with one another by exchanging gluons. Leptons do not participate in this interaction since they do not carry colour and are therefore unable to emit or to absorb gluons. The gluons carry colour themselves. Hence they act as their own source:

in complete analogy with gravity. The equations of motion for the gluons must contain self-interaction, i.e. are non-linear. The gluon cloud that surrounds a quark carries colour. As it is the case for gravity the cloud amplifies the effect of the source. If the quark is red the cloud will on average also be red. The bare colour of the quark is smaller than the total colour of the object. A test particle that penetrates the cloud will sense a smaller interaction if it comes close to the bare quark than if it passes at large impact parameter

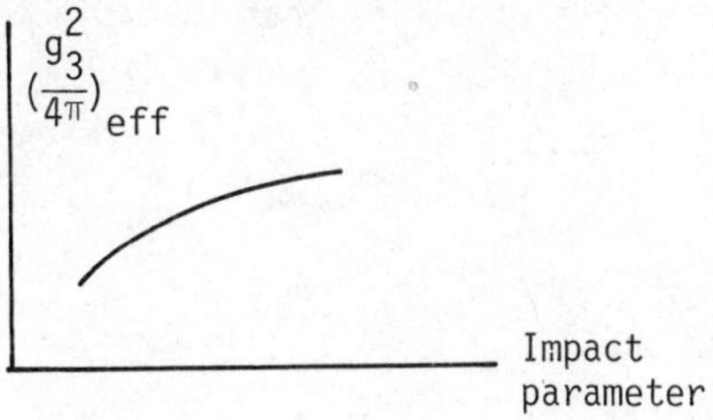

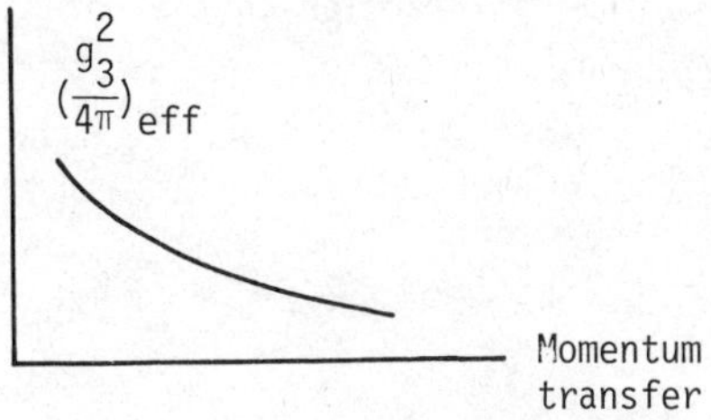

In "normal" field theories like QED the opposite is true: if a bare electron is put into the vacuum it surrounds itself with a cloud that predominantly contains positrons: the vacuum shields the charge of the bare particle

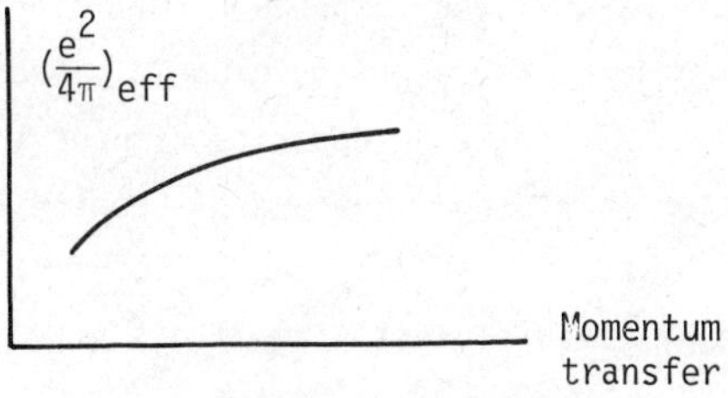

(The photons contained in the cloud do not carry charge; their presence does not affect the charge distribution and therefore does not show up in the effective strength of the interaction between a test particle and the electron.)

There is a shielding effect due to the quarks contained in the cloud also in QCD. In fact if there are too many ($>$ 16) quark flavours then the shielding effect even overcomes the amplification due to the gluons.

In contrast to gravity QCD is a renormalizable theory. The dependence of the coupling constant on momentum transfer can be described quantitatively within perturbation theory. The coupling constant of QCD is renormalized by the interaction which it generates. It does not make sense to give a value to the strength g_3 of the interaction unless one specifies the momenta of the particles that participate in the interaction. This problem arises in any renormalizable theory. The coupling constant $\alpha = e^2/4\pi$ of QED e.g. is subject to a similar renormalization. In that case one may however adopt a convenient physical renormalization prescription by identifying α with the value of a physical cross section like compton scattering at threshold. This procedure does not work in QCD since one is not able to quantitatively relate physically accessible low energy parameters to the coupling constant of the theory. At this time quantitative statements are only possible about high energy processes - for these there is no natural value of the momenta at which to specify g_3. This problem is not a serious one because one has control over the dependence of the strength of the interaction on the momentum q of the particles that participate in this interaction:

$$\frac{g_3^2}{4\pi} = \frac{12\,\pi}{N_3 \ln \dfrac{q^2}{(\Lambda_3)^2}} \tag{4.1}$$

The number N_3 stands for

$$N_3 = 33 - 2\,N_q$$

where N_q is the number of quark flavours that are sufficiently light to be produced by an interaction at momentum q. For a purely gluonic cloud $N_3 = 33$; if u, d and s quarks contribute then $N_3 = 27$. Note the opposite sign of the gluon and quark contributions to N_3 (amplification/shielding). Instead of specifying the value of g_3 at some given momentum one may equivalently specify the value of Λ_3. (Formula (4.1) is valid for sufficiently large q, i.e. sufficiently small g_3).

There is at least a qualitative test of the validity of (4.1). It predicts that for deep inelastic electron scattering on protons the interaction between the quarks should become negligible if one probes with sufficiently high momentum transfer. The quarks should then behave as if they were free (asymptotic freedom, Bjorken scaling) with small and calculable "radiative" corrections due to the interaction mediated by gluons. The qualitative prediction concerning the way in which these gluon corrections to free quark behaviour should show up is verified by the data. One may even extract a rough measurement of the value of Λ_3 from these data: $\Lambda_3 \simeq 500 \pm 200$ MeV.

The corresponding values of $g_3^2/4\pi$ are plotted as a function of q in Fig. 1. For $q = 3$ GeV, $N_q = 3$ we have $g_3^2/4\pi = 0.4$; the value then very slowly decreases with growing momentum to reach $g_3^2/4\pi \simeq 1/45$ at $q \simeq 10^{16}$ GeV (allowing for 6 quarks). If we wanted the strong interactions to become as weak as the electromagnetic ones are at low energy $g_3^2/4\pi = 1/137$, we would have to go to outrageous values of q of the order of 10^{53} GeV !

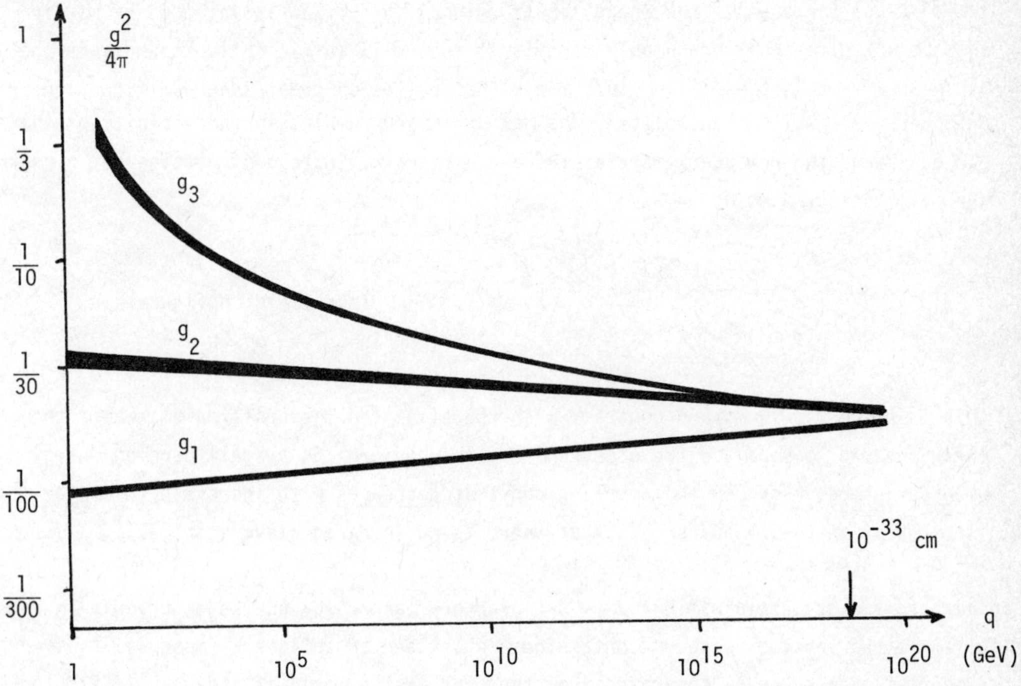

Fig. 1: Coupling constants of SU(3)xSU(2)xU(1) as a function of the momentum scale. The width of the bands indicates the uncertainties in Λ_3 and in $\sin^2\theta_W$.

5. CONFINEMENT

The same property of QCD that explains why the quarks in a proton appear to behave as if there was essentially no binding - provided one only probes the proton with sufficiently high momentum transfer - may also explain why quarks have not been seen as isolated objects. The formula (4.1) shows that near $q \simeq \Lambda_3$ the coupling constant should become very large. Of course the perturbation calculation on which this formula is based then becomes invalid. The perturbation theory picture that takes free quarks as a zero order approximation to the real world miserably falls at low energies. This should be so if quarks do not occur as isolated on-shell particles. Perturbation theory itself suggests that the force between quarks becomes strong if they are far apart - the assumption that this property of perturbation theory is indeed a property of QCD is referred to as <u>infrared slavery</u>. The picture associated with this notion is that if one attempts to separate the quark from the antiquark in a meson then one has to build up a gluon field that requires more and more energy as the distance increases - the energy in the gluon field of an isolated quark may in fact be infinite. What presumably happens if one pulls sufficiently strongly on the two quarks is that the strong gluon field that builds up creates an additional quark-antiquark pair which then splits - the new antiquark remains in the vicinity of the old quark and the new quark marries the old antiquark. Instead of getting out a quark one will get out a meson.

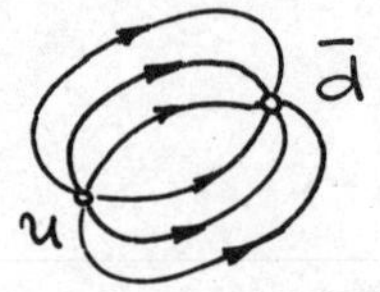

Gluon field configuration in the π^+ meson

This is of course in marked contrast with the situation for positronium, where the energy needed to separate the electron from the positron is a small binding energy of order $\alpha^2 m_e c^2$. For QED the coupling constant <u>decreases</u> with increasing distance - perturbation theory guarantees that there is no infrared slavery in QED.

Infrared slavery explains in a very satisfactory manner why quarks occur only in bags of three (baryons) or in bags containing a quark and an antiquark (mesons). A given number of quarks can be separated from the rest of the world if and only if its total colour is zero such that there is no long range gluon field to be built up. Configurations which are colour neutral are invariant under colour transformations, i.e. are colour singlets. The simplest colour singlets that we can construct out of the quark fields (which transform according to the triplet representation of $SU(3)_{colour}$) are states containing a quark-antiquark pair (symmetric state with equal probabilities for a red-antired, green-antigreen and blue-antiblue pair) or states involving three quarks (one of each colour). A state containing a single quark or two quarks is not colour neutral and therefore requires a long range gluon field. A state containing four quarks and one antiquark also gives rise to a singlet configuration; it may be interpreted as a two particle state containing one baryon and one meson.

The same property of QCD that explains free quark behaviour at high energies (scaling) thus also gives a natural (although at this time only qualitative) explanation of <u>quark confinement</u>. A quantitative solution of the infrared problem in QCD should allow us to calculate the size of a meson or a baryon, to determine the energy of the ground state, the excitation energies (slope of Regge trajectory) etc. in terms of the single parameter Λ_3 that characterizes the strength of the strong interaction. We will come back to this problem in section 10.

6. FLAVOUR AND CHIRAL SYMMETRY

Once the strong interactions are switched on the symmetry group U(N) of free dust ceases to be a symmetry of the Lagrangian. QCD distinguishes between leptons, quarks and antiquarks. It does not hoever distinguish the different flavours of the quarks as it affects u, d, c, s, ... in exactly the same manner. We denote the set of all elements of U(N) that commute with the colour subgroup as the flavour group. This definition implies that the flavour generators do not carry colour - flavour transformations must take leptons into leptons and cannot mix quark fields with antiquark fields. For N_ℓ two component lepton fields, N_q two component quark fields and N_q antiquark fields the flavour group has the structure $U(N_\ell)xU(N_q)xU(N_q)$. The factor $U(N_\ell)$ represents the symmetry group of the leptons which are described by free massless fields at this stage. The chiral factor $U(N_q)xU(N_q)$ acts on the quark fields; it allows to independently transform the lefthanded quarks among themselves and the righthanded quarks among themselves (the righthanded antiquarks follow the lefthanded quarks).

Formally, the QCD Lagrangian is invariant under the full flavour group. Accordingly the N_q^2 vector currents $V_\mu^i(x)$ and the N_q^2 axial currents $A_\mu^i(x)$ associated with the chiral group are formally conserved. In our notation the vector current $V_\mu^{1+i2}(x)$ e.g. is given by the sum $\psi_u^+ \sigma_\mu \psi_d + \psi_{\bar{d}}^+ \sigma_\mu \psi_{\bar{u}}$, the axial current $A_\mu^{1+i2}(x)$ by the difference $\psi_u^+ \sigma_\mu \psi_d - \psi_{\bar{d}}^+ \sigma_\mu \psi_{\bar{u}}$; a trace over the colour indices of the quark fields is understood. It is however dangerous to draw conclusions from formal manipulations with a Lagrangian that requires counter terms, renormalization and all that. An analysis of the Ward identities satisfied by the vector and axial currents in fact reveals that the flavour singlet axial current has an anomaly and fails to be conserved. The true global symmetry of the quark-gluon Lagrangian is not $U(N_q)xU(N_q)$, but $SU(N_q)xSU(N_q)xU(1)$. There are N_q^2 conserved vector charges:

$$T_i^V = \int d^3x\, V_i^o(x); \; [T_i^V, H_{QCD}] = 0; \; i = 1, \ldots, N_q^2 \tag{6.1}$$

but only $N_q^2 - 1$ conserved axial charges

$$T_i^A = \int d^3x\, A_i^o(x); \; [T_i^A, H_{QCD}] = 0; \; 1 = 1, \ldots, N_q^2 - 1 \tag{6.2}$$

Note that quark mass terms in the Lagrangian would break this symmetry. If all quarks would e.g. be given the same mass term, the symmetry group would reduce to $U(N_q)$, if they were given different masses, only U(1)xU(1)x... would be left. It is a crucial feature of the standard model that the quark masses are not put into the Lagrangian by hand, but arise as a self-energy effect due to the electroweak interactions. The violations of exact flavour symmetry that we observe in the particle spectrum are due to QFD, not to QCD. Since we did not yet switch on the electroweak interactions we may at this stage enjoy the full chiral symmetry of massless quarks.

This high degree of symmetry has the following consequences for the hadron spectrum. Since the vector charges commute with the Hamiltonian, the state $T_i^V\,|f>$ has the same energy as the state $|f>$. The spectrum therefore consists of degenerate multiplets of $SU(N_q)$. In the three quark sector the ground state multiplet consists of P, N, Λ, Σ, Ξ, some charmed baryons and baryons containing the quark responsible for the upsilon - all of them at the same mass at this stage. In the quark-antiquark sector the pseudoscalar multiplet is lowest in energy: π, K, η, D, F, η_c .... As mentioned above the fact that these multiplets are not exactly degenerate in mass is the merit of QFD, not of QCD. In fact there is more symmetry. The axial charges commute with the Hamiltonian, too: $T_i^A|f>$ has the same energy as $|f>$. Since the axial charges carry negative parity we should find a state of opposite parity, mass degenerate with any given state. There seems to be no trace of this symmetry in the spectrum of the baryons and mesons ! Nambu and Jona-Lasinio have given a beautiful explanation for this apparent discrepancy between theory and experiment: they have pointed out that for some dynamical reason the lowest state of the theory, the vacuum, may happen not to be symmetric under T_i^A:

$$T_i^A\,|0> \neq 0$$

This implies that we must have $N_q^2 - 1$ states with the quantum numbers of $T_i^A\,|0>$: no energy, no momentum, no angular momentum and negative parity (spontaneously broken symmetry, "Goldstone-Theorem"). Indeed the 8 lightest hadrons (π, K,η) are 0^- particles. If we anyway have to have them at equal mass it is as good an approximation to the true situation if we identify them (and their charmed, top and bottom partners) with the Goldstone bosons of the spontaneously broken symmetry generated by the axial charges. Then there is no need for a parity partner of the proton: the state $T_i^A|P>$ simply consists of a proton and a pseudoscalar meson knocked out from the vacuum by T_i^A.

It still remains to be clarified why the symmetric Fock space vacuum that perturbation theory uses as a starting point to solve QCD is not stable; states with lower energy than the dressed perturbation theory vacuum remain yet to be found. Since confinement already suggests quite a drastic failure of perturbation theory in the low energy region it should however not come as a surprise that the structure of the state which is in the extreme infrared, the vacuum, differs from what perturbation theory would suggest.

To summarize the picture at this stage: The quarks and gluons form colourless droplets. Droplets that differ only in the flavour of the quarks that they contain, not in their state of internal motion, have the same mass. There are $N_q^2 - 1$ massless pseudoscalar mesons. The leptons are still moving around freely waiting for flavour to be appreciated. The quantity Λ_3 which determines the QCD coupling constant is the only parameter of the theory.

7. EQUIVALENCE PRINCIPLE OF THE ELECTROWEAK INTERACTIONS

As described in the last section the QCD Lagrangian possesses a global flavour symmetry group with the structure $U(N_\ell)xSU(N_q)xSU(N_q)xU(1)$. The standard model is based on the assumption that a particular subgroup of this symmetry, a subgroup with the structure SU(2)xU(1) is indeed not only a global, but a local symmetry. The SU(2) factor acts exclusively on the lefthanded dust particles (and on the righthanded antiparticles); it leaves the righthanded particles untouched. A lefthanded up-quark e.g. may be transformed into a lefthanded downquark and vice versa ($c \leftrightarrow s$, $t \leftrightarrow b$, $e \leftrightarrow \nu_e$, ...). The U(1) factor only affects the phases of the various fields without changing their flavour. The equivalence principle of the electroweak interactions thus states that what is called left-up in Peking may be called left-down in Moscow; what is however considered right-up in Santiago is called right-up everywhere. Again the equivalence principle fixes the couplings uniquely in terms of the two coupling constants g_2 and g_1 associated with SU(2) and U(1) respectively. At low energies the values of these coupling constants are

$$\frac{g_2^2}{4\pi} = \frac{1}{29} \qquad \frac{g_1^2}{4\pi} = \frac{1}{108}$$

(The sum of the two denominators is the inverse fine structure constant, the ratio $29/137 = 0.21$ is $\sin^2\theta_W$).

Since the 3 gauge fields of SU(2) carry the charge to which they are coupled (non-abelian gauge field theory), the vacuum amplifies this part of the electroweak interaction. Perturbation theory describes this amplification through a formula analogous to (4.1):

$$\frac{g_2^2}{4\pi} = \frac{12\pi}{N_2 \ln \frac{q^2}{\Lambda_2^2}}$$

with $N_2 = 22$ if there is no shielding from quarks and leptons and $N_2 = 22 - 4G$ if G generations of fermions participate in the cloud, each generation consisting of two quark triplets, a charged lepton and a lefthanded neutrino. (First generation: u, d, e, ν_e; second generation: c, s, μ, ν_μ; third generation: t, b, τ, ν_τ). The dependence of $g_2^2/4\pi$ on momentum is depicted in Fig. 1 for 3 generations of fermions, G = 3. The value of Λ_2 may be read off from $g_2^2/4\pi \simeq 1/29$ for $q \simeq 10$ GeV (the energy has to be sufficiently large to have all three generations participate in the cloud): $\Lambda_2 \simeq 10^{-23}$ GeV. Since there are only 3 gauge fields in SU(2) as compared to 8 in SU(3) the amplification effect is less pronounced here. This is responsible for the fact that the two curves cross at a momentum of the order of 10^{16} GeV.

The strength of the U(1) coupling is not amplified by the vacuum, because the corresponding gauge field does not carry the charge to which it is coupled (abelian gauge field theory like QED). In this case only the shielding effect due to the fermions in the cloud of a bare charge is present. Perturbation theory describes this effect as

$$\frac{g_1^2}{4\pi} = \frac{12\pi}{N_1 \ln \frac{q^2}{\Lambda_1^2}}$$

With $N_1 = 0 - \frac{5}{3} \times 4$ G. The dependence of $g_1^2/4\pi$ on q is also shown in Fig. 1 for G = 3. The value $\frac{g_1^2}{4\pi} = \frac{1}{108}$ for q = 10 GeV implies $\Lambda_1 \simeq 10^{45}$ GeV.

In contrast to g_2 and g_3 the normalization of the U(1) coupling constant g_1 is not fixed by the geometry. In unified theories for which the fermions are indeed organized in generations with the content described above a more natural normalization is to replace $\frac{g_1^2}{4\pi}$ by $\frac{g_1^{*2}}{4\pi} = \frac{5}{3}\frac{g_1^2}{4\pi}$. (Compare the values of N_1, N_2, N_3: $N_1 = -\frac{5}{3} \cdot 4$ G; $N_2 = 22 - 4$ G, $N_3 = 33 - 4$ G). It is very interesting to observe that this renormalization of g_1 shifts the U(1) curve up in Fig. 1 such that the three coupling constants do obtain roughly the same strength $\frac{g^2}{4\pi} \simeq \frac{1}{45}$ at $q \simeq 10^{16}$ GeV. This may be a sign of a common origin of the three interactions. Note that the unification scale is not very far from the Planck mass $M_P = (\hbar c/G_{New})^{1/2} = 1{,}22 \cdot 10^{19}$ GeV, which characterizes the scale of gravity.

8. SPONTANEOUS SYMMETRY BREAKDOWN

Despite the deep relationship between QCD and QFD there is a very crucial difference: colour seems to be confined, flavour certainly is not. In the standard model this fundamental difference is associated with the symmetry properties of the vacuum. The Lagrangian of QCD+QFD is invariant under SU(3)xSU(2)xU(1), the state of lowest energy only has the symmetry SU(3)xU(1).

There are many examples in solid state theory which illustrate the fact that the ground state of a system need not have all the symmetries of the Hamiltonian. If a symmetry of the Hamiltonian is not a symmetry of the ground state (spontaneously broken symmetry) then it is very difficult to see a trace of that symmetry in the phenomenological properties of the system. (Example: magnet, the symmetry under rotations is lost in the spontaneously magnetized state - measurable quantities depend on the direction of the magnetization). The hidden symmetry only manifests itself indirectly in the appearance of modes for which there is no restoring force (in the example given long wave length spin waves have zero frequency).

It is crucial for QFD to be a realistic theory that the local SU(2)-symmetry is broken spontaneously by the vacuum. If $SU(2)_{flavour}$ was as perfect a symmetry as we think $SU(3)_{colour}$ is, then:

- the weak interaction would have long range like the strong interaction between two colours or the electromagnetic interaction between two charges (gauge invariance in QED requires the photon to be massless, i.e. the interaction to have long range);
- quarks and leptons would be strictly massless ($SU(2)_{flavour}$ only affects the left-handed electron; a mass term produces left-right transitions and hence violates this symmetry).

There are specific models involving Higgs fields which demonstrate that what is known to happen in solid state theory may also happen in a relativistic quantum field theory. In these models there indeed are asymmetric states with lower energy than the symmetric Fock space vacuum. In these models a W-particle moving through the asymmetric vacuum surrounds itself with a cloud that senses the asymmetry - it picks up self-energy and moves at $v < c$. In fact every field occurring in the Lagrangian acquires a mass, unless it has some special excuse not to do so. The point is that in field theory the interactions have to be arranged in a very special manner if one wants to prevent the fields from picking up self-energies, not if one wants them to become massive. The chiral symmetry of QFD provides for such a special arrangement only if the vacuum is symmetric. Among the four gauge fields of SU(2)xU(1) only the photon is excused: for this particular field the U(1)-symmetry of the ground state (gauge invariance of QED) guarantees that it does remain massless. The quarks and lepton fields are exposed to the same self-energy effects. It is a very satisfactory feature of the standard model that it does not ascribe the actual values of the quark and lepton masses to terms that have to be inserted into the Lagrangian by hand, but blames them on the asymmetries of the vacuum. These asymmetries depend on the details of the spontaneous symmetry breakdown. There is no reason why the various fermions should get the same mass. We need a better understanding of the way in which the symmetry is broken to calculate these masses in terms of the basic constants appearing in the Lagrangian.

9. QUARK MASSES

The propagation properties of the fermion fields may be characterized by a general mass term of the form

$$\psi^T \varepsilon m \psi + \text{h.c.} \tag{9.1}$$

(ε is the antisymmetric 2x2 matrix, $\varepsilon = i\sigma_2$). Since the vacuum is assumed to be SU(3)-symmetric, the mass matrix m does not connect quarks with leptons. Furthermore, U(1)-symmetry implies that quarks of charge $+\frac{2}{3}$ $(-\frac{1}{3})$ are only connected with antiquarks of charge $-\frac{2}{3}$ $(+\frac{1}{3})$. In the quark sector the mass term is therefore described by two 3x3 matrices:

$$\begin{pmatrix} m_{\bar{u}u} & m_{\bar{u}c} & m_{\bar{u}t} \\ m_{\bar{c}u} & m_{\bar{c}c} & m_{\bar{c}t} \\ m_{\bar{t}u} & m_{\bar{t}c} & m_{\bar{t}t} \end{pmatrix} , \quad \begin{pmatrix} m_{\bar{d}d} & m_{\bar{d}s} & m_{\bar{d}b} \\ m_{\bar{s}d} & m_{\bar{s}s} & m_{\bar{s}b} \\ m_{\bar{b}d} & m_{\bar{b}s} & m_{\bar{b}b} \end{pmatrix}$$

The mass matrix allows us to distinguish the three quarks of charge $+\frac{2}{3}$ from one another: u, c and t are the eigenstates of the first one of these two matrices. With an appropriate change of basis the mass term may be written as

$$m_u \; \psi_{\bar{u}}^T \; \varepsilon \; \psi_u + m_d \; \psi_{\bar{d}}^T \; \varepsilon \; \psi_d + \ldots + \text{h.c.}$$

or, equivalently, in the more familiar notation which collects the two component fields $(\psi_u(x), \; \varepsilon\, \psi_{\bar{u}}^*(x))$ in the four component spinor u(x):

$$m_u \bar{u}u + m_d \bar{d}d + m_c \bar{c}c + m_s \bar{s}s + m_t \bar{t}t + m_b \bar{b}b \qquad (9.2)$$

Note that this diagonalization interferes with the weak couplings. In the new basis the emission of a W^+ by an up-quark will not always produce a down-quark, but some linear combination of d, s, b described by Cabibbo angles (the mass matrix determines these mixing angles).

I emphasize that although the origin of the fermion masses is assumed to be in the weak and electromagnetic interactions it is by no means necessary that the quark and lepton masses are small compared to the scale Λ_3 of the strong interactions. Spontaneous breakdown involves new mass scales whose magnitude depends on the mechanism responsible for the breakdown. (In a Higgs model the parameters characterizing the Higgs Lagrangian are relevant.) The weak and electromagnetic interactions lead to two classes of effects in the world of hadrons:

- there is a small direct perturbation due to weak boson exchange. This direct effect is responsible for the weak and electromagnetic decays of hadrons, for the electromagnetic self energy of the proton etc.
- there is a much more sizeable indirect effect on the spectrum of the baryons and mesons: the quark masses. To a very good approximation one may neglect the direct effects of the weak and electromagnetic interactions on the hadron spectrum. Only if we also neglect the indirect effects, viz. the quark masses, then we obtain the very coarse picture described in section 6: massless pseudoscalar mesons, no mass difference between P, N, Λ, Σ, Ξ, ... etc.

As far as the strong interaction goes the net effect of spontaneously broken QFD is that it supplies the QCD Lagrangian with a quark mass term. The values of the quark masses m_u, m_d, m_s, m_c, ... reflect the mechanism of symmetry breakdown for which we only have provisional models. Up-quark is the name of the lightest quark with charge $+\frac{2}{3}$, c is the name of the next heavier one with this charge etc. It is plausible that if one gives the quarks a mass the states containing quarks become more heavy. The heavier the quark the heavier the meson or baryon that contains it: π, P, N should become the lightest mesons and baryons because they only contain the quarks u and d and are in the ground state of internal motion. In the limit of very large quark masses we expect that the mass of a baryon approaches the sum $m_1 + m_2 + m_3$ of the masses of its quarks - the effect of the gluon field energy and internal motion should be comparatively small, of relative order v^2/c^2. In the opposite extreme, if we neglect the quark masses, then the baryon ground state will be a degenerate state of mass M_0, the partners of the proton differing from it only in the flavour of the quarks then get the same mass. A simple formula that interpolates between these two extremes is

$$M = M_o + m_1 + m_2 + m_3 \tag{9.3}$$

There is a slightly more learned justification of this simple formula for the dependence of a baryon on the mass of its quarks. The quark mass term may be considered as a perturbation of the QCD Lagrangian. Writing $H = H_o + H_1$ with H_1 representing the quark mass term, the first order shift in the energy levels is given by

$$E = \langle H_1 \rangle = \int d^3x \; \langle m_u \bar{u}u + m_d \bar{d}d + m_s \bar{s}s + \ldots \rangle \tag{9.4}$$

If we assume that the quarks move around nonrelativistically, there should be only a small difference between $\bar{u}u$ and $\bar{u}\gamma^o u$. The value of the integral $\int d^3x \bar{u}\gamma^o u$ is however known: it counts the number of up-quarks (minus the number of anti-up-quarks) in the state in question. The energy shift therefore becomes $\Delta E = m_1 + m_2 + m_3$. Adding this to the expectation value of H_o leads to (9.3).

It is a simple matter to check whether the formula (9.3) works. It predicts

$$\begin{aligned} M_P &= M_o + 2\, m_u + m_d \\ M_N &= M_o + m_u + 2\, m_d \\ M_\Lambda &= M_{\Sigma o} = M_o + m_u + m_d + m_s \\ M_{\Xi^-} &= M_o + m_d + 2\, m_s \end{aligned} \tag{9.5}$$

etc. The formula thus fails to explain the mass difference between $\Lambda(1115)$ and $\Sigma^o(1192)$ - this deficiency is an indication of the order of magnitude by which it fails in general. To estimate $m_s - m_u$ from the baryon spectrum we use the masses of Ξ^- and P:

$$m_s - m_u = \frac{1}{2}\,(M_{\Xi^-} - M_P) = 190 \text{ MeV}$$

The difference between m_u and m_d must be smaller. To explain why the neutron is somewhat heavier than the proton we need

$$m_d - m_u \simeq 2 \text{ MeV}$$

(This difference is reduced by the direct effects of the electromagnetic interaction which are however too weak to invert the sign of the proton-neutron mass difference.)

The formula (9.5) may also be applied (with a different value for M_o) to the decuplet of baryon resonances. Here it works perfectly (the familiar equal spacing rule) and gives $m_s - m_u$ = 145 MeV. For the mesons the analogous formula reads

$$M = M_o + m_1 + m_2 \qquad (9.6)$$

Applying it to the vector mesons ρ, ω, K* and ϕ it again works quite well, however with a somewhat smaller value for $m_s - m_u$:

$$m_s - m_u = 120 \text{ MeV}$$

From the mass of the ψ we get the estimate

$$m_c - m_u = \frac{1}{2}(M_\psi - M_\rho) = 1160 \text{ MeV}$$

With this input we may then estimate the mass of the D* and F* mesons (consisting of $c\bar{u}$ and $c\bar{s}$ respectively). We get M_{D*} = 1940 MeV to be compared with the experimental value 2006 MeV and M_{F*} = 2060 MeV to be compared with (2140 $\pm$ 60) MeV. In view of the coarse interpolation between the relativistic regime (ρ meson) and the non-relativistic regime (ψ) the agreement is better than we have any right to expect.

The formula (9.6) does not work at all for the pseudoscalar mesons - as indeed it should not. The pseudoscalar mesons play a special role - they become massless Goldstone bosons if the quark masses tend to zero: the quantity M_o vanishes. The formula (9.6) would imply that the mass of the pion is simply given by the sum of the masses of its quarks with no effect from binding. There is a general argument that for small quark masses the mass of the pseudoscalar mesons M should not tend to zero in proportion to $m_1 + m_2$, but in proportion to $\sqrt{(m_1 + m_2)}$:

$$M^2 = M_1(m_1 + m_2) \qquad (m_1, m_2 \rightarrow 0) \qquad (9.7)$$

The simplest interpolation between this behaviour and the large quark mass limit ($M = m_1 + m_2$ + finite binding effects) is

$$M^2 = M_1 (m_1 + m_2) + (m_1 + m_2)^2 \qquad (9.8)$$

This formula does work for the pseudoscalar mesons. It allows us to extract absolute values for m_u, m_d, m_s, ..., not only quark mass differences. The numerical analysis - which I invite you to do-it-yourself - shows that for the light quarks m_u, m_d, m_s the term $(m_1 + m_2)^2$ is small compared to the term $M_1(m_1 + m_2)$. We may therefore use (9.7) as a first approximation and get

$$\frac{m_u + m_d}{m_u + m_s} = \frac{M^2}{M_K^2} \rightarrow \frac{\frac{1}{2}(m_u + m_d)}{m_s} \simeq \frac{1}{24} \tag{9.9}$$

i.e. the s quark is about 24 times as heavy as the u or d quarks. Putting the various estimates together we obtain a coherent picture with the following orders of magnitude [4]:

$$m_u = 4\ \text{MeV},\ m_d = 7\ \text{MeV},\ m_s = 150\ \text{MeV},\ m_c = 1200\ \text{MeV},\ m_b = 4400\ \text{MeV} \tag{9.10}$$

The quantity $(m_d - m_u)/(m_s - \frac{1}{2}(m_u + m_d))$ measures the strength of the isospin breaking effects in comparison with SU(3) breaking. This quantity may be evaluated rather accurately from the observed pattern of baryon and meson masses without making use of the coarse arguments described above. A recent analysis [15] gives

$$\frac{m_d - m_u}{m_s - \frac{1}{2}(m_u + m_d)} = 0.021 \pm 0.002$$

The absolute values of the quark masses are more difficult to pin down. It is interesting to observe that the values (9.10) agree quite well with the old SU(6) formula ($F_\rho \simeq \sqrt{2} \cdot F_\pi$)

$$\frac{m_u + m_d}{2} = \frac{m_\pi^2\, F_\pi}{3\, m_\rho\, F_\pi} = 5.4\ \text{MeV}$$

I should however point out that this value for the light quark mass is not entirely uncontroversial. Part of the problem derives from the fact that the quark masses are of course subject to renormalization. If one quotes values rather than only ratios one is implicitly choosing a renormalization convention. The scattering of the quark mass values found in the literature [16] can however not be blamed on this ambiguity. The root of the discrepancies between the various theoretical "measurements" is the question whether (9.7) describes the chiral limit correctly, with a sizeable value for M_1 determined by the strong interaction scale Λ_3. If this is the case then (9.9) follows to within SU(3) breaking effects. Note that what we are talking about here are the quark mass terms in the Lagrangian ("current" quark masses). These masses describe the inertia of a quark against hard kicks. If one instead probes their reaction to soft kicks for which the cloud of soft gluons surrounding the quarks in a bound state has enough time to readjust one does not measure this quantity but measures a constituent mass of order Λ_3 that includes the inertia of the gluon cloud which the quarks have to drag along in this state. (In some cases the quark mass is simply confused with the quark momentum, the quark wave length is mistaken for the quark Compton wave length).

It is an amusing exercise to work out the masses of the D or F mesons (the pseudoscalar particles containing one charmed quark) or the mass of the pseudoscalar and vector particles that contain the b quark plus one of the ordinary u, d, s, c quarks etc. within the simple scheme described above. For the particles that have already been seen experimentally the mass values agree to within about 100 MeV. The mass M_0 and M_1 for the various multiplets turn out to be of order 1 GeV - these quantities have nothing to do with the perturbation caused by QFD, they are determined by QCD alone. Hopefully we will some day learn how to calculate them in terms of the QCD coupling constant $\Lambda_3 \simeq 500$ MeV.

10. BOUND STATES

After this readers' digest [5)] of the ingredients involved in a theory of the hadrons I return to the issue of confinement. Perturbation theory splits the Hamiltonian into a part that describes free gluons and free quarks and an interaction term

$$H_{QCD} = H_{free} + H_{int} \tag{10.1}$$

At short distances (high momenta) H_{free} indeed dominates the matrix elements of H_{QCD} (asymptotic freedom, g_{eff} is small) and it makes sense to treat H_{int} as a perturbation. In a meson the quarks are however moving only at modest momenta and the cloud surrounding these quarks involves soft gluons. In this situation H_{int} has the same weight as H_{free} (g_{eff} is large), perturbation theory fails. If one treats H_{int} as a perturbation nevertheless, one is punished with infrared singularities.

To arrive at a quantitative description of mesons or baryons we need a better splitting:

$$H_{QCD} = H_o + H_{int}' \tag{10.2}$$

where H_o represents a dressed form of H_{free} that includes the collective effects of the soft gluons surrounding the bare quarks or bare gluons. More specifically, H_o should be chosen in such a manner that it dominates the matrix elements of H_{QCD} at small as well as at large momenta. If one succeeds in finding such an improved zero order approximation to QCD then the full theory may be recovered by treating H_{int}' as a perturbation. Of course, the splitting is not unique - what counts is the total Hamiltonian. Different zero order approximations merely lead to a different remainder H_{int}'. In particular, every Hamiltonian H_o that has the same behaviour at short distances as H_{free} will work equally well in the region where standard perturbation theory works; what one is looking for is a deformation of H_{free} that allows one to extend this region.

To be more specific, let us consider a meson of momentum p. I denote the corresponding state vector by $|p, n\rangle$ where n stands for the remaining quantum numbers such as spin, spin direction, mass, charge etc. In QCD this state is expected to have a nonvanishing overlap with the two quark sector. The corresponding matrix element

$$\langle 0|T\bar{q}^{\beta}(y)\, q^{\alpha}(x)\,|\,p,\, n\rangle = \psi(x,y)^{\alpha\beta} \tag{10.3}$$

represents a wave function describing the quark distribution in the meson. Since gauge invariance with respect to the colour group is the central feature of QCD it is important to look at the gauge transformation properties of this wave function. Both $|0\rangle$ and $|p, n\rangle$ are assumed to be invariant under the colour group (colour singlets). The quantity $\bar{q}(y)\, q(x)$ is not - it transforms into $\bar{q}(y)\, U^+(y)\, U(x)\, q(x)$. This implies that the wave function as defined in (10.3) depends on the gauge in which one chooses to work. To get a meaningful quantity one changes the definition of the wave function by allowing for a gluon factor

$$\langle 0|T\bar{q}(y)^{\beta}\, G(y, x)\, q^{\alpha}(x)\, |p, n\rangle = \psi(x, y)^{\alpha\beta}$$

$$G(y, x) = P \exp ig \int_x^y dx^{\mu'}\, B_{\mu}(x') \tag{10.4}$$

P denotes ordering along the path of integration; a trace over the colour indices is understood: for fixed quark flavours ψ is a 4 x 4 matrix.

In the nonrelativistic quark model the spectrum of mesons is accounted for by only counting the quark degrees of freedom. In this picture the gluon degrees of freedom are considered to be frozen in - the gluon cloud follows the motion of the quarks. In QCD one expects however that sooner or later gluonic excitations should start to show up in the meson spectrum, even if there is no experimental evidence for these excitations at this time.

One way to describe gluonic excitations is to allow the integration in (10.4) to run along a curved path connecting x with y. If one knew how the wave function changes from path to path one would presumably be able to distinguish the various modes of excitation of the gluonic cloud. An alternative, intuitively perhaps more plausible method to cope with excited gluonic degrees of freedom (in particular, in a non-relativistic picture of the bound state) is to consider wave functions associated with operators of the type

$$\bar{q}(y)\, G(y, z)\, B_{\mu\nu}(z)\, G(z, x)\, q(x); \; \bar{q}\, G\, B_{\mu\nu}\, G\, B_{\rho\sigma}\, G\, q, \; \ldots$$

where $B_{\mu\nu}(z)$ stands for the gluon field strength. In this language a meson for which the gluon degrees of freedom are not frozen in would be interpreted as a bound state containing two constituent quarks and one or several constituent gluons. One also expects that purely gluonic states containing e.g. two constituent gluons should exist. In the picture I am advocating here these gluonic states should be described through wave functions of the form

$$<0|\ T\ B_{\mu\nu}(x)\ G(x,\ y)\ B_{\rho\sigma}(y)\ G(y,\ x)\ |p,\ n> = \psi_{\mu\nu\rho\sigma}(x,\ y)$$

a trace over colour again being understood. (The axial anomaly indicates that wave functions of this sort may have sizeable values even for mesons such as the η' which in the nonrelativistic quark model is described as a bound $q\bar{q}$ pair.)

In the following we assume that excitations of the gluonic degrees of freedom do not play a decisive role for the lowest lying mesons. The success of the nonrelativistic quark model suggests that for the lowest states it makes sense to assume the gluon cloud to be determined by the quark configuration. We do not try to answer the question how the gluon cloud looks like in these states. Accordingly we identify the path of integration that occurs in the definition of the wave function with the straight line from x to y, such that the wave function $\psi^{\alpha\beta}(x,\ y)$ indeed only involves the degrees of freedom of the two quarks.

An important consequence of this identification of the wave function with a matrix element of local fields is the spectrum condition: not every function of x and y can represent a matrix element of the type (10.4). To state the spectrum condition, let us for the moment take spinless quarks described by scalar fields A(x), B(y):

$$\psi(x,\ y) = <0|\ T\ A(x)\ B(y)\ |p,\ n>$$

and let us also suppose that the meson does not carry spin. In this case Lorentz invariance implies that the wave function is of the form

$$\psi(x,\ y) = e^{-\frac{i}{2}p(x+y)}\ \phi(pz,\ z^2)$$

with $z = x - y$. The spectrum condition states that for $z^2 < 0$ the Fourier transform of ϕ with respect to pz vanishes outside the interval $-\frac{1}{2} \leq \xi \leq \frac{1}{2}$:

$$\phi(pz,\ z^2) = \int_{-\frac{1}{2}}^{+\frac{1}{2}} d\xi e^{i\xi pz}\ \Phi(\xi,\ z^2)$$

(This property follows from the fact that A(x) and B(y) commute at spacelike separation.)

The spectral condition has a simple interpretation in the infinite momentum frame: it states that the two quarks move in the same direction as the meson. It also holds for spinning quarks; the gluon gauge factor in (10.4) does not upset it as long as the integration runs along the straight line from x to y.

11. BETHE-SALPETER-EQUATION

The standard analysis of bound state wave functions is based on the BS framework. One considers the Green's function

$$S(xy|x'y') = \frac{1}{3}\{<0|T\bar{q}(y)\ G(y,x)\ q(x)\ \bar{q}(x')\ G(x',y')\ q(y')|0> - <0|T\bar{q}(y)\ G\ q(x)\ |\ 0>\ <0\ |\ T\ \bar{q}(x')\ G\ q(y')|0>\} \tag{11.1}$$

The second term in this definition subtracts those diagrams for which the points x, y are disconnected from x', y'; the factor $\frac{1}{3}$ compensates the trace over colour. In the free quark limit S reduces to a product of two free quark propagators:

$$S_{free}\ (xy|x'y') = S(x - x') \times S(y' - y) \tag{11.2}$$

The differential equation

$$(-i\gamma^{\mu}\ \partial^{x}_{\mu} + m)\ x(i\gamma^{\mu}\ \partial^{y}_{\mu} + m)\ S_{free}\ (xy|x'y') = \delta^4(x - x')\ \delta^4(y - y')$$

shows that if one interprets the Green's function as an operator acting on 4 x 4 matrix functions $\psi^{\alpha\beta}(xy)$ of two variables then it has an inverse given by

$$S^{-1}_{free}\ (xy|x'y') = (-i\gamma^{\mu}\ \partial^{x}_{\mu} + m) \times (i\gamma^{\mu}\partial^{y}_{\mu} + m)\ \delta^4(x - x')\ \delta^4(y - y')$$

The BS equation is based on the assumption that the full Green's function also has an inverse (a property that holds in perturbation theory). The BS-kernel $V(xy|x'y')$ may then be defined as

$$S^{-1} = S^{-1}_{free} + V \tag{11.3}$$

or, equivalently,

$$S = S_{free} - S_{free}\ V\ S \tag{11.4}$$

(products of two operators involve an integration over two coordinates as well as a sum over two Dirac indices.)

The eigenvalue equation for bound state wave functions may be obtained as follows. Translation invariance allows one to represent S as

$$S(xy|x'y') = \int d^4p \, e^{-\frac{i}{2} p(x+y-x'-y')} \, S(z|p|z') \tag{11.5}$$

with $z = x - y$, $z' = x' - y'$. If there is a bound state at $p^2 = m^2$ then the quantity $S(z|p|z')$ develops a pole at this value of p^2:

$$S(z|p|z') \sim \frac{1}{m^2-p^2} \, \phi(z) \, \overline{\phi}(z)$$

The residue of the pole is related to the wave function of the bound state in question:

$$\psi(x, y) = e^{-\frac{i}{2} p(x+y)} \, \phi(x - y) \tag{11.6}$$

To exploit this information in (11.4) one takes the Fourier transform and looks at the pole at $p^2 = m^2$. In order for the equation to hold near $p^2 = m^2$ the wave function of the bound state must satisfy

$$\psi = - S_{free} \, V \, \psi$$

or, multiplying with S_{free}^{-1}

$$(- i\gamma^{\mu} \partial_{\mu}^{x} + m) \, \psi(x, y) \, (i\gamma^{\mu} \overleftarrow{\partial}_{\mu}^{y} + m) + \int V(xy|x'y') \, d^4x' \, d^4y' \, \psi(x'y') = 0 \tag{11.7}$$

If the BS kernel V is given then this equation amounts to an eigenvalue problem that fixes the bound state wave functions as well as the mass of the corresponding mesons.

As such (11.7) is an exact equation accounting for the full QCD Hamiltonian. It merely performs the splitting $H_{QCD} = H_{free} + H_{int}$ on the level of the meson wave functions. In contrast to H_{int} the kernel V is however not given explicitly - this is the price to pay for reducing the problem to the quark degrees of freedom. It is here that one would run into a problem if it should turn out e.g. that gluonic excitations play a decisive role even for the lowest lying states: a two-quark-kernel V exists also in that case, but it would presumably be impossible to calculate it without studying these excitations in detail, i.e. without going beyond the $q\bar{q}$ sector.

In perturbation theory V can be calculated as a power series in the coupling constant. For QCD the lowest order contribution to V (exchange of one gluon) is:

$$V(xy|x'y') = \frac{4}{3} g_3^2 \, \gamma^{\mu} \times \gamma^{\nu} \, D_{\mu\nu}(x - y) \, \delta^4(x - x') \, \delta^4(y - y') \tag{11.8}$$

where $D_{\mu\nu}(z)$ is the gluon propagator. In addition, the lowest order expression for V contains a quark self-energy contribution. One may separate the diagrams contributing to V into those that represent $q\bar{q}$ interactions, i.e. connect the quark line with the antiquark line ("one particle irreducible diagrams") and those that represent self-energy effects. The self-energy effects may be summed up by replacing the free

Green's function S_{free} by a product of two dressed quark propagators. Accordingly, the product of the two Dirac operators in (11.7) must then be replaced by the product of two dressed inverse quark propagators. It is however quite questionable whether a splitting into one-particle irreducible and reducible diagrams is a significant step in an analysis of the confinement problem. The splitting does make sense if the quarks are essentially moving freely; in the presence of a soft gluon cloud the propagation properties of one of the quarks are presumably quite dependent on the location of the other quark. Whether one uses free quark propagators as a zero order approximation or uses a product of two dressed quark propagators - a zero order approximation that factorizes presumably has little chance to be a good starting point.

12. VERY HEAVY QUARKS

The bound state problem of QED - positronium - is very well described by the lowest order expression for the one-particle irreducible part of V, eq. (11.8) ($\frac{4}{3} g_3^2 \to e^2$). If one solves the bound state equation (11.7) to all orders in e with this input ($\equiv$ ladder approximation) one obtains an excellent approximation both for the spectrum and for the wave functions of the system. If the same approximation is used for bound quark states one of course finds the same spectrum, in particular, one finds that the system can be ionized by supplying it with an energy of order $m\, g_3^4$.

If the quarks are very heavy, then this description of the bound state spectrum may be quite decent at least for the lowest lying states for the following reason. For heavy quarks the Bohr orbits of the positronium-like bound states are small: the radius is of order $(m\, g^2)^{-1}$. The average momentum of the quarks is large, of order $m\, g^2$; the average momentum of the gluons whose exchange provides the binding force is also of this order. Hence one expects that the strength of the QCD coupling constant g_3 should be given by the perturbation theory formula (4.1) evaluated at a momentum scale of this order. If m is sufficently large this leads to a small effective value of g_3; in this situation a perturbative treatment of the interaction is self-consistent. Only high excited states, in particular, states that fail to be bound in this approximation will explore the behaviour of the $q\bar{q}$ potential at large distances - for these the approximation of course breaks down. If the perturbative treatment of the interaction applies, then the velocity of the quarks is of order $g^2 \cdot c$, i.e. the motion is nonrelativistic. In this case it is appropriate to take the nonrelativistic limit of the bound state equation [6)]. In this limit the potential (11.8) takes the form of an instantaneous interaction at a distance: only the value of the wave function at $x^0 = y^0 = t$ matters. Furthermore, the wave function decomposes into large and small components. Only 4 of the 16 components of ψ are large - we denote these large components by $\chi(t, \vec{x}, \vec{y})$. The 12 small components are dependent quantities that may be expressed in terms of χ. In lowest order the bound state equation (11.7) then reduces to the Schrödinger equation

$$\frac{1}{i}\frac{\partial\chi}{\partial t} + (m_1 - \frac{\Delta^x}{2m_1})\chi + (m_2 - \frac{\Delta^y}{2m_2})\chi - \frac{4}{3}\frac{g_3^2}{4\pi}\frac{1}{|\vec{x}-\vec{y}|}\chi = 0 \qquad (12.1)$$

The eigenvalue spectrum of this equation is given by

$$M_n = m_1 + m_2 - (\frac{4}{3}\frac{g_3^2}{4\pi})^2 \frac{m_1 m_2}{m_1+m_2}\frac{1}{2n^2} \; ; \; n = 1, 2, \ldots \qquad (12.2)$$

Most papers on the $q\bar{q}$ bound state problem are based on this approximation. Two effects are discussed to improve the approximation:

(1) Corrections of order $\frac{v^2}{c^2} \sim g^4$. The fine- and hyperfine-splittings of positronium belong to this category. These effects follow unambiguously from the BS equation based on the one-gluon exchange approximation. In the mass spectrum these effects give rise to contributions of order $m\, g^8$.

(2) To account for the failure of the 1/r potential at large distances one adds a long-range potential to the nonrelativistic Hamiltonian (12.1). The most popular potential is motivated by a classical picture of the gluon cloud. One assumes that if the quarks are far from one another then the cloud takes the shape of a tube of finite diameter. The energy of this field configuration is proportional to the length of the tube. Accordingly the confining potential is taken to be linear in $|\vec{x} - \vec{y}|$. In the simplest version one also takes it to be independent of the spin direction of the quarks and of their flavour. A review of nonrelativistic $q\bar{q}$ bound state models may be found in ref. 7.

The nonrelativistic approach to the bound state problem is limited to sufficiently heavy quarks. A (current) quark mass of the order of 5 MeV is bound to move at a velocity comparable to c. Even if the effective dynamical mass of the quark (constituent mass) should turn out to be considerably larger than 5 MeV due to the fact that the quark has to drag along a cloud of soft gluons it would come as a surprise if it should be possible to describe the pion within a nonrelativistic scheme - recall that the pion itself moves at v = c in the chiral limit.

Even for charmonium the success of the nonrelativistic models is not overwhelming. As of a couple of months ago the main problem was the following. The pseudoscalar partners (↑↓) of the lowest lying spin one (↑↑) state $\psi(M_{\psi} = 3.097$ GeV) and of its radial excitation $\psi'(M_{\psi'} = 3.686$ GeV) were observed at a surprisingly low mass $M_{\eta_c} = 2.83$, $M_{\eta_c'} = 3.45$) whereas the standard nonrelativistic model predicted these states to occur less than 100 MeV below ψ, ψ'. (In the nonrelativistic model hyperfine splittings are considerably smaller than the splittings due to a different orbital state of motion, i.e. different principal quantum number n). Even if one is able to cook up suitable nonrelativistic models of the confining potential that do produce sufficiently large hyperfine splittings it is very difficult if not impossible to understand why the rates for the transitions $\psi \to \eta_c + \gamma, \psi' \to \eta_{c'} + \gamma$ are so small.

Meanwhile the experimentalists seem to have lost their pseudoscalar states. At the moment the size of the hyperfine splitting is an open question. If the states η_c, η_c' should finally be rediscovered close to ψ, ψ' and if the corresponding transition rates should turn out to be in reasonable agreement with the expectations based on the standard nonrelativistic model then one would conclude that this model indeed provides a picture of the charmonium states that is basically sound.

A further problem the nonrelativistic model has to cope with is the fact that the prediction for the splitting of the $b\bar{b}$ states turned out to be too small. Some of the proposals that were made to remedy this defect seem to be rather far-fetched.

In the remainder of these lectures I describe an attempt at a relativistic treatment of the long range force between quarks [8)]. It is remarkable that within this framework large hyperfine splittings arise in a natural manner as a consequence of the requirement that the long range force is described by a local potential. It is a crucial feature of the model described below that the interaction modifies the connection between the large and the small components of the wave function (interaction effects of order v/c). The standard prescription that expresses the transition rate for $\psi \to \eta_c + \gamma$ in terms of the quark magnetic moment is based on the assumption that the interaction does not affect the current to order v/c. The model thus offers an explanation for the failure of the nonrelativistic predictions for this rate. As a further plus I mention that the model does predict the proper splitting between Y and Y'.

The model poses a number of problems, too. Just to mention one: it does not suffice, of course, to explain why a given calculation of transition rates fails - one should replace it by a better calculation. At the moment we are however not in a position to do this. I will come back to some other open questions at the end.

13. RELATIVISTIC BOUND STATES

The BS equation (11.7) does not suppose that the motion of the system is nonrelativistic. Indeed, one may try to solve the BS equation as it stands for the one-gluon exchange potential (11.8), suitably modified at large distances. There is however a basic problem with this equation that arises as soon as one leaves the safe grounds of perturbation theory that guarantees nonrelativistic internal motion. The problem has to do with the appearance of two time variables. The equation involves, in addition to the time $t = \frac{1}{2}(x^o + y^o)$ also a relative time variable $z^o = x^o - y^o$ that plays an important role if one is not working in the nonrelativistic, instantaneous approximation. The trouble can be seen even if one switches the interaction off all together. In this limit it is clear that one is free to specify the wave function at any fixed time t as an arbitrary function of $\vec{x}$, $\vec{y}$ as well as of the relative time z^o. The BS equation then determines how these initial values evolve with t. The variable z^o plays a role similar to $\vec{x}$, $\vec{y}$: we have a system with 7 degrees of freedom rather than only 6 as would be appropriate for two particles. (In other words, most of the solutions of the BS equation for V = 0 are not describing two free quarks). The disease also manifests itself if the potential is switched on: again, z^o plays the role of an independent degree of freedom. A complete set of eigenstates of the BS equation contains states that involve excitations of the orbital motion in $\vec{x}$, $\vec{y}$ as well as excitations in relative time. One is free to prescribe the wave function as a function of $\vec{x}$, $\vec{y}$ and z^o at some fixed time t and to decompose it into the complete set of eigenstates. In particular, one may choose the wave function to vanish in the vicinity of $x^o = y^o$, (say $|x^o - y^o| < T$), but to have nonzero values as soon as x^o is sufficiently different from y^o. In view of the field equations of QCD this is strange. The field equations specify the time evolution of the state $\bar{q}(y)\, G(y, x)\, q(x)|0\rangle$ both with respect to x^o and with respect to y^o. A proper initial value problem for the wave function should involve only the value of $\psi(x, y)$ at a given time x^o and a given time y^o. The dependence of ψ on $x^o - y^o$ should be determined by QCD. In fact, just as one derives the BS equation one may also project the field equations on the two quark sector:

$$\gamma^\mu \partial^x_\mu \psi(xy) = \int W(xy|x'y')\, dx'\, dy'\, \psi(x'y')$$
$$\psi(xy)\, \gamma^\mu \overleftarrow{\partial}^y_\mu = \int \hat{W}(xy|x'y')\, dx'\, dy'\, \psi(x'y') \qquad (13.1)$$

These relations are also exact. In perturbation theory, W and $\hat{W}$ may be calculated in a power series of g just as the BS kernel can be calculated order by order. If one switches the interaction off, these equations do however not allow unphysical solutions in contrast to the situation for the BS equation. Accordingly, we may expect that if we use some approximate expression for the kernels W, $\tilde{W}$ and solve the pair of equations (13.1) there will be no unphysical excitations in relative time.

The heart of the problem is again to find an approximate expression for the kernels W, $\tilde{W}$. Our model is characterized by the requirement that the contributions of the long range force to these kernels is local:

$$W(xy|x'y') \sim W(x - y)\, \delta^4(x - x')\, \delta^4(y - y') \tag{13.2}$$

and similarly for $\tilde{W}$. I will give a more precise formulation of this assumption in section 15. It turns out that the locality requirement is extremely strong - it essentially fixes the form of the long range interaction. To motivate the locality requirement I briefly discuss some aspects of quantum mechanics at infinite momentum (quantum mechanics on a null plane). The model can very well be formulated without invoking null planes; the excursion is however extremely instructive.

14. EXCURSION: INITIAL VALUES ON A NULL PLANE

In this section I again consider scalar quarks, i.e. a single component wave function $\psi(x, y)$. In this case the Poincaré generators take the form

$$\begin{aligned} P_\mu \psi &= i(\partial^x_\mu + \partial^y_\mu)\, \psi \\ M_{\mu\nu} \psi &= i(x_\mu\, \partial^x_\nu - x_\nu\, \partial^x_\mu)\, \psi + i\, (y_\mu\, \partial^y_\nu - y_\nu\, \partial^y_\mu) \end{aligned} \tag{14.1}$$

Mass and spin of the meson are related to the eigenvalues of these operators.

If the quarks are free the wave function obeys the wave equations:

$$(\Box^x + m_1^2)\, \psi = 0 \qquad (\Box^y + m_2^2)\, \psi = 0 \tag{14.2}$$

If one prescribes the wave function $\psi(x, y)$ on the null plane $x^- \equiv x^0 - x^3 = 0$, $y^- \equiv (y^0 - y^3) = 0$, then the wave equations determine it all over Minkowski space. In fact, they may be solved for ∂^x_-, ∂^y_-:

$$\partial^x_-\, \psi = \frac{1}{2}\, (\partial^x_+)^{-1}\, \{(\partial^x_1)^2 + (\partial^x_2)^2 - m_1^2\}\, \psi$$

The operator on the right hand side requires only knowledge of the wave function on the plane $x^- = \text{const}$.

The Hamiltonian of the system generates the translations in x^-, y^-:

$$P_-^{free}\,\psi = \frac{i}{2}(\partial_+^x)^{-1}\{\Delta_T^x - m_1^2\}\,\psi + \frac{i}{2}(\partial_+^y)^{-1}\{\Delta_T^y - m_2^2\}\,\psi \tag{14.3}$$

This shows that the Hamiltonian of two free quarks is a local differential operator in the two transverse directions of the null plane; it is not local with respect to the lightlike directions x^+, y^+ $[(\partial_+^x)^{-1}$ is an integral operator$]$. The fact that P_-^{free} is a local differential operator in the transverse directions reflects causality: the value of the wave function at "time" $x^- = y^- = \varepsilon$ depends only on those initial data on the plane $x^- = y^- = 0$ that can be reached by moving backwards with less than the speed of light. In the limit $\varepsilon \to 0$ only the data at points with the same transverse coordinates matter.

Let us now switch on the interaction by adding an interaction term to P_-^{free}. We postulate that the interaction does not destroy this causality property: even in the presence of interaction the value of the wave function at $x^- = y^- = \varepsilon$ is assumed to depend only on those initial data at $x^- = y^- = 0$ that are in the causal past of the point in question. More precisely, we assume that P_- as well as the other generators of the Poincaré group are differential operators in the transverse directions of at most second order (the dependence of these operators on x^+, y^+ is left unspecified). There is an equivalent, covariant formulation of this requirement. We have shown in ref. 8 that the Poincaré generators are local differential operators in the transverse directions if and only if ψ obeys a pair of covariant second order wave equations that may be written in the form $(z = x - y)$

$$\psi(x, y) = e^{-\frac{i}{2}p(x+y)}\,\phi(z)$$

$$\{\Box - u_2 D^2 - u_1 D - u_o - n\,p^2\}\,\phi = 0 \tag{14.4}$$

$$\{p\partial - v_2 D^2 - v_1 D - v_o\}\,\phi = 0$$

The operator D stands for $D = z^\mu \partial_\mu$, the "potentials" u, v, n are functions of z^2 and pz. They may involve z^2 in an arbitrary manner, but can be at most quadratic in pz. Free quarks are of course a special case. In the above notation they are described by $(m_1 = m_2 = m)$:

$$(\Box + m^2 - \frac{1}{4} p^2)\, \phi = 0 \qquad p\partial\phi = 0 \tag{14.5}$$

The relativistic harmonic oscillator

$$\{\Box - \lambda^2 z^2 + m^2 - \frac{1}{4} p^2\}\, \phi = 0 \qquad \{p\partial - \lambda\, pz\}\, \phi = 0 \tag{14.6}$$

is another example of a system of this type. The oscillator may equivalently be described by two equations of the type (13.1)

$$\begin{aligned} \Box_x \psi &= (\lambda^2 z^2 - m^2 - i\,\lambda\, Pz)\, \psi \\ \Box_y \psi &= (\lambda^2 z^2 - m^2 + i\,\lambda\, Pz)\, \psi \end{aligned} \tag{14.7}$$

In the language used in the last section, the harmonic oscillator is described by field equation kernels W, $\tilde{W}$ that are local in the sense (13.2). (The occurence of a term linear in Pz requires a contribution proportional to the first derivative of the wave function in addition to a strictly local term).

One might expect that by choosing different potentials u, v, n in (14.4) one may obtain a large variety of systems that all obey the causality requirement formulated above. This expectation is false for the following reason. It is in general a crime to impose more than one equation on a single unknown. If one imposes two differential equations on the single wave function $\psi(x, y)$ with random coefficients the only solution will be $\psi = 0$. In order for this not to happen the two differential equations have to satisfy an integrability condition which imposes very strong restrictions on the potentials. We have shown in ref. 8 that these integrability conditions in fact determine the potentials up to a few constants (the general solution is a generalized oscillator). To understand how this comes about let us consider the following particular case. Suppose that the field equation kernels are strictly local:

$$\Box_x \psi = W(z)\, \psi \; ; \qquad \Box_y \psi = \tilde{W}(z)\, \psi \tag{14.8}$$

The integrability condition is obtained by looking at the commutator of the two differential operators:

$$[\Box_x - W, \quad \Box_y - \tilde{W}] = -4\, \{\tilde{W}'\, z^\mu\, \partial^x_\mu + W'\, z^\mu\, \partial^y_\mu\} + 4\, \{z^2\, W - z^2\, \tilde{W}\}'' \tag{14.9}$$

(Lorentz invariance implies that W, $\tilde{W}$ depend only on z^2; W' denotes the derivative of W with respect to z^2). Applied to a solution of the two wave equations the commutator vanishes. Hence ψ must be annihilated by the right hand side of (14.9). This imposes a new differential equation on ψ . We require that the system of wave equations has sufficiently many solutions in the following sense: it should admit arbitrary initial values on the plane $x^- = y^- = 0$. The two wave equations (14.8) do not impose any restriction on these initial value data; they instead determine the time evolution. The new constraint does however impose a restriction unless $W' = \tilde{W}' = 0$ i.e. unless the two particles are free. In order to have an interaction without dropping the locality requirement we have to allow the potentials W, $\tilde{W}$ to contain terms proportional to the first or to the second derivative of ψ [see (14.7)]. Even with this general form of the locality requirement the class of admissible potentials is very narrow.

Why do we not find, say, the Coulomb potential as one of the possibilities ? Is the locality requirement which I have shown above to be related to causality a necessary condition ? Are we saying that one-photon or one-gluon exchange violates causality ? The answer to these questions is the following: Field theory is manifestly causal on the space of all states. Projecting the system on the two quark sector of Fock space one in general has to live with interaction kernels that are not manifestly causal. This is related to the fact that by specifying the wave function at some given time one is by no means making sure that there will be exactly two bare quarks in this state. The state will have Fock space matrix elements with two pairs of quarks or no quarks, but several gluons etc. These components do not show up in the wave function. If time evolves these components will however produce transitions to the two quark sector of Fock space and show up as contributions to the wave function that are not necessarily in a causal relation to the two-quark component at an earlier time. Positronium theory keeps track of these other sectors of the Fock space of QED (vacuum polarization). QED is of course causal, the equations of motion for the wave functions of positronium are not. Only the zero order approximation to positronium - free leptons - is causal even on the two lepton sector. We are merely pointing out that there is a narrow class of long range interactions which retain this feature of the standard zero order approximation. We do not expect the full $q\bar{q}$ interaction to be described by this approximation. Gluon exchange, which is known to dominate the interaction at short distances, must be treated as a perturbation of this approximate description much as one treats photon exchange in positronium.

In my opinion it is important to have a zero order approximation with decent properties. To my knowledge none of the alternative guesses at an approximate relativistic description of the long range forces between quarks that can be found in the literature avoids the problem of unphysical modes associated with relative time. I am convinced that the $q\bar{q}$ bound state spectrum of QCD does not contain modes of this sort.

Whether QCD cures the disease by producing a soft gluon cloud that gives rise to a causal effective $q\bar{q}$ interaction (as we assume it to be the case) or whether QCD solves the problem in a more subtle manner is an open question.

15. HARMONIC CONFINEMENT

Applying the above ideas to spin $\frac{1}{2}$ quarks we postulate that for large separations of the two quarks the kernels W, $\tilde{W}$ in (13.1) are approximately local

$$W(xy|x'y') = W_o\, \delta(x - x')\, \delta(y - y') + W_1(xy|x'y') \qquad \psi = \psi_o + \psi_1 + \dots$$

$$\gamma^\mu\, \partial^x_\mu\, \psi_o(xy) = W_o\, \psi_o(xy);\; \partial^y_\mu\, \psi_o(xy)\, \gamma^\mu = \tilde{W}_o\, \psi_o(xy) \tag{15.1}$$

We allow W_o and $\tilde{W}_o$ not only to depend on z = x - y, but also to involve the differential operator $P_\mu z^\mu$. Since we are dealing with first order equations here we assume that the kernels are at most linear in Pz. We do not impose any restrictions on the manner in which W_o or $\tilde{W}_o$ depend on z or on the spin:

$$W_o\, \psi = w_1\, \psi + w_2\, \not{z}\, \psi + w_3\, \gamma^\mu\, \psi\, \gamma_\mu + \dots$$

The potentials w_1, w_2, ... are functions of z^2 and Pz, arbitrary in z^2 but linear in Pz.

Again the consistency of the two wave equations requires an integrability condition to be satisfied. This condition imposes strong restrictions in the form of differential equations to be satisfied by the potentials w_1, w_2 ... . The general charge conjugation symmetric solution[8)] of these differential equations involves only three constants:

$$\begin{aligned}
&\gamma^\mu\, \partial^x_\mu\, \psi = -\lambda\, \gamma_5\, \psi\, \gamma_5\, \not{z} - i\, A\, \psi + \mu\, Pz\, P_+\, \psi \\
&\partial^y_\mu\, \psi\, \gamma^\mu = \lambda\, \not{z}\, \gamma_5\, \psi\, \gamma_5 + i\, A\, \psi + \mu\, Pz\, P_+\, \psi \\
&A\, \psi = (m - 2\,\mu)\, P_+\, \psi + \kappa\, P_-\, \psi - \frac{1}{2}\, \mu\, P_+\, \gamma^\nu\, \psi\, \gamma_\nu \\
&P_\pm\, \psi = \frac{1}{2}\, (\psi \pm \gamma_5\, \psi\, \gamma_5)
\end{aligned} \tag{15.2}$$

(To be precise there are two different solutions.The second one is however the adjoint of the one given). The three independent constants are m, κ and μ; the parameter λ is related to these by $\lambda = \kappa\mu$.

If we set $\mu = 0$, $\kappa = m$ then we recover the free Dirac equations. If λ is different from zero then the wave equations involve a potential that grows linearly with z; the potential couples to the spin direction in a very particular manner. (Note that a linear potential in the Dirac equation amounts to a harmonic potential in the corresponding Klein-Gordon equation).

The system has the following properties:

(1) The eigenfunctions are of the form $[z^2 = (z^o)^2 - (\vec{z})^2]$:

$$\psi(x, y) = e^{-\frac{1}{2} ip(x+y)} e^{\frac{1}{2}\lambda z^2} P(z)$$

where P(z) is a polynomial. This shows that the model satisfies the spectrum condition discussed in section 10.

(2) The eigenvalues of $P^2 = M^2$ are equally spaced:

$$M^2 = 4\kappa m + 8\lambda n, \quad n = 0, 1, 2 \ldots$$

(3) There are no timelike excitations. The initial values of the wave function say at $x^o = y^o = 0$ determine it all over Minkowski space. This implies of course that the set of eigenstates is not a complete set on Minkowski space.

(4) Only 4 of the 16 components of ψ are independent. As a consequence of this property and of (3) the spectrum of the model has the same degrees of freedom as the nonrelativistic quark model. To every orbital wave function (characterized by an orbital angular momentum ℓ and a radial quantum number k = 0, 1, 2, ...) there are 4 states differing in the orientation of the quark spins. For one of these states the quark spins are antiparallel (s = 0), the other three have total quark spin s = 1. The total angular momentum is given by $j = \ell$ for s = 0 and $j = \ell + 1$, or $j = \ell$ or $j = \ell - 1$ for s = 1. In terms of these quantum numbers the mass of the state is given by

$$M^2 = 4\kappa m + 8\lambda(\ell + 2k + s) \tag{15.3}$$

(5) For fixed quark spin and radial quantum number we have linear Regge trajectories parametrized by ℓ. For the lowest states the spectrum is shown in Fig. 2.

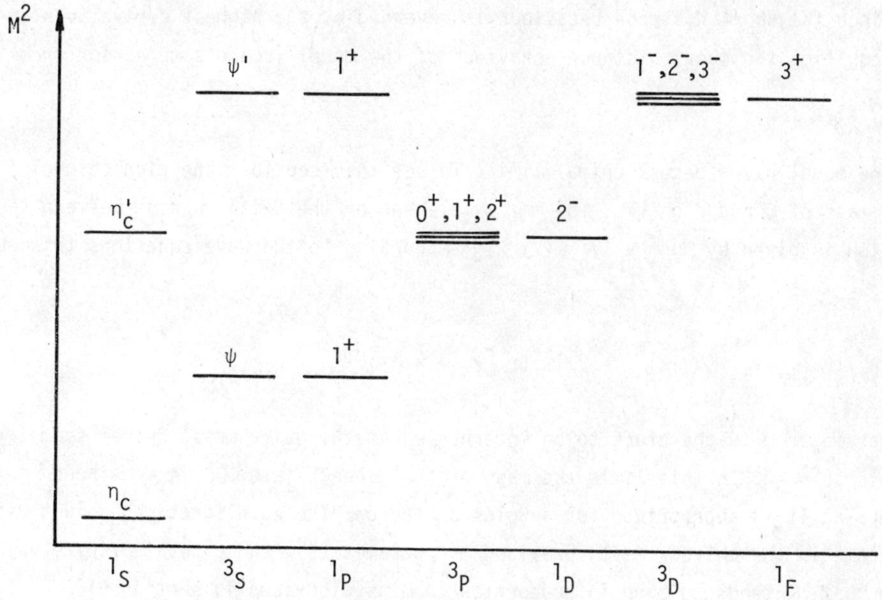

Fig. 2: Spectrum of the relativistic oscillator.

(6) One of the main features of the model is a strong hyperfine splitting. [Compare (15.3); the spacing between s = 1 and s = 0 is the same as the spacing between successive states on a Regge trajectory, say between $\ell = 1$ and $\ell = 0$.] This is in marked contrast to positronium where the levels analogous to ψ and η_c are essentially degenerate, the splitting being hyperfine. The same remarks apply to the radial (η_c') and angular excitations of η_c. The strong spin-spin interaction manifests itself in the relation

$$M_\psi^2 - M_{\eta_c}^2 = \frac{1}{2}(M_{\psi'}^2 - M_\psi^2)$$

(7) The qualitative features of the spectrum are best seen in the nonrelativistic limit of the model, $\lambda \ll \kappa m$. In this limit the system may be described by the Hamiltonian

$$H = -\frac{1}{2m}\Delta_x - \frac{1}{2m}\Delta_y + \frac{\lambda^2}{m}(\vec{x} - \vec{y})^2 + \frac{2\lambda}{m}\vec{s}_1 \cdot \vec{s}_2 \tag{15.4}$$

which explicitly shows that the strength of the spin-spin interaction is governed by the spring constant of the oscillator.

(8) Since the short distance behaviour is governed by the highest derivatives in the wave equation the short distance behaviour of the model is the same as for free quarks.

(9) The model has a decent chiral limit. To see this consider the pion channel for equal mass of u and d quarks. The matrix element of the axial current between vacuum and pion is given by $\mathrm{tr}\left[\gamma_\mu \gamma_5 \psi(x, x)\right]$. According to the wave equations this quantity obeys

$$\partial_x^\mu \, \mathrm{tr}\left[\gamma_\mu \gamma_5 \psi(x, x)\right] = 2\, im \, \mathrm{tr}\left[\gamma_5 \psi(x, x)\right]$$

The parameter m is therefore to be interpreted as the quark mass. Chiral symmetry amounts to $m = 0$. In this limit the mass of the ground state $M^2 = 4\kappa m$ indeed tends to zero as it is appropriate for a Goldstone boson. The wave functions do not disintegrate in the chiral limit, provided the parameters κ and μ retain finite values. (Note that M_π^2 tends to zero in proportion to m as discussed in section 9).

It should be clear that any model with a decent chiral limit ($M_\pi = 0$, $M_\rho \neq 0$) must involve spin-spin interactions with a strength comparable to the force responsible for the Regge trajectories (M_ρ^2 and the slope $|\alpha'|^{-1}$ are of the same order of magnitude).

16. SHORT DISTANCE CONTRIBUTIONS TO THE $q\bar{q}$ INTERACTION

As pointed out in section 14 a causal effective $q\bar{q}$ interaction can at most account for the long range force due to soft gluons. What about the remainder ? In the following I argue that perturbation theory allows one to make reliable statements about the short distance behaviour of the interaction. It is possible to calculate the remainder at least at short distances and thus to improve the accuracy of the approximation [9].

Asymptotic freedom states that the strength g_3 of the strong interaction is weak if the quarks come close to one another. If g_3 is small then perturbation theory is reliable and the lowest order graphs will dominate. The lowest order graph contributing to the $q\bar{q}$ interaction is one-gluon exchange. This contribution which behaves like 1/r should therefore dominate the $q\bar{q}$ interaction at short distances. The harmonic long range interaction which we have been describing in the last section is soft at short distances and may be neglected in comparison with the one-gluon exchange contribution if r is small. On the other hand, with a fixed value of g_3 one-gluon exchange is negligible in comparison with the harmonic long range potential for large values of r. We may therefore improve the range of validity of the approximation by adding the two contributions.

Treating one-gluon exchange as a perturbation we have worked out the corresponding lowest order shift in the energy levels. In this approximation one only needs to calculate the expectation values of the one-gluon exchange potential in the unperturbed oscillator wave functions, a calculation that can be done analytically. The net effect is a shift of the levels in the direction of the positronium spectrum, as was to be expected. One obtains an excellent fit to the well established states of charmonium. With this fit one then predicts the masses of the remaining states, in particular of the pseudoscalar states η_c, η_c'. The prediction is within 30 MeV of the experimental values published some time ago as 2.83 GeV and 3.454 GeV respectively. Furthermore, the model requires a state at 3.6 GeV with $j^{PC} = 2^{-+}$ for which there is (was ?) also some experimental evidence. The prediction of the model concerning these states are firm in the sense that if these states are not found within say 60 MeV of the old published values, the model fails. We do not have adjustable parameters that would e.g. allow us to shift η_c , η_c' up to the vicinity of ψ, ψ'.

As discussed in section 12 the effects of the soft gluon cloud should become less and less important the heavier the quarks. For sufficiently large quark mass the low lying bound states should to a good approximation be described by the one-gluon exchange potential alone. In this situation one can of course not treat one-gluon exchange as a perturbation. We have therefore iterated the one-gluon exchange diagram in the presence of the harmonic long range potential, a calculation that can only be done numerically[14]. Quite apart from providing us with predictions for heavy quark systems this analysis also allows us to check the stability of the first order calculation[10]. The main result in this respect is that the levels are rather stable. The values of the three parameters that occur in a fit to the spectrum ($M_0^2 = 4\kappa m$, λ and $g_3^2/4\pi$) do however change considerably. In particular, the value needed for $g_3^2/4\pi$ in the first order calculation for charmonium is 0.55; the higher order ladder graphs reduce this value to 0.4.

It is interesting to compare the values of $g_3^2/4\pi$ with the perturbation theory formula (4.1). To make this comparison, one has to find the momentum scale of the one-gluon exchange interaction in charmonium. What should count here is the average momentum of the exchanged gluons or the average momentum of the quarks. At least qualtitatively, the momentum distribution of the quarks is given by the Fourier transform of the square of the wave function. If one estimates $g_3^2/4\pi$ on this basis then one generally finds rather large values $\gtrsim 0.5$. (The perturbation theory formula gives $g_3^2/4\pi = 1, 0.5, 0.4, 0.3, 0.2$ for $q = 1, 2, 3, 5, 10$ GeV respectively).

A different estimate is obtained from the decay $\psi \to 3$ gluons, which leads to $g_3^2/4\pi \simeq 0.2$ for the strength of the interaction responsible for this decay. If $\Lambda_3 = 500$ MeV then it takes a momentum scale of the order of 10 GeV to produce such a small value ! Even if Λ_3 should turn out to be only 250 MeV, one would still need a momentum scale of order 5 GeV which is hard to justify in a decay for which each one of the gluons on the average only gets 1 GeV. I conclude that within the uncertainties involved the value $g_3^2/4\pi = 0.4$ is tolerable; smaller values such as the value extracted from the decay $\psi \to 3$ gluons are hard to reconcile with the direct "measurements" of the strong interaction coupling constant based on deep inelastic scattering.

17. OPEN PROBLEMS

One of the open problems concerns the high excited states. The unperturbed oscillator contains two degenerate 1^{--} states at every second level (the two states corresponding to S and D waves respectively). The consecutive levels have equal spacing in M^2. The one gluon exchange corrections do not affect this equal spacing rule very strongly, but produce a small splitting between the two degenerate states [14]. (In the case of $\psi(3.685)$ and $\psi^*(3.77)$ the splitting does have the proper magnitude). The experimental situation concerning the 1^{--} states in the continuum is not very clear, but there seems to be rather good evidence for a state at 4.03 - whereas the equal spacing rule would place the two degenerate states at 4.19. The one-gluon exchange correction moves them to 4.18 and 4.24 respectively. The higher state is expected to have smaller width; it may correspond to the peak seen at 4.16. We thus have a sizeable discrepancy with the data above threshold. The same phenomenon occurs in the Y spectrum for which the analogous state Y" is below threshold. Again the model predictions are too high.

The model does appear to give a decent extrapolation from charmonium spectroscopy to Y spectroscopy. The extrapolation is based on the following input. The ocillator involves three constants μ, κ, m. As such the model does not tell us how these constants depend on quark flavour. From QCD one expects however that the properties of the $q\bar{q}$ bound states involve only one flavour dependent parameter: the quark mass. The only other scale that should enter the problem is the universal scale Λ_3 of the strong interactions. We should therefore be able to express the two quantities μ and κ as a function of the quark mass m. Concerning μ the simplest assumption is that it is flavour independent. The quantity κ cannot be flavour independent, because in the nonrelativistic limit κ becomes equal to m up to effects of the order of the binding energy. In the other extreme, when m tends to zero (chiral limit) κ retains a finite value κ_o. The simplest interpolation between these two extremes is

$$\kappa = \kappa_o + m$$

This implies that the parameter $\lambda = \mu\kappa$ grows with the quark mass in such a manner that the frequency of the oscillator $w = \frac{2\lambda}{m}$ tends to a flavour independent value for sufficiently large quark mass. Since λ measures the inverse slope of the Regge trajectories an increasing value of λ leads to a slope that decreases with the mass of the quark. There is good experimental evidence for this effect. Comparing the slope of the I = 1 mesons with the slope of charmonium we get $[\hat{m} = \frac{1}{2}(m_u + m_d)]$

$$\lambda_{\bar{c}c} \simeq 2\,\lambda_{\bar{u}d} \rightarrow \kappa_o + m_c = 2\,(\kappa_o + \hat{m}).$$

Since the light quark mass $\hat{m}$ is negligible this leads to $\kappa_o = m_c = 1200$ MeV. From the value of the slope we then get $\mu = 120$ MeV. Furthermore, we may calculate the light quark mass from the relation $M_\pi^2 = 4\,\kappa_o\,\hat{m}$ with the result

$$\frac{1}{2}(m_u + m_d) = 4 \text{ MeV}$$

in reasonable agreement with the estimate given in section 9. (Note, however that this calculation is very crude as it ignores gluon exchange corrections all together.)

Once κ_o and μ are determined, Y spectroscopy involves a single unknown, the mass of the b quark. (The strength g_3 of one-gluon exchange is scaled up from charmonium to Y in the standard manner.) The experimental value of the Y mass may be used to determine m_b. One may then calculate[14] the splitting Y' - Y and obtains 563 MeV to be compared with the experimental value 556 MeV. (As mentioned above, the prediction for the state Y" is however too high: $M_{Y''}$ = 10.52 whereas, experimentally $M_{Y''}$ = 10.38).

As far as the charge conjugation symmetric bound states go the model does seem to offer a simple description of the flavour asymmetries caused by the quark masses. There is however the following problem: It is not possible to describe bound states containing quarks of unequal mass in terms of a strictly local effective $q\bar{q}$ interaction[11]. If one perturbs the symmetric oscillator by giving the quarks a mass difference, locality is lost. This does not mean that for asymmetric mesons such as K or D the long range interaction cannot be approximated by a local potential. It does imply however, that there necessarily are nondominating contributions on the level of mass terms (neither dominating at small nor dominating at large distances) that fail to be local. This may cast a shadow of doubt on the hyperfine splitting. In fact, the dominating long range force only implies that all spin orientations have the same Regge slope. The value of the hyperfine splitting itself is a question that does not concern the slope but the intercept of the various trajectories. This intercept is affected by mass terms. If we allow suitable nonlocal contributions on the level of mass terms we can of course choose the hyperfine splitting at will. In our opinion this is however not the proper conclusion to draw. Since a large hyperfine splitting is crucial for a decent chiral limit we do think that the model provides for an adequate zero order approximation to pure QCD (massless quarks). The flavour asymmetries produced by QFD via quark mass terms however perturb the system and give rise to a nonlocal contribution to the potential that has to be treated in a perturbative manner in the same way as the short distance contributions which in our language are also non-local. From this point of view the proposal given above for the flavour dependence of κ, μ amounts to the observation that for $m_1 = m_2$ even the mass terms remain local provided the range λ of the long range force is renormalized (λ grows with the quark mass). Phenomenologically, the slopes of the Regge trajectories do vary strongly with the mass of the quarks ($M^2_{\Upsilon'} - M^2_{\Upsilon} = 10.8$, $M^2_{\psi'} - M^2_{\psi} = 4$, $M^2_{\rho'} - M^2_{\rho} = 1.8$). One may however maintain the view that this is only a low energy effect [12]. Whether the range of the long range force in QCD is flavour independent or is subject to renormalization that depends on the mass of the quark is an open question.

A further problem that we yet have to learn how to cope with is the calculation of transition rates. In particular, to work out the rate of the transitions into e^+e^-, we need a proper normalization of the wave functions. In the case of the BS equation the eigenstates, including the states involving relative time oscillations, are complete on Minkowski space. This property may be used to construct the norm in an unambiguous manner. In our model the solutions of the wave equations are only complete on an initial surface, say $x^0 = y^0 = 0$. The construction of conserved currents and of the corresponding norm is not a simple matter.

The transition rates for photon emission or pion emission pose a similar problem. In this connection it is important to realize that it is inconsistent to use the standard minimal substitution in the wave equations - they fail to be compatible in the presence of external fields. Even in the nonrelativistic limit the prescription is not unambiguous because the connection between the small and the large components of the wave function involves interaction effects of order v/c. The magnetic dipole transitions are matrix elements of this order and it is therefore not permissible to ignore the interaction effects in a calculation of the corresponding rates.

It would of course be of interest to generalize the model to the baryons. Some attempts at formulating wave equations for three body systems can be found in the literature[13]. To my knowledge all of these models are sick in one way or another (relative time oscillations or failure to reduce to free particles when the coupling is turned off).

The main problem of course is the question whether QCD cares about producing a long range interaction that remains causal when projected onto the quark degrees of freedom. If this question has an affirmative answer it should be possible to establish the correct interface between the long range forces and the short distance contributions of perturbation theory.

FOOTNOTES AND REFERENCES

1) Some of the material presented in the first part of these lecture notes is adapted from the Proceedings of the Spring School on Weak Interactions and Gauge Theories, SIN 1978.

2) The analysis of the effect of gravitational self-interaction on the orbits of the planets was carried out in collaboration with P. Minkowski.

3) For a quantitative discussion of asymptotic freedom in a quantum theory of gravity see E.S. Fradkin and G.A. Vilkovisky, preprint Lebedev Institute and Bern, 1976.

4) H. Leutwyler, Phys.Letters 48B, 431 (1973);Nucl.Phys. 76B, 413 (1974); Proceedings of the Topical Seminar on Deep Inelastic and Inclusive Processes, Suchumi, 1974.

J. Gasser and H. Leutwyler, Nucl.Phys. 94B, 269 (1975).

S. Weinberg, Transactions of the New York Academy of Sciences, Series II, 185 (1977).

5) An excellent review, containing many references to the original literature is H. Joos and M. Böhm, Eichtheorien der schwachen, elektromagnetischen und starken Wechselwirkung, DESY preprint 78/27.

6) See e.g. L.D. Landau and E.M. Lifshitz, Vol. IVa.

7) K. Gottfried, Proceedings of the Int. Symposium on Lepton and Photon Interaction at High Energies, Hamburg 1977.
J.D.Jackson, Proceedings of the European Conference on Particle Physics, Budapest 1977.
M. Kramer and H. Krasemann, Schladming Lectures 1979.

8) H. Leutwlyer and J. Stern, Phys.Letters 73B, 75 (1978); Ann.Phys.(NY) 112, 94 (1978); Nucl.Phys. 133B, 115 (1978) and Orsay preprint IPNO/TH 78-44.

9) J. Jersak, H. Leutwyler and J. Stern, Phys.Letters 77B, 399 (1978).

10) J. Jersak and D. Rein, preprint TH Aachen 1979.

11) S. Mallik, Nucl.Phys., to be published.

12) P. Minkowski, preprint Bern 1979.

13) See e.g. T. Takabayashi, Prog.Theor.Phys. 58, 1229 (1977) and preprints Nagoya University.

14) J. Gasser, J. Jersak, H. Leutwyler and J. Stern, to be published.

15) J. Gasser & A. Zepeda, to be published; P.Minkowski & A.Zepeda, Preprint Univ.Bern 1979.

16) M.D. Scadron and H.F. Jones, Phys. Rev. D10, 967 (1974); J.F. Gunion, P.C. McNamee and M.D. Scadron, Nucl.Phys. B123, 445 (1977); H. Sazdjian and J. Stern, Nucl.Phys. B129, 319 (1977); R.L. Jaffe, Oxford preprint 1979.

High Energy Behavior of Nonabelian Gauge Theories

J.Bartels

II.Institut für Theoretische Physik, Universität Hamburg

Abstract:

The high energy behavior (in the Regge limit) of nonabelian gauge theories is reviewed. After a general remark concerning the question to what extent the Regge limit can be approached within perturbation theory, we first review the reggeization of elementary particles within nonabelian gauge theories. Then the derivation of a unitary high energy description of a massive (= spontaneously broken) nonabelian gauge model is described, which results in a complete reggeon calculus. There is strong evidence that the zero mass limit of this reggeon calculus exists, thus giving rise to the hope that the Regge behavior in pure Yang-Mills theories (QCD) can be reached in this way. In the final part of these lectures two possible strategies for solving this reggeon calculus (both for the massive and the massless case) are outlined. One of them leads to a geometrical picture in which the distribution of the wee partons obeys a diffusion law. The other one makes contact with reggeon field theory and predicts that QCD in the high energy limit is decribed by critical reggeon field theory.

I. Introduction

These lectures intend to give a review of our present understanding of the Regge limit of nonabelian gauge theories, in particular QCD. Since cross sections are large in this kinematic regime, high energy physicists have always been interested in understanding the dynamics behind it (especially the nature of the Pomeron), but a theoretical description which is based on an underlying quantum field theory is still missing. Most previous attempts to understand the Regge limit within a field theory have been based on perturbation theory, and the main difficulty (besides the question which field theory to choose) was that the number of Feymann diagrams that could be handled always turned out to be too small Now QCD is believed to be the right theory of strong interactions, and we are asked to understand its behavior in the Regge limit. Can we hope that the conventional approach, i.e. the start from perturbation theory, might be successful for this theory? Let me say a few words about this general question, before I come to details. The point I would like to make is that there are good reasons to believe that perturbation theory is a useful starting point, because the Regge limit is not far from that kinematic region in which perturbation theory works[1] (hard scattering processes). But, on the other hand, the Regge limit is also sensitive to certain features which are commonly referred to as nonperturbative.

Let us start with the optimistic part of the argument and consider elastic forward scattering of a very heavy photon off a nucleon. This is the process measured in deep inelastic leptoproduction (Fig.1), and the standard argument about light cone dominance tells us that in the Bjorken limit ($-q^2 \to \infty$, $s \sim -q^2(\frac{1}{x} - 1) \to \infty$, x fixed) one probes the short distance structure of the nucleon target: if y_1 and y_2 are the two space-time points where incoming and outgoing photons couple to the nucleon, then $(y_1 - y_2)^2 \lesssim -1/q^2$ .Within QCD the property of asymptotic freedom then allows to use perturbation theory for this short distance process. Either by means of the operator expansion and renormalization group techniques or, equivalently, by extracting and summing leading logarithmus of Feynmann diagrams, one can calculate the q^2 - dependence, i.e. the change of the cross section when we move closer and closer to the light-cone $(y_1 - y_2)^2 = 0$. We now imagine that at some large value of q^2 we take a different limit: keeping now q^2 fixed and taking $x \to 0$, we reach the Regge limit $\cos\theta_t \sim s/-q^2 \to \infty$. By choosing q^2 large enough, our investigation of the Regge limit can be carried out very close to the light cone, but once q^2 is kept fixed we always stay away from it by some finite distance. In terms of QCD Feynmann diagrams it is not difficult to see that those diagrams (Fig.1) which govern the leading q^2-behavior of the Bjorken limit cannot be expected to correctly also describe the region of very small x. The tower diagrams of Fig.1 do not contain "final state interactions" of the produced quarks and gluons and, hence, cannot satisfy unitarity which is known to be important in the Regge limit ($x \to o$ limit). If one wants to investigate the Regge limit within this perturbative approach, it is, therefore, necessary first to find all Feynmann diagrams (beyond those of Fig.1), which are required by unitary for yielding a sensible $x \to 0$ behavior, then to compute their behavior in the limit $x \to 0$.

We conclude from this that, since the Regge limit, i.e. the Pomeron, can be investigated very close to the light cone where the (effective) coupling constant is small and perturbation theory works, perturbation theory may be a good starting point also for studying the Pomeron. The problem then consists of two major parts: first one has to decide which terms in the perturbation expansion (Feynmann diagrams) have to be taken into account. Because of unitarity which is crucial for the Pomeron physics, these terms will not be the same as those which govern the Bjorken limit. Secondly , one has to find a method for summing them up. As I will make clear later, this part of the problem will require new techniques.

But as I have already indicated before, the Pomeron is also sensitive to certain features of long distance physics ("confinement dynamics"), which implies that at some stage nonperturbative aspects might have to enter the calculations. In the elastic scattering process of a very energetic hadron (say, in the rest frame of the target) the projectile appears as a composite system of partons which are spread out in impact parameter space. The probability of finding a slow parton at distance b is given by the impact parameter transform of the elastic scattering amplitude:

$$\frac{1}{s} T(s, b^2) = \frac{1}{2\pi s} \int d^2k_\perp \, e^{-i\vec{k}_\perp \vec{b}} \, T(s, k_\perp^2 = -t). \qquad (1.1)$$

The hadron radius is defined as:

$$\langle b^2 \rangle = \frac{1}{s} \int d^2b \;\; b^2 \, T(s, b^2) \qquad (1.2)$$

and, in general, it will depend on the energy s. It might, again, be useful to relate this to the hard scattering process in the Bjorken limit. In the deep inelastic scattering process the photon couples just to those constituents of the hadron which carry the fraction x of the hadron momentum (x is the Bjorken scaling variable). When approaching the Regge limit $x \to o$, these constituents are more and more wee: the Pomeron feels the distribution of the wee partons inside the hadron. When the energy increases, i.e. the incoming hadron becomes more energetic, more decay processes are necessary before a fast parton slows down and eventually creates wee partons, and this may occupy a larger region in impact parameter space. As a result, the radius $\langle b^2 \rangle$ may grow as a function of s. In order to estimate how fast this growth could be in a realistic model, it may be useful to recall the multiperipheral model where

$$\frac{1}{s} T(s, b^2) = \frac{const}{\alpha' \ln s} e^{-b^2/4\alpha' \ln s} \tag{1.3}$$

and

$$\langle b^2 \rangle = const \cdot \alpha' \ln s \tag{1.4}$$

(α' is the Pomeron slope). Lowest order perturbation theory (Fig.2) in a field theory with massless vector particles [2], on the other hand, leads to

$$\frac{1}{s} T(s, b^2) \underset{|b| \to \infty}{\sim} |b|^{-4} \tag{1.5}$$

$$\langle b^2 \rangle = \infty . \tag{1.6}$$

This indicates that only after summing many more diagrams one may hope to come somewhat close to (1.3), (1.4): the quantity $\langle b^2 \rangle$ can serve as a guide in estimating to what extent the use of perturbation theory alone is sufficient to "confine" the wee partons inside the fast hadron, and the fact that it is infinite in lowest order perturbation theory of QCD indicates how difficult it may be to obtain a correct theory of the Pomeron.

Before I can start to describe how well understood the Pomeron is within nonabelian gauge theories (and this understanding is almost entirely based on perturbation theory), I have to mention the other approach towards a theory of the Pomeron, namely reggeon field theory (RFT). As it is well known[3], physics of the Pomeron is most easily be discussed in terms of singularities in the angular momentum plane, and the interaction of moving pole and cut singularities has been formulated by Gribov in his reggeon calculus (or reggeon field theory). The rules of this formalism follow from certain analyticity properties of the S-matrix (existence of partial wave continuation,and t-channel unitarity equations) and are expected to hold in field theories that contain moving Regge singularities. The values of the parameters of reggeon field theory (intercepts, slopes, and interaction vertices), however, are not very much constrained from these analyticity arguments alone, and as long as RFT has not been considered in the context of a specific underlying field theory, they have been choosen freely. As the most interesting case, the Pomeron with intercept one has been studied extensively, and the best-known result is the critical Pomeron theory with

$$\sigma_{total} \sim (\ln s)^{-\gamma} \quad , \quad \gamma \sim 0.2 \quad . \tag{1.7}$$

Since this solution has also been shown to be consistent with the most restrictive constraints imposed by s-channel unitarity, it is an excellent candidate for a theory of strong interactions at high energies. More recently [4], also the case of the Pomeron intercept being above one has been investigated. Apart from the question how presently available energy ranges fit into these Pomeron field theories, the outstanding theoretical problem remains the derivation of the Pomeron parameters from an underlying field theory, such as QCD. It is not unexpected that these features of angular momentum theory will play an important role in analyzing the high energy behavior of nonabelian gauge theories.

After this introduction I can begin with a brief outline of the program of my talk. The aim is a review of what at present we know about the Regge limit (i.e. the Pomeron) of nonabelian gauge theories, and since the problem has not yet been solved completely, I shall attempt to describe both was has been achieved so far and what seem to be the main strategies for the future. Most of the existing calculations are determined to find the high energy behavior of QCD, the theory of (confined) quarks and gluons, but mainly because of the infrared problems, they start from spontaneously broken gauge theories. The mass of the vector particles then is considered to be an infrared cutoff which at the end of the calculations is taken to zero, hoping that in this limit one reaches QCD. As I have said already, all this will be based on perturbation theory.

In the first two sections of my talk I shall outline our present understanding of what the formal behavior of massive (=spontaneously broken) nonabelian gauge theories is in the Regge limit. First I shall discuss the question of reggeization of elementary particles in these theories which divides the gauge theories into two classes: those where (at least) all vector particles reggeize and those where some of them don't. Then I shall describe (for a simple model) how this property of reggeization is seen to lead to a full reggeon calculus: this follows from the requirement of having (asymptotic) unitarity in both s and t channel, and the elements of the reggeon calculus are calculable in the limit of small coupling constant. Although such a reggeon calculus is of interest by itself, I shall consider it mainly as an intermediate step on the way towards finding the high energy behavior of massless Yang-Mills theories. This then requires a study of the zero mass limit of the reggeon calculus, and I shall briefly discuss what we know about this limit. As to the question how the use of perturbation theory may be extended into the Regge limit, this first part of my talk then basically selects all those terms in the perturbation expansion which have to be taken into account for a reliable high energy description: the selection criterion is unitarity, and the Feynmann diagrams which are included in the reggeon calculus are just enough to satisfy unitarity.

Section IV deals with the question of how to solve this reggeon calculus, i.e. how to perform the summation of all the terms that we have decided to keep. First I shall briefly scetch an approach which, although it has not been pushed very far yet, has the advantage of asking directly for the distribution of the wee partons in impact parameter space. It allows a rather direct control over $\langle b^2 \rangle$ and, according to what has been said in the introduction, over the validity of the use of perturbation theory. Moreover, this technique seems to be applicable also to QED, where a high energy description which takes full account of unitarity is still missing. Within this approach one sees the possibility that, after summing all terms that have been obtained in the first part, the distribution of wee partrons may come close to the multiperipheral picture. Then I shall describe how one might use the full apparatus of reggeon field theory, in particular its phase structure as a function of the bare Pomeron intercept, in order to determine the high energy behavior of massless Yang-Mills theories. Under the assumption that confining QCD can be obtained as the zero mass limit of spontaneously broken gauge theories with a modified $i\epsilon$ -prescription, this approach predicts critical high energy behavior for QCD, i.e. $\sigma_{total} \sim (\ln s)^{-\gamma}$.

II. Reggeization in Yang-Mills theories

Let us first consider gauge theories quite in general and ask which of them can be expected to have a "good" high energy behavior. The requirements one would like to impose on a realistic theory are the existence of moving Regge singularities and analyticity of the scattering amplitudes in the complex angular momentum plane. In particular, one would not like to have fixed singularities of the Kronecker delta function type which seem to exist if the theory contains nonreggeizing particles. This leads us to the question of reggeization in Yang-Mills theories.

It might be useful to recall what reggeization of a particle in a given field theory means. Suppose the theory contains a particle with spin J_0 and mass μ. The exchange of this particle in lowest order perturbation theory (Fig.3) yields the following contribution to the t-channel partial wave:

$$T(J,t) = const \cdot \delta_{JJ_0}, \tag{2.1}$$

which is nonanalytic in J. Higher order diagrams for the same amplitude then can have two possible effects: either they leave the lowest order term (2.1) unchanged and simply add some new contributions:

$$T(J,t) = const \cdot \delta_{JJ_0} \quad + \text{ terms analytic near } J_0 . \tag{2.2}$$

In this case the particle stays elementary and leads to a nonanalytic term in the partial wave amplitude. Alternatively, the higher order contributions remove the δ-function in (2.1), for example:

$$T(J,t) = \frac{\alpha(t) - J_0}{\alpha(t) - J} \cdot const \tag{2.3}$$

$$\alpha(t) = J_0 + (t-\mu^2)\cdot\beta(t) \tag{2.4}$$

($\beta(t)$ is proportional to the coupling constant of the theory and vanishes in lowest order perturbation theory. Eq. (2.3) then reduces to (2.1)). In this case the particle is said to reggeize: it lies on the trajectory (2.4), and the partial wave (2.3) is analytic in J. Experience with strong interaction physics clearly favors this second alternative: there is no evidence that singularities of the type (2.1) should be present. Therefore, one should look for theories in which, if possible, <u>all</u> particles reggeize.

Is there a simple way to decide whether, in a given theory, a particle reggeizes or not? The safest way, of course, is the explicit calculation: one computes the next-to-lowest order term of the partial wave and compares with the power series expansion of (2.3). This is the method by which, in the early sixties, Gell-Mann et.al.[5] found that the fermion in massive QED lies on a Regge trajectory. Based on these calculations the same authors derived certain criteria which must be satisfied if a particle is to reggeize. One of them implies that the theory must contain particles of spin one (or higher spin). This excludes scalar theories such as φ^3 or φ^4 (although these theories may still contain moving Regge singularities). Another criterium requires certain factorization properties of the Born amplitudes. Later on, Mandelstam[6] gave counting arguments which say under what conditions a particle <u>must</u> necessarily reggeize. All those methods, when applied to (massive) QED, agree in that the fermion reggeizes but the photon does not. The boson in scalar QED has also been found[7] to reggeize: this result came out only after direct calculation of Feynmann diagramms up to eighth order, and it illustrates that factorization and counting arguments [8] have to be applied with great care.

Theories in which also the vector particle reggeizes must be of the non-abelian type. To be more specific let us consider models of the following kind:

$$\mathcal{L} = -\frac{1}{4} F^a_{\mu\nu} F^{a\mu\nu} + \frac{1}{2} (\mathcal{D}_\mu \phi)^2 - V(\phi) + \text{spinor part} \tag{2.5}$$

$$F^a_{\mu\nu} = \partial_\mu A^a_\nu - \partial_\nu A^a_\mu + g f^{abc} A^b_\mu A^c_\nu \tag{2.6}$$

$$\mathcal{D}_\mu \phi = (\partial_\mu - ig A^a_\mu T^a)\phi \tag{2.7}$$

$$[T^a, T^b] = i f^{abc} T^c . \tag{2.8}$$

The potential $V(\phi)$ is invariant under the gauge group G and has its minimum at some nonzero value $\langle\phi\rangle \neq 0$. The pattern of spontaneous symmetry breaking may be rather complicated (for general symmetry breaking schemes see, for example, Refs.9 and 1o), and the resulting particle spectrum may mask the original gauge group G. Is there a <u>simple</u> criterium which tells us under which conditions the vector particles reggeize? We first answer the question for two popular models: (i) the

Higgs SU(2) model and (ii)the Weinberg-Salam model, and then state the result for the general case.

For the first case the gauge group G is SU(2) , and the scalar field comes in two SU(2) doublets. With the scalar potential

$$V(\phi) = -\frac{1}{2}\mu^2\phi^2 + \frac{\lambda}{4}(\phi^2)^2 \tag{2.9}$$

the Higg's mechanism makes all three vector particles massive:

$$M^2 = g^2 \mu^2/\lambda \ , \tag{2.10}$$

and leaves one (massive) scalar particle. (Generalization to SU(n) is made in the following way [11]: one starts from the gauge group U(n), adds n complex fundamental representations of scalar fields which make all n^2 vector particles massive, and then restricts oneself to the SU(n) subgroup of U(n)). Using the factorization criterion of the Born amplitudes, Grisaru et al.[11] found that both the fermions and the vector particles of this model reggeize. For the scalars the situation is still somewhat unclear: recently [8] it has been pointed out that it may reggeize, but in a more complicated way, similarly to the boson in scalar QED. This is contrary to the former belief [12] that reggeization of the scalar particle can occur only for special values of the parameters of the theory (masses and coupling constants). Presumably, only calculations similar to those of Ref.7 will settle this point.

For the Weinberg-Salam model with gauge group G = SU(2) $\times\, \mathcal{U}(1)$ one adds one doublet of complex scalar fields:

$$\mathcal{L} = -\frac{1}{4} F^a_{\mu\nu} F^{a\mu\nu} - \frac{1}{4} B_{\mu\nu} B^{\mu\nu} + \frac{1}{2}(\mathcal{D}_\mu\phi)^+(\mathcal{D}_\mu\phi) + V(\phi)$$
$$+ \mathcal{L}(\text{spinors}) \tag{2.11}$$

$$B_{\mu\nu} = \partial_\mu B_\nu - \partial_\nu B_\mu \tag{2.12}$$

$$V(\phi) = -\mu^2\phi^+\phi + \lambda(\phi\phi^+)^2 \ . \tag{2.13}$$

As a result of the Higg's mechanism one has the three massive vector bosons:

$$W_\mu^\pm = \frac{1}{\sqrt{2}} \left(A_\mu^1 \mp i A_\mu^2 \right) \tag{2.14}$$

$$Z_\mu = \frac{-g A_\mu^3 + g' B_\mu}{\sqrt{g^2 + g'^2}} \tag{2.15}$$

$$g' = g \cdot \tan \Theta_W \tag{2.16}$$

and the massless photon:

$$A_\mu = \frac{g B_\mu + g' A_\mu^3}{\sqrt{g^2 + g'^2}} \quad . \tag{2.17}$$

With the same arguments which have been used for the Higg's model one finds[13] that in the Weinberg-Salam model only the W-bosons reggeize whereas the Z and the photon don't. Obviously, it is the U(1) subgroup of G which destroys the reggeization: the W's being purely made out of the nonabelian A-fields still reggeize. The Z and the photon, on the other hand, contain the U(1)-type B-field, which destroys the reggeization.

The general connection between the structur of the gauge group G and the reggeization of the vector particles has recently been investigated by two groups[13) 14)]. It turns out that in order to make all vector particles reggeize G must be simple or semisimple. If G does not have this property - in particular if it has an abelian invariant subgroup - , some of the gauge particles lie on Regge trajectories, but others do not reggeize. The Higgs model with G= SU(2) and the Weinberg-Salam model with G = SU(2) x U (1) are examples for these two types of models. It is important to note that this result on the reggeization depends on the gauge group G but not on the way in which the Higgs scalars enter: this may introduce additional (global) symmetries into the theory, which manifest themselves in the mass spectrum of the vector particles but are independent of G. Finally, if after invoking the Higgs mechanism some vector particles are left massless, their trajectory functions (if they reggeize) have to be regularized by some infrared cutoff (cf.

(2.4)).

The most important implication of this result concerns the reggeization of the photon within grand unification schemes. In one of the most popular versions[15], weak, electromagnetic, and strong interactions are embedded into a SU(5) gauge theory in which, according to the result stated above, all vector particles reggeize:

$$SU(2) \times U(1) \times SU(3)_{color} \subset SU(5) \quad . \tag{2.18}$$

Such a scheme, for the first time, would allow to get rid of the undesired Kronecker-type partial wave singularities connected with the abelian photon, which according to our present understanding of this problem persist in QED and also the Weinberg-Salam model.

III. Construction of a (asymptotically) unitary S-matrix in a massive Yang-Mills theory. The zero-mass limit.

In this section I come to the longest part of my talk. Concentrating on the SU(2) - Higg's model which has been introduced in the previous section, I would like to describe how one can construct a high energy description of this model which satisfies (in an asymptotic sense) unitarity in both the direct and the crossed channel. The mass of the vector particle mainly serves as a convenient way to avoid the problems connected with massless particles and will be kept different from zero until the end of this section where I will mention what is known about the zero mass limit. The logic of this approach for investigating the high-energy behavior of massless Yang-Mills theories is illustrated in Fig.4a and b: in the massive case one studies the high energy behavior of processes with the (massive) vector bosons as external particles. In order to be able to take the zero mass limit, one has to replace the external particles by appropriate hadron wave function models. The gluon "soup" exchanged between these hadrons (the box of Fig.4b) is taken to be the zero mass limit of the box of Fig.4a. Within the line of arguments set up in the introduction the requirement of having for the massive case unitarity serves as a guide in selecting those terms in perturbation theory which one has to sum up for having a reliable description of the Regge limit of massless gauge theories.

After having established that the massive vector particles of the Higgs model reggeize, it seems very natural to expect that the full high energy description of this field theory should come in form of a complete reggeon calculus: the three reggeons (i.e. the reggeized vector particles) interact with each other through all possible (momentum dependent) interaction vertices which are allowed by signature conservation. This will, in fact, be the result of this section: starting from the requirement of both s and t-channel unitarity, a full reggeon calculus emerges. The method I am going to describe also allows, at least in principle, to compute the elements of this reggeon calculus.

Before I am going into more detail, a few words should be said about the method of calculation. So far the problem of finding a reliable description for the high energy behavior of a field theory has not been solved, and this failure has, at least in part, to do with the calculational technique. For each order of perturbation theory the high energy behavior of a scattering amplitude, say for the $2 \to 2$ process, can be written as:

$$g^{2n} s \left[(\ln s)^{n-1} f_{n-1}(t) + (\ln s)^{n-2} f_{n-2}(t) + \ldots + \ln s\, f_1(t) + f_0(t) \right] + O(s^0) + O(s^{-1}) + \ldots \quad . \tag{3.1}$$

Conventionally, the leading term of this expansion, $f_{n-1}(t)$, is found by writing down all Feynmann diagrams of this order perturbation theory, and, by means of a clever parametrization (Sudakov variables, infinite momentum variables, α-parameter representation), extratracting the highest power of lns. Summation over all orders in g then yields the leading -logarithmic approximation (LLA). As it turns out, however, in both QED and nonabelian vector theories this approximation violates the Froissart bound and, hence, is inacceptable. Because of the tremendous technical complications, the nonleading terms in (3.1) $f_{n-2}, \ldots, f_0$ can be computed so far only for very few special cases (for recent progress in this direction see Ref.16),but not for the vector theories we are interested in. In order to make further progress it seems, therefore, necessary to look for other methods of computation.

Since the main defect of the LLA was the violation of the Froissart bound, i.e. the lack of unitarity, the first goal must be restauration of unitarity. This suggests to use unitarity for the construction of the amplitudes from the very beginning: the Lagrangian is used only for determining vertices in the tree approximation. Amplitudes and all higher order corrections are then found by means of dispersion relations, i.e. by using our knowledge about the analytic structure of multiparticle amplitudes in the Regge limit. This garantees s-channel unitarity and, thanks to the reggeization of the vector particle, also t-channel unitarity in terms of partial wave unitarity. In terms of the expansion (3.1), presumably only parts of the nonleading coefficient functions $f_{n-2}, \ldots$ can be found in this manner: those which are necessary for achieving unitarity. In other words, what one obtains is likely to be the small- g approximation of the unitary S-matrix (which does not agree with the LLA). Whether this approximation is sufficient to give a reliable high energy theory can, at earliest, be answered after the summation of all these terms has been carried out. This problem will be subject of the next part of my lectures.

The presentation of how the construction of the unitary S-matrix works will be organized in the following way. Since extensive use will be made of the analytic structure of multiparticle amplitudes in the Regge limit, I start (part A) with a short review of those features which will be needed in the following. Then (part B) the construction of $T_{n\to m}$ in the LLA will be described. In part C this approximation will be unitarized, leading to the full reggeon calculus. In the final part D I shall discuss features of the zero mass limit.

A. Analytic structure of multiparticle amplitudes at high energies.

Throughout this section I shall take a very pragmatic attitude: rather than describing any proofs or derivations, I shall restrict myself to listing those results which will be needed in the following. Those who wish to learn more details I refer to the lectures of Stapp and White[17] in the Les Houches Summer School 1975 and to Refs. 18-2o.

Let me first recall a few well known facts about the $2\to 2$ amplitude at high energies:

(i) $T_{2\to 2}$ satisfies a dispersion relation in s with right and left hand cuts. For a theory with vector particles one needs two subtractions.

(ii) Real and imaginary part of the amplitude are connected via the signature factor:

$$T_{2\to 2} = \frac{1}{2\pi i}\int dj\; s^{j}\left(-i + \frac{\cos\pi j + \tau}{\sin\pi j}\right) F(j,t) \quad . \qquad (3.2)$$

The partial wave $F(j,t)$ is real in the physical region of the s-channel.

(iii) t-channel unitarity comes in form of partial wave unitarity relations. Starting from the normal t-channel unitarity equations, for example:

$$\mathrm{disc}_t\, T = \qquad + \ldots \qquad t \geq 16\mu^2 , \qquad (3.3)$$

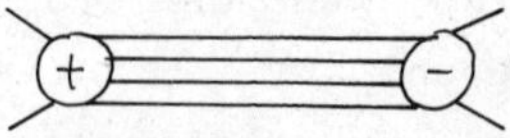

one projects out the partial waves $F(j,t)$, assumes that the two pairs of intermediate state particles couple together to moving Regge poles and continues down to $t < 0$:

$\mathrm{disc}_j\, F(j,t) =$ 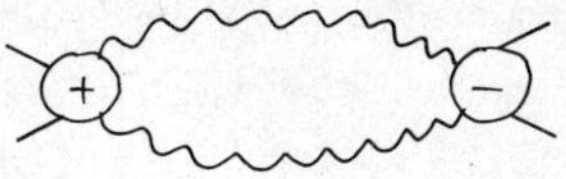 $+ \ldots$ (3.4)

("+" and "-" now refer to the j-variable). Together with

disc_j 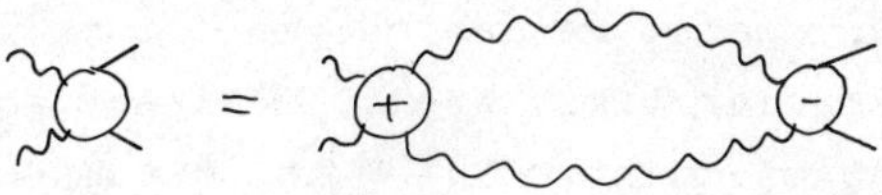 (3.5)

and

disc_j 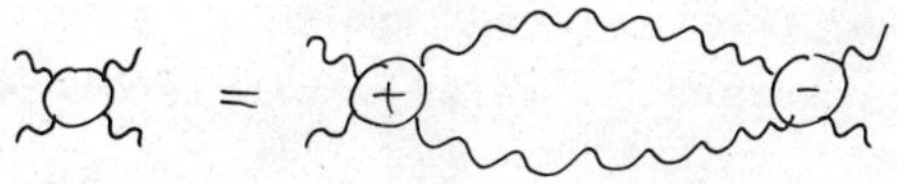 , (3.6)

one has a coupled set of "partial wave" unitarity equations: this is the form in which t-channel unitarity enters the region $s \to \infty,\ t \leq 0$. In order to obtain a solution to these unitarity equations, it is sufficient that the partial wave function F(j,t) takes the form of a reggeon calculus. It is important to note that at this stage the parameters of the reggeon calculus (trajectory function, interaction vertices) may be rather general: t-channel unitarity alone requires only that the formal rules of the reggeon calculus are satisfied (presence of signature factors etc.).

We now come to the simplest case of an inelastic amplitude: $T_{2\to 3}$ in the double Regge region (Fig.5). Before statements analogous to (i)-(iii) of $T_{2\to 2}$ can be made for $T_{2\to 3}$, we have to use one of the key results on the analytic structure of multiparticle amplitudes. It says that the amplitude $T_{2\to 3}$ in the double Regge limit splits into two parts, one having energy discontinuities only in the variables s and s_{ab}, the other one in s and s_{bc} (see Fig.6). In the partial wave representation one has:

$$T_{2\to 3} = \frac{1}{(2\pi i)^2} \iint dj_1\, dj_2 \Big[s^{j_2}\, s_{ab}^{j_1 - j_2}\, \xi_{j_2}\, \xi_{j_1 j_2}\, F_R + s^{j_1}\, s_{bc}^{j_2 - j_1}\, \xi_{j_1}\, \xi_{j_1 j_2}\, F_L \Big] \qquad (3.7)$$

$$\xi_j = \frac{e^{-i\pi j} + \tau}{\sin \pi j}$$

$$\xi_{jj'} = \frac{e^{-i\pi (j-j')} + \tau\tau'}{\sin \pi (j-j')} \quad . \qquad (3.8)$$

This decomposition is necessary for both s and t-channel unitarity. In the s-channel the Steinmann relations forbid simultaneous discontinuities in energy variables of overlapping channels (in the present case, the (ab) and(bc) channels are mutually overlapping). This problem is avoided when $T_{2\to 3}$ is written as in (3.7). Modern dispersion theory also proves that both pieces in (3.7) (or Fig.6) have only normal threshold singularities in their respective energy variables: more complicated singularities, such as Landau singularities, are subdominant in the double Regge limit. From the t-channel point of view the decomposition (3.7) is important, because only in this representation the partial waves F_L and F_R are real, i.e. free from internal phase factors. For both F_L and F_R a reggeon calculus exists [20] which satisfies t-channel partial wave unitarity.

We are now in the position to list the properties analogous to (i)-(iii). Once the decomposition (3.7) (or Fig.6) of $T_{2\to 3}$ has been made, we have:

(i) each of the two terms satisfies a double dispersion relation (with both right and left hand cuts and the appropriate number of subtractions).
(ii) Real and imaginary parts a related through the signature factors (3.8).
(iii) For each partial wave a reggeon calculus exists which satisfies t-channel partial wave unitarity.

The generalization to more general multiparticle amplitudes is now rather straightforward. The crucial step in each case is that one first has to find the necessary decomposition of the amplitude, before one is able to write a multiple dispersion relation for the amplitude or a reggeon calculus for the partial wave functions.

In Fig.7 this decomposition is illustrated for the two sixpoint amplitudes $T_{2\to 4}$ and $T_{3\to 3}$:it holds in the kinematic region where all energy variables are as large as possible and the momentum transfers and Toller angles kept fixed. A move detailed discussion of these amplitudes (in particular certain subtleties connected with the last two terms in the decomposition of $T_{2\to 4}$ and $T_{3\to 3}$) can be found in Ref. 21.

As it can be seen from these few examples, the number of terms in the decomposition grows rather fast as the number of external particles increases. In practical calculations, however, it seems not necessary to go beyond the sixpoint amplitude: it is believed that these amplitudes already contain all the essential complications of the analytic structure of multiparticle amplitudes (note that some of these complications do not yet show up in $T_{2\to 3}$: the five point amplitude is still "too simple"). Once the correct expression for these amplitudes has been found, it seems possible to generalize to higher order amplitudes.

B. $T_{n\to m}$ in the leading logarithmic approximation

The construction of the multiparticle amplitudes $T_{n\to m}$ in the LLA which will be described in the following has first been started by the Leningrad group [22] and then been carried through independently in Refs.23 and 21. A summary of the Leningrad school calculations can be found in Ref.24. I will not have the time to present calculations in detail but will concentrate on making the logic as clear as possible. I will use the notations of Ref.21, and more details can be looked up there. The starting point is the computation of tree graphs for $T_{n\to m}$ which, in the language of dispersion relations, serve as subtraction constants. For illustration consider the 2-> 2 vector scattering amplitude in lowest order perturbation theory. There are seven Feynmann diagrams (Fig.8), and one finds that some of them individually have a bad high energy behavior (e.g. they grow like s^2). However, when the sum is taken over all diagrams, these unwanted terms cancel and the final result has the appealing form:

$$2g^2 s \frac{1}{t-M^2} \cdot \text{helicity matrices} \cdot \text{group structure} , \qquad (3.9)$$

where the helicity matrices are constant (independent of s). As a graphical notation for (3.9) we use the diagram on the rhs of Fig.8. The fact that all (but one) Feynmann diagrams of this order are necessary for obtaining (3.9) illustrates the extensive cancellations between different contributions of pertubation theory which are typical for vector theories. For the next amplitude, $T_{2\to 3}$, the number of Feynmann diagrams, which have to be taken into account in order to find the correct high energy behavior of the tree approximation, is already much larger. But the result is again simple (Fig.9):

$$2g^3 s \frac{1}{t_1-M^2} \vec{\Gamma}(q_1,-q_2) \frac{1}{t_2-M^2} \cdot \text{helicity matrices} \cdot \text{group structure} , \qquad (3.10)$$

where the three component vector

$$\vec{\Gamma}(q_1,-q_2) = (\Gamma_\sigma e_\sigma^1, \Gamma_\sigma e_\sigma^2, \Gamma_\sigma e_\sigma^3) \qquad (3.11)$$

stands for the production vertex labelling the polarizations of the produced vector particle. It has a nontrivial dependence upon the momenta q_1, q_2 ($q_1^2=t_1$, $q_2^2=t_2$) and the Toller variable $\eta = s_{ab} \cdot s_{bc}/s$. For $T_{2\to 4}$ the result is shown in Fig.1o: it takes the form of a multiperipheral production amplitude, the production vertex being given by (3.11)[21]. $T_{3\to 3}$ is obtained from $T_{2\to 4}$ by crossing one of the produced particles.

An elegant method for computing these tree approximations for general $T_{n\to m}$ has been suggested by Lipatov [23]. Writing down a t-channel dispersion relation (without subtraction constants), only the particle pole contributes to the tree approximation of $T_{2\to 2}$, and for this only the on-shell vertex functions have to be computed. The result agrees with (3.9). For $T_{2\to 3}$ a double dispersion relation in t_1 and t_2 is needed which has to be saturated by the pole contributions. The only new element is the production vertex whose off-shell continuation follows from direct computation and the requirement of gauge invariance:

$$\Gamma_\sigma(q_1,-q_2)\cdot(q_1-q_2)^\sigma = 0 \quad . \qquad (3.12)$$

Proceeding in this way it is possible to verify the results of Figs.9, 1o and to show that for general $T_{n\to m}$ the tree approximations always have this multiperipheral structure. Within this approach the Lagrangian is needed only for the calculation of vertices in the tree approximation: tree amplitudes are built up by means of t-channel dispersion relations.

In the next step these tree approximations will be "dressed", and this is done by using s-channel dispersion relations plus unitarity. The amplitudes $T_{n\to m}$ in the LLA are then built up order by order perturbation theory. As a result of this "dressing" procedure, the elementary exchanges of the tree approximation will be reggeized.

Let me illustrate how this happens. To order g^4 one has the one loop contribution to $T_{2\to 2}$. For this amplitude the dispersion relation is:

$$T(s,t) = a(t) + b(t)\cdot s + \frac{s^2}{\pi}\int_{s_0}^{\infty} ds' \frac{\mathrm{disc}\, T(s',t)}{s'^2 (s'-s)} + \text{left hand cut} . \qquad (3.13)$$

The discontinuity follows from unitarity which, in this order of perturbation theory, has only the two-particle intermediate state:

$\mathrm{disc}\, T =$ 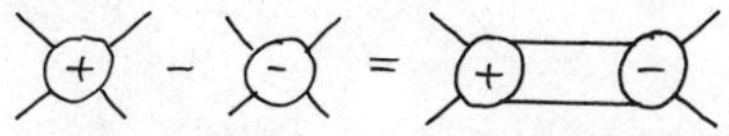 . (3.14)

On the rhs of this equation, only the $2 \to 2$ amplitude to order g^2 - which is the tree approximation - is needed: this we know from the first step of our calculations. Thus eqs. (3.13), (3.14) are sufficient to determine the leading term of $T_{2\to 2}$ in fourth order prturbation theory: the integral in (3.13) goes as $s \cdot \ln s$ and, hence, dominates over the subtraction constants. The result is the term of the order g^4 of the reggeizing vector exchange:

$$T_{2\to 2} = -g^2 s^{\alpha(t)} \frac{e^{-i\pi\alpha(t)} - 1}{t - M^2} \cdot \text{helicity factors} \cdot \text{group structure} , \qquad (3.15)$$

where

$$\alpha(t) = 1 + (t - M^2) g^2 \int \frac{d^2k}{(2\pi)^3} \cdot \frac{1}{k_\perp^2 + M^2} \cdot \frac{1}{(q-k)_\perp^2 + M^2} , \quad q_\perp^2 = -t . \qquad (3.16)$$

Comparison with the tree approximation (3.9) shows that the elementary exchange of (3.9) has been replaced by the reggeizing vector exchange.

In order g^5 we have to calculate the one loop correction to the $2 \to 3$ amplitude. In principle, we could proceed in the same way as we did for $T_{2\to 2}$: one uses the decomposition of Fig.6 and writes down a double dispersion relation for each of the two terms, including the right number of subtraction terms. The various discontinuities and double discontinuities are computed via unitarity, for example:

$\mathrm{disc}_{s_{ab}} T_{2\to 3} =$ 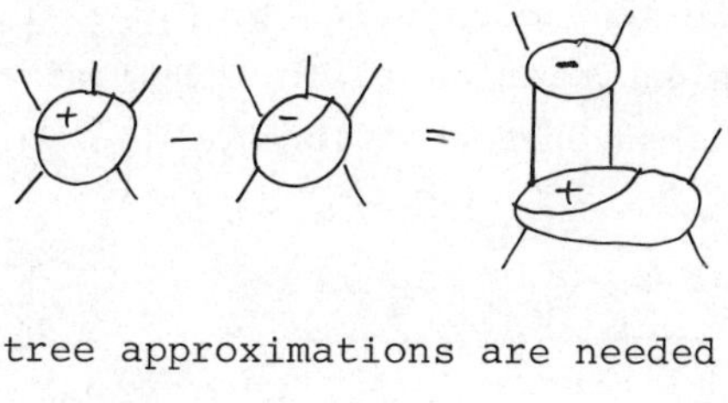 . (3.17)

On the rhs, only tree approximations are needed in this order of g. Let me, however, shortcut these calculations a little bit. I directly use the ansatz (3.7) and, anticipating the result that the singularities in j_1 and j_2 will be the poles belonging to the reggeizing vector particle, I simply write:

$$T_{2\to 3} = s^{\alpha_2} s_{ab}^{\alpha_1-\alpha_2} \xi_{\alpha_2} \xi_{\alpha_1-\alpha_2} F_R + s^{\alpha_1} s_{bc}^{\alpha_2-\alpha_1} \xi_{\alpha_1} \xi_{\alpha_2-\alpha_1} F_L \qquad (3.18)$$

(the functions $\alpha_i = \alpha(t_i)$ should, of course, be the same as in (3.16)). The unknown quantities are now the coefficient functions F_L and F_R. F_R, for example, is determined by taking the s_{ab} -discontinuity of eq.(3.18), expanding in powers of g, and comparing the term g^5 with the rhs of eq.(3.17). A consistency check can be made by taking the s-discontinuity of (3.18) and comparing it with the result of evaluating the unitarity equation which yields the s -discontinuity: both F_R and F_L in (3.18) are already fixed by the s_{ab} and s_{bc} discontinuities, resp., and no further freedom is left. What we have found in this way is that, up to this order of perturbation theory, $T_{2\to 3}$ is given by the exchange, in both the t_1 and t_2 channel, of the reggeized vector particle. This has to be compared with the tree approximation (Fig.9) where the exchanges are the elementary vector particles. In order to make this comparison more explicit (and also for later convenience), we rewrite eq. (3.18). Using the results for F_L and F_R [21], (3.18) can be written:

$$T_{2\to 3} = 2 g^3 s \, \frac{s_{ab}^{\alpha_1 - 1}}{t_1 - M^2} \, \vec{\Gamma}(q_1, q_2) \, \frac{s_{bc}^{\alpha_2 - 1}}{t_2 - M^2} \times \text{helicity matrices} \times$$

$$\times \text{group structure} \qquad (3.19)$$

(where terms of the order $g^5 \ln \frac{s_{ab} s_{bc}}{s}$ have been neglected). Eq. (3.19) is the form in which the double Regge exchange amplitude (Fig.11) would conventionally be represented: it is equivalent to (3.18), but it looses the information about the analytic structure in the energy variables.

In order g^6 two contributions have to be calculated: the two-loop correction of $T_{2\to 2}$ and the one loop correction to $T_{2\to 4}$ ($T_{3\to 3}$). For $T_{2\to 2}$ we again use the dispersion relation (3.13). The unitarity equation for the discontinuity now has several contributions:

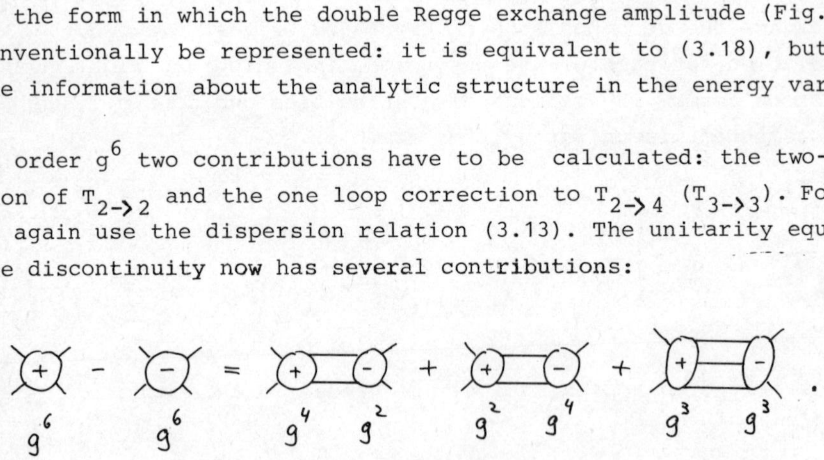

(3.20)

On the rhs of this equation, all amplitudes are known from previous steps: $T_{2\to 2}$ in order g^2 and g^4, and $T_{2\to 3}$ in the tree approximation. Inserting the result into the dispersion integral, we correctly reproduce the coefficient proportional to g^6 of the amplitude (3.15).

The one-loop contribution to $T_{2\to 4}$ (and $T_{3\to 3}$) is obtained in the same way as for $T_{2\to 3}$: to proceed most generally, one makes the decomposition (Fig.7) and writes a multiple dispersion relation for each term (with the right number of subtractions). Then unitarity equations are used for computing, in the given order, single and multiple discontinuities. But we again shortcut this procedure and make the ansatz analogous to (3.18). There are now five unknown coefficient functions which can be determined from the (single) discontinuities in the five subenergy variables. The discontinuity across the total energy s again serves as a consistency check. The result for $T_{2\to 4}$ can be written (Fig.12):

$$T_{2\to 4} = 2g^4 s \, \frac{s_{ab}^{\alpha_1 - 1}}{t_1 - M^2} \, \vec{\Gamma}(q_1, -q_2) \frac{s_{bc}^{\alpha_2 - 1}}{t_2 - M^2} \, \vec{\Gamma}(q_2, -q_3) \frac{s_{cd}^{\alpha_3 - 1}}{t_3 - M^2} \times \text{helicity terms} \times \text{group structure} \qquad (3.21)$$

and what we have just computed is the term g^6 in the power series expansion of this equation. This result (Fig.12) is the "dressed" generalization of the tree approximation in Fig. 10.

This procedure of calculating order by order perturbation theory all multiparticle amplitudes $T_{n\to m}$ in the LLA can be continued up to arbitrarily high order (in Ref.21, this has been done up to the order g^8). Let me, however, stop already here and state the general result. For the four, five, and sixpoint amplitudes we have found that the $T_{n\to m}$ have the simple multiregge form with only pole exchanges, and one should expect that this holds for general $T_{n\to m}$ (Fig.13). This then generalizes the reggeization of the vector particle, as it was found already by Grisaru et al.[11] for $T_{2\to 2}$ on the level of the Born approximation. A nontrivial feature of this result is the fact that no Regge cuts appear: signature conservation rules would very well allow for two Regge cuts in the central rapidity gap of $T_{2\to 4}$ (Fig.12), but as a result of some subtle cancellations[21] these cut contributions drop out for the LLA. As we shall see later, such cut contributions will, however, come in when we go beyond this leading logarithmic approximation, requiring full s-channel unitarity.

One may ask how well justified our extrapolation from $T_{2\to 2}$, $T_{2\to 3}$,

$T_{2\to 4}$, $T_{3\to 3}$ to general $T_{n\to m}$ was. As a "proof" for the correctness of this generalization one can perform a consistency check and test the unitarity content of the $T_{n\to m}$:unitarity puts nonlinear constraints on the elements of the set $T_{n\to m}$ which, on the level of the LLA, must be satisfied if our result is correct. For the simplest case, $T_{2\to 2}$, it can, in fact, be shown [25] that the elements $T_{2\to 2}$ and $T_{2\to n}$ satisfy the "bootstrap" equation:

$\mathrm{disc}_s\, T_{2\to 2} =$

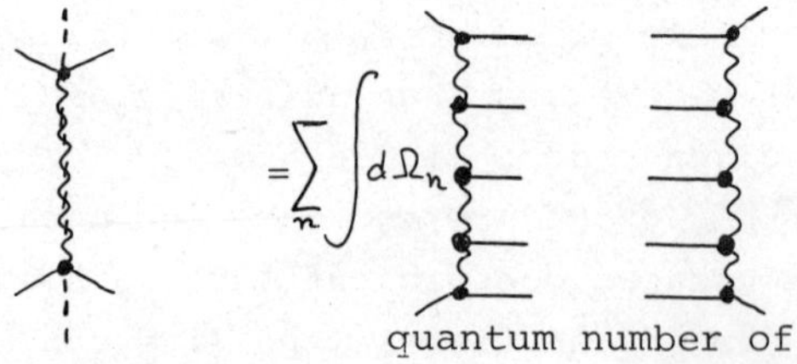

(3.22)

quantum number of the vector particle

When "squaring", on the rhs of this equation, the $T_{2\to n}$ amplitudes, one has to take that quantum number configuration which corresponds to the exchange of the vector particle: from Fig.13 it is clear, that in the LLA <u>all</u> t-channels carry the quantum number of the vector particle and, in particular, there is no vacuum quantum number exchange yet. (For our SU(2) model we have, after the symmetry breaking due to the Higg's mechanism, a global SU(2) symmetry. If we call this symmetry, for the time being, isospin, than the vector particle carries the quantum number I= 1. On the rhs of eq. (3.22) we then have the possibilities I = o, 1, 2, and it is the I = 1 configuration that we must take).

For $T_{2\to 3}$ we have three constraints given by unitarity. In Ref.21 it is shown that the following relations hold:

$\mathrm{disc}_{s_{bc}}\, T_{2\to 3} =$

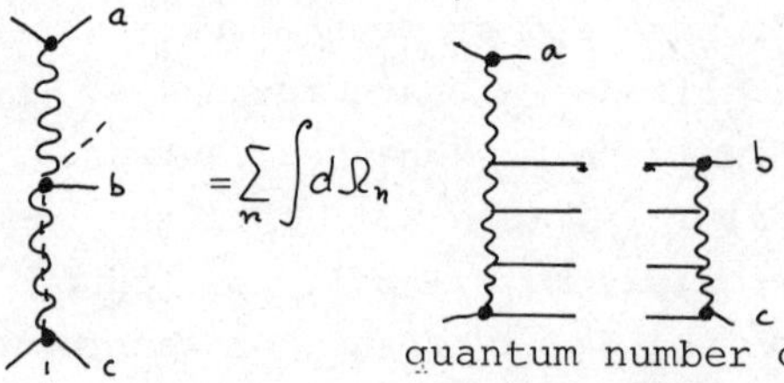

(3.23)

quantum number of vector particle

and

$\mathrm{disc}_s\, T_{2\to 3} =$

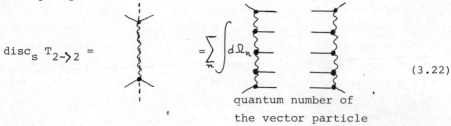

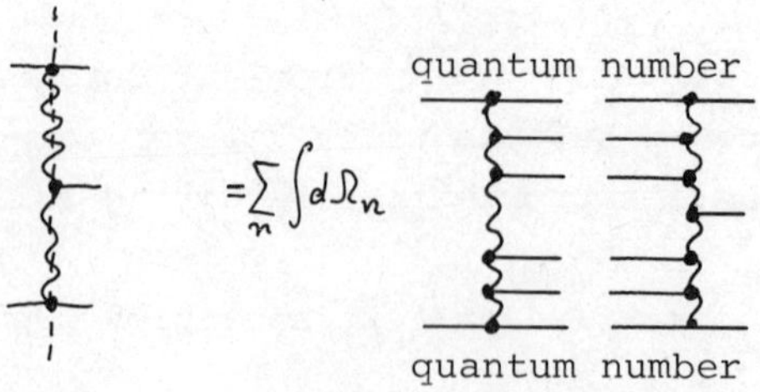

(3.24)

quantum number of vector particle

quantum number of vector particle

On the rhs of these equations the t-channel quantum numbers again have to be that of the vector particle. In the same way it can be shown that for $T_{2\to 4}$ and $T_{3\to 3}$ unitarity holds for all energy variables.

In order to summarize these unitarity properties of the $T_{n\to m}$ in the LLA I use a matrix notation. Let T be the matrix whose elements are the $T_{n\to m}$:

$$T = \begin{pmatrix} T_{2\to 2} & T_{2\to 3} & \cdots \\ T_{3\to 2} & T_{3\to 3} & \cdots \\ \cdots & \cdots & \cdots \end{pmatrix}, \qquad (3.25)$$

and let the subscript "1" remind us that we are dealing with the LLA. Then eqs. (3.22) - (3.24) are elements of the following matrix equation:

$$T^{(1)} - T^{(1)^+} = 2i \cdot T^{(1)} \cdot T^{(1)^+} \Big|_{\text{quantum number restricted}} \qquad (3.26)$$

On the rhs, all t-channels must have the quantum number of the vector particle This restriction signals that $T^{(1)}$ is not yet completely unitary: to find the missing pieces will be the task of the following subsection.

C. Unitarization

In order to find a T-matrix which is fully unitary, i.e. satisfies eq. (3.26) without any restrictions on the rhs, we make the ansatz:

$$T = \sum_n T^{(n)} \qquad (3.27)$$

with all the $T^{(n)}$ being matrices of the form (3.25).The expansion parameter is the following. $T^{(1)}$ is the LLA which has been obtained in the previous part; in the sense of the expansion in eq.(3.1) it represents the sum of the leading terms $f_{n-1}\ (\ln s)^{n-1}$. $T^{(2)}$ is the sum of the next-to-leading terms, but as it was said in the beginning of this section, only those parts of the f_{n-2} will be found which are required by unitarity.Similarly, $T^{(3)}$ corresponds to the f_{n-3}, etc.

To begin with the elements of $T^{(2)}$, we recall that $T^{(1)}$ had nothing but the quantum number of the vector particle in all exchange cahnnels. This was because the leading power of lns in each order of perturbation theory always belongs to odd signature (the expansion of the signature phase factor ($e^{-i\pi\alpha(t)} \pm 1$) in powers of g^2 starts with the constant -2 for odd signature, but with $O(g^2)$ for even signature). The requirement that the amplitude is odd under s <-> u crossing projects out the quantum number of the vector particle. $T^{(2)}$ therefore must contain even signature exchanges, in particular the Pomeron. The easiest way to find these amplitudes is via unitarity:

$$T^{(2)} - T^{(2)+} = 2\, T^{(1)}\, T^{(1)+} \Big|_{\text{even signature}} \tag{3.28}$$

This <u>defines</u> $T^{(2)}$: on the rhs, at least <u>one</u> t-channel must have even signature, otherwise we would be back at eq. (3.26) and nothing new would have been found. Eq. (3.28) defines discontinuities; for obtaining the full amplitudes one uses the Sommerfeld-Watson representations, e.g. (3.7). The simplest example is $T_{2\to 2}^{(2)}$. From (3.28) we have:

$$\mathrm{disc}_s\, T_{2\to 2}^{(2)} = \sum_n \int d\Omega_n \, \left| T_{2\to n} \right|^2_{\text{even signature}} \; . \tag{3.29}$$

This determines the partial wave of $T_{2\to 2}$ and is illustrated in Fig.14. For one of the even signature channels, the Pomeron, the leading singularity in the j-plane comes out as a fixed cut [25)] to the right of j=1: it violates the Froissart bound and also dominates the LLA (3.15), both being a clear indication that the expansion (3.27) cannot be truncated after the first or the second term.

In case of the 2- 3 amplitude the signature degree of freedom allows for three amplitudes contributing to $T^{(2)}$: the configurations (τ_1, τ_2) = (-,+), (+,-), (+,+). For the first case a closer look at the signature factors (i.e. counting powers of g^2 in eq. (3.81)) shows that, out of the two terms in the decomposition of $T_{2\to 3}$(eq.(3.7) or Fig.6), the second one which has the s_{bc}-discontinnity dominates over the first one. Hence, the amplitude can be constructed out of the s_{bc} - discontinuity alone:

$$\mathrm{disc}_{s_{bc}}\, T_{2\to 3}^{(2)} = \sum_n \int d\Omega_n \; T_{2\to n+1} \; T_{2\to n}^{+} \Big|_{\text{even signature}} \; . \tag{3.30}$$

In terms of reggeon diagrams, this equation is illustrated in Fig. 15. For the signature configuration (+ , +), the g^2-expansion of the signature factors implies that the amplitude is proportional to its s-discontinuity:

$$T^{(2)}_{2\to 3} \sim disc_s\, T^{(2)}_{2\to 3} = \sum_n \int d\Omega_n\, T_{2\to n}\, T^{*}_{n\to 3} \Big|_{even\ signature} \qquad (3.31)$$

The reggeon diagrams for this amplitude are shown in Fig. 16.

The construction of $T_{2\to 4}$, $T_{3\to 3}$ proceeds in the same way. For the various signature configurations (-, -, +), (-, +, -), (+,-,-), (+,+,-) (-,+,+) (note that (+, -, +) does not belong to $T^{(2)}$ but to $T^{(3)}$) , it is always sufficient to compute single discontinuities, and what one obtains are diagrams similar to Figs. 14-16. The following pattern then emerges for the elements of $T^{(2)}$: whereas the elements of $T^{(1)}$ (Fig.13) have always just one reggeon in the t-channel, those of $T^{(2)}$ (Figs.14-16) can have either one (for odd signature exchange) or two reggeons (for even signature exchange) in each t-channel. Since the rules, according to which the reggeon diagrams of Figs.14-16 are constructed, agree with the general reggeon calculus for inelastic production amplitudes[2o)], the elements of $T^{(2)}$ also satisfy t-channel unitarity. The elements of this reggeon calculus (Fig.17) are obtained from the defining equations (3.28), (3.29) (analytic expressions will be given in Ref.26). To complete the construction of $T^{(2)}$, let me mention that also certain nonleading terms obtained from expanding the signature factors of the elements of $T^{(1)}$ have to be counted as elements of $T^{(2)}$.

The construction of $T^{(3)}$, $T^{(4)}$... essentially repeats the steps which have lead to $T^{(1)}$ and $T^{(2)}$. At the level of $T^{(3)}$ new contributions to the partial waves with only odd signature exchanges appear: this are reggeon diagrams involving the higher order 1-3 reggeon vertex (Fig.18) or the one reggeon + three reggeons $\to$ particle production vertex (Fig. 19). Compared to the diagrams of $T^{(1)}$ (Fig.(3)), these new contributions have two more powers of g^2 (or, in other words, are down by two powers of lns). The lowest order (in powers of g^2) contributions to Figs.18 and 19 are shown in Fig.2o : these diagrams, having only elementary exchanges, are of the order g^6s and g^7s, respectively, and contain no logarithm of any energy variable. Furthermore, they are real. This implies that they cannot be obtained by just iterating s - channel unitarity (in the language of dispersion relations, they are subtraction constants), but they must be computed by hand: going back to the Lagrangian, one has to use methods which are similar to those which were used for the tree approximations at the level of $T^{(1)}$. Details on this will be found in Ref. 27. To find the elements of $T^{(3)}$, one proceeds very much in the same way as we did for $T^{(1)}$: the lowest order elements (Fig.2o) play the same role as the tree approximations, and $T_{2\to 2}$, $T_{2\to 3}$, ... are computed order by order perturbation theory by computing the energy discontinuities from unitarity equations (cf (3.14), (3.17), (3.2o)). On the rhs of these equations, a careful counting of powers of g^2 is needed, and contributions from $T^{(1)}$, $T^{(2)}$, and $T^{(3)}$ have to be taken into account. As a result of these calculations the elementary exchanges of Fig.2o are "dressed", i.e. they are reggeized, and they also interact via the quartic reggeon vertex which was found in the previous step. The unitarity content of $T^{(3)}$ is the following:

$$T^{(3)} - T^{(3)\dagger} = 2i \left[T^{(1)} T^{(3)\dagger} + T^{(2)} T^{(2)\dagger} + T^{(3)} T^{(1)\dagger} \right]_{\text{odd signature}} . \qquad (3.32)$$

This is the analogue of eq. (3.26) for $T^{(1)}$. As in the case of $T^{(2)}$, also nonleading terms of $T^{(1)}$ and $T^{(2)}$ have to be counted as elements of $T^{(3)}$: they are obtained from expanding the signature factors in powers of g^2.

$T^{(4)}$ can be obtained from s-channel unitarity without computing new subtraction constants:

$$T^{(4)} - T^{(4)\dagger} = 2i \left[T^{(1)} T^{(3)\dagger} + T^{(2)} T^{(2)\dagger} + T^{(3)} T^{(1)\dagger} \right]_{\text{even signature}} . \qquad (3.33)$$

In the matrixelements on the rhs, at least one t-channel must have even signature. Otherwise we would be back at (3.32). Eq. (3.33) is the analogue of (3.28) for $T^{(2)}$. The reggeon diagrams of $T^{(4)}$ contain up to four reggeons in the t-channels.

Repeating these steps, higher and higher $T^{(n)}$ are obtained: at each step the (maximal) number of reggeons in an exchange channel increases by one, and new elements (vertices with a nontrivial momentum dependence) appear. The result for T is a complete reggeon calculus, with the reggeizing vector particle being the reggeon and having (infinitely many) selfinteration vertices. In principle all these vertices are calculable, but so far only a few of them are known, and to find a simple expression for the most general n->m reggeon vertex remains a subject of future work.

The fact that the result of our unitarization procedure comes in form of a complete reggeon calculus was to be expected as soon as the reggeization of the vector particle had been established. For future investigations it might, however, be useful to mention that, by a slight rearrangement in the expansion of T, a more physical picture of the (elastic) scattering process can be obtained. The idea is simply to reexpand each reggeon diagram in the expansion

$$T_{2\to 2} = \sum_n T^{(n)}_{2\to 2} \tag{3.34}$$

in powers of $g^2/(j-1)$ (note that each reggeon line by itself represents a power series in this parameter: $[j-1-(\alpha(t)-1)]^{-1} = [j-1]^{-1} \cdot \sum_m \left(\frac{\alpha-1}{j-1}\right)^m$ with $\alpha-1 = O(g^2)$:

$$F^{(n)}_{2\to 2} = \sum_m \left(\frac{g^2}{j-1}\right)^m t^{(n)}_m \tag{3.35}$$

($F^{(n)}_{2\to 2}$ is the partial wave of the amplitude $T^{(n)}_{2\to 2}$). Since the variables j-1 and lns are conjugate to each other (cf. eq.(3.2)), (3.35)leads to an expansion of $T_{2\to 2}$ in powers of g^2 lns and has a physical interpretation close to that of the well-known multiperipheral model (for a description of these ideas see Ref.27). The term proportional to $(g^2 \ln s)^m$ represents the following subprocess of elastic scattering: out of the incoming fast hadron which is a composite system of virtual constituents (partons), some parton has initiated a m-step cascading decay. At the end of this decay slow partons (wee partons) have been produced which can interact with the target at rest. This process is illustrated in Fig.21. In the introduction I raised the question whether the hadron radius can be made finite: this means that we are interested in the distribution of these wee partons in impact parameter space. As I will explain a little later, the expansion (3.35) may be a better starting point for investigating this question than the reggeon calculus representation of the T-matrix that was obtained in the first instance. This is the reason why I mentioned this second form of representing the matrix T.

D. The zero mass limit

In the first three parts of this section I have been dealing with the question of how to select those terms in the perturbation expansion which are needed for obtaining a reliable high energy description. All this was done for the massive SU(2) Higg's model, but our final aim is the pure Yang-Mills case. We therefore have to investigate how our T-matrix behaves under the limit which takes us from the Higg's model to the pure Yang-Mills case.

The T-matrix whose construction I have outlined before depends only on the two parameters g (the gauge coupling) and the mass of the vector particle $M^2 = g^2\mu^2/\lambda$ (cf. eq.(2.1o)), but not on the Higg's parameters μ and λ seperately. For our purposes it is, therefore, sufficient to demand that $M^2 \to o$, g staying fixed. A brief investigation of low order perturbation theory shows that, in order to decouple the Higgs sector from the gauge particles, on should take λ and μ to infinity such that $\mu^2/\lambda \to 0$. For the time being, I shall concentrate on the question how our T-matrix behaves when M^2 is taken to zero. But it seems to me that a more accurate study of the transition from the Higgs model to the pure Yang-Mills case would be very desirable.

First it is necessary to replace the external states which so far have been taken to be massive vectors and scalars. It is wellknown from QED calculations[28] that the simplest case of a high energy scattering amplitude which is finite in the zero mass limit of the photon is that of elastic photon-photon scattering (or elastic photon-electron scattering) the incoming photon dissociates into a electron-positron pair which interacts with the target via photon exchanges (note that elastic electron electron scattering via multiphoton exchange is not infrared finite). Th can easily be generalized to the nonabelian case[24] (Fig. 22a): replace the external photons by hadrons, say vector mesons with some wave functions, and take the fermions to be quarks. It can then be shown[26] for $T^{(2)}_{2\to 2}$, the first term in the expansion (3.27) which contributes to elast scattering, that in the zero quantum number exchange channel the limit $M^2 \to o$ exists and is finite to all orders of g^2.For higher terms in (3.27 $T^{(4)}_{2\to 2}$ etc., this can be shown [26], so far, only for important subsets of terms (for example those shown in Fig.22b); but from the results of studying infrared singularities in hard scattering processes[29] it seems likely that in the vacuum quantum number (color zero) channel infrared singularities should always cancel.

Let me assume that this, in fact, is true for all the $T^{(n)}$ in (3.27). Then our present situation can be described as follows (Fig.23).Starting again from the deep inelastic region where the use of perturbation theory (and in this case even the summation of only leading logarithms) rests on a safe ground, we now have isolated those Feynmann diagrams which have to be summed when the Regge limit is taken (q^2 fixed and $x \to o$). They are obtained as the zero mass limit of our T-matrix which is coupled to the quark loop as external source.

I finish this long section with a few comments on other approaches to the same problem. When describing the derivation of the LLA, I have restricted myself to that method which, as I believe, is most suitable for achieving unitarity: the use of the analytic structure of multiparticle amplitudes together with unitarity. Other groups of authors[30)31)] have followed the more conventional method of investigating Feynmann integrals and extracting the leading term by use of a clever choice of integration variables. This approach has so far been restricted to the 2->2 amplitude (with one exception[32)]) in the LLA and one step beyond (in our notation: $T^{(2)}_{2\to 2}$). Wherever a comparison can be made, the results of the different approaches agree. As to the next logical step, namely the unitarization of the LLA, Refs.33 and 34 claim that the fully unitary S-matrix takes a simple eikonal form, both for QED and the nonabelian case. However, when reexpanding this eikonal representation, it appears that pieces are missing which are necessary for having s and t-channel unitarity. From the s-channel point of view, subchannel unitarity is not satisfied, i.e. rescattering contributions to inelastic production amplitudes are missing. T-channel unitarity (partial wave unitarity) requires that the lowest order g^2-expansion coefficients of the 3-reggeon, 5-reggeon,.... cut diagrams are real. Hence, they cannot be obtained from iterating s-channel unitarity alone, as it is done in the eikonal expression of Ref.34.

A very different approach has been taken in Ref.35. For the case of quark-quark scattering, the leading infrared divergent terms are isolated by means of the equations of Cornwall, Tictopoulos and Korthaus-Altes, de Rafael, and then the behavior of these terms in the Regge limit is studied. The result is a fixed cut singularity at j=1. Compared to the procedures which I have been describing so far, this amounts to taking the two limits (Regge limit $s \to \infty$ and zero mass limit $M^2 \to o$) in the reverse order. As it has been shown by Bronzan and Sugar[36)], these two limits do not commute: the terms found in Ref.35 (first $M^2 \to o$, then $s \to \infty$) form a subset of those obtained from the other approach (first $s \to \infty$, then $M^2 \to o$) and, hence, do not seem to satisfy unitarity.

IV. Summation of the diagrams

I now come to the final part of my talk: how can one try to sum all the contributions that have been obtained in the previous section? Let me recall the two quantities we wanted to concentrate on: the s - dependence of the total cross section as the most important observable, and the hadronic radius $\langle b^2 \rangle$ as a test for the reliability of the calculations. Unfortunately, I will not be able yet to give you final answers. The task of summing all these contributions of T (or, at least, of extracting the relevant information about the leading s-behavior) requires new techniques, and all I can do is to outline the main ideas and mention those results which we already have. For the investigation of the two quantities σ_{total} and $\langle b^2 \rangle$ two different approaches seem to emerge: the first one starts from the reggeon calculus representation of the S-matrix and than makes use of the phase structure of reggeon field theory which has been studied within the last few years. For a study of the parton distribution in b - space, on the other hand, the power series (3.35) seems to be a good starting point, and I would like to begin with this approach first.

To be specific, let us consider the model illustrated in Fig. 22 (elastic scattering of two q-$\bar{q}$ bound states via gluon exchanges), assuming that the zero mass limit exists for all $T^{(n)}_{2\to 2}$ in the vacuum exchange channel. As explained before, the expansion in powers of $g^2 \ln s$ (c.f. (3.35)) can be related to the parton picture: each power of $g^2 \ln s$ stands for a change in rapidity by one unit, and the term $(g^2 \ln s)^m \, t_m$ corresponds to a m-step decay of some fast constituent into slower ones such that at the end wee partons emerge. For the rest frame of the target, this situation is illustrated in Fig.21, for the CM-system in Fig. 24: each horizontal line denotes a point in rapidity (i.e.all points on one line have the same rapidity), but each vertex has its own impact parameter coordinate. The (statistical) distribution of all these points in impact parameter space defines the extension of the hadron. For the lowest approximation to the elastic scattering process, $T^{(2)}_{2\to 2}$, there is a direct analogy to Figs.21 (or 24): two successive steps in the expansion (3.35) are connected by a single two-dimensional transverse momentum integration, i.e. $t^{(2)}_{m+1} = K \cdot t^{(2)}_m$ with K being an integral operator (the explicit form of this recursion relation can be found in Ref. 25) or $t^{(2)}_{m+1} = K \times K \times \ldots K \times t^{(2)}_1$. The lines in Fig. 24a then denote the flow of transverse momentum in $T^{(2)}_{2\to 2}$. At the level of $T^{(4)}_{2\to 2}$, the momentum flow becomes more complicated (Fig.24b): between two rungs there may be more than two vertical lines, since the Kernel in $t^{(4)}_{m+1} = K \cdot t^{(4)}_{m}$ involves more than one $k_\perp$ -integral. With increasing n the number of vertical lines (i.e. the number of $K_\perp$- integrations in $T^{(n)}_{2\to 2}$) increases, giving rise to more and more interaction between partons of different rapidity.

In order to study the b-distribution of the partons in Fig.24 we shall investigate how the leading j-plane singularity of the partial wave in (3.35) is generated (i.e. we study the behavior of the expansion near the rightmost value of j for which the series diverges). For this we use an observation made by Kuraev et al.[25] for the case of $T^{(2)}_{2\to 2}$: the divergence of the expansion

$$F^{(2)}_{2\to 2} = \sum_m \left(\frac{g^2}{j-1}\right)^m t^{(2)}_m \tag{4.1}$$

comes from a specific region of phase space of the $k_\perp$ - integrations in Fig. 24a. Each $k_\perp$ -integration is perfectly finite, but for large m (which is the number of rungs or cells in Fig.24a) the dominant region of integration in those cells which are far away from the external particles moves more and more towards large $k_\perp$ -values:

$$\langle \ln k_\perp^2 \rangle = c_2 \cdot \sqrt{m} \quad , \tag{4.2}$$

where c_2 is a computable number (note that this type of growing transverse momentum is quite different from that found in hard scattering processes). Eq.(4.2) means that the average value of $\ln k_\perp^2$ obeys a diffusion law as a function of the number of steps. Once $k_\perp^2$ is large, the resulting singularity will not depend on finite quantities such as the mass M^2 or the momentum transfer $q^2 = -t$: this explains its nature of being a fixed cut. The j-value j_c for which this singularity arises lies to the right of j = 1 and leads to a total cross section which grows like $\sigma_{total} \sim s^{j_c - 1}$ (Fig.25a). Since with (4.2) also $\langle k_\perp^2 \rangle$ grows, as the number of steps m increases, the variable b^2 conjugate to $k_\perp^2$, which stands for the distance in impact parameter between neighboring vertices in Fig. 24, becomes smaller and smaller, and the parton distribution inside the upper (or lower) hadron is of the form shown in Fig. 25b. As to the zero mass limit $M^2 \to 0$, the most interesting point is that of $q^2 = -t = o$ [24]: the large b-behavior comes from the small $q_\perp$ -region. For $M^2 = o$, $q^2 = o$ the diffusion picture of $\ln k_\perp^2$

still holds, but $\langle \ln k_\perp^2 \rangle$ now moves in both positive and negative direction:

$$\langle \ln k_\perp^2 \rangle = \pm c_2 \sqrt{m} \tag{4.3}$$

(c_2 being independent of M^2 is the same as in the massive case (4.21)). In b-space the large negative values of $\langle \ln k_\perp^2 \rangle$,i.e. the small values of $k_\perp^2$, allow for longer and longer steplengths in b-space, and the radius $\langle b^2 \rangle$ grows too fast (as a power of s). This implies that, at the level of the approximation $T^{(2)}$, (a) the radius $\langle b^2 \rangle$ is too large, and (b) the limit $M^2 \to o$, although it exists order by order perturbation theory, is discontinous for the leading s-behavior. The last point has been made explicit[24)] by solving at t= o the integral equations of $T^{(2)}_{2\to 2}$ for $M^2 \neq o$ and for $M^2 = o$: there is a jump in the s-behavior of σ_{total} at the point ($M^2 = o$, $t = o$), compared to $M^2 \neq o$ and / or $t \neq o$.

As a guideline to what the situation in a realistic high energy theory should be, it might be usefull to recall a few features of the multiperipheral model. Writing the amplitude in the form (4.1), one find that $t_m \sim [\beta(t)]^m$ with $\beta(t)$ being the integral in equation (3.16), and the resulting j-plane singularity is a moving pole. Since the $k_\perp$ -integrations in $[\beta(t)]^m$ are always superconvergent, and their mean välues do not depend on m at all, the average steplength in b space is constant, and we have the well known random walk picture in b-space with $\langle b^2 \rangle \sim \alpha' \ln s$. This suggests that in our nonabelian gauge theory model we should look for a mechanism which stops the growth of $\langle k_\perp^2 \rangle$ as a function of the number of steps.

Let me briefly outline[37)] how the presence of the higher $T^{(n)}$ could lead to a change in the right direction (as long as one does not know the form of general $T^{(n)}$ in full detail I can describe this only qualitatively). A simple dimensional argument for the general n->m reggeon vertex shows that the diffusion law (4.2) for $\ln k_\perp^2$ will always hold, provided the limit $M^2 \to o$ is finite. The only new feature compared to $T^{(2)}_{2\to 2}$ is that the momentum integration between two steps in Fig. 24b now may consist of two or more $k_\perp$ -loops, and the variable which grows is the mean value of these $k_\perp$'s :

$$\ln k_\perp^2 = \frac{1}{n} \left[\ln k_{\perp 1}^2 + \dots + \ln k_{\perp n}^2 \right] \sim c_n \sqrt{m} \tag{4.3}$$

or

$$k_\perp^2 = \sqrt{k_{1\perp}^2 \cdot k_{2\perp}^2 \cdot \ldots \cdot k_{n\perp}^2} \quad . \tag{4.4}$$

The numbers c_n in (4.3), belonging to the approximation $T^{(n)}$, will be different from c_2 in (4.2): if the growth of $\ln k_\perp^2$ should come to a stop, we must have $c_n \to o$ as $n \to \infty$. The situation of the n variables $\ln k_{i\perp}^2$, whose "center of mass" coordinate obeys the diffusion law, resembles that of the one-dimensional motion of n atoms, moving in a potential which depends only on the relative distance of the atoms from each other, but not on the center of mass position. In such a case the center of mass coordinate obeys the diffusion law, and depending on whether the relative forces between the atoms are attractive or repulsive the diffusion will be slower or faster than in the absence of those forces. The crucial observation now is that, if the forces are sufficiently attractive, the motion of the center of mass can come to stop when the number of atoms becomes infinite. Applying these ideas to our $T^{(n)}$ we see that if the number of $k_{\perp i}$ -variables in (4.4) becomes very large - i.e. in Fig. 24 there is more and more interaction between different horizontal lines, each of which represents a certain rapidity in the "gluon cloud" around the incoming hadron - the growth of $k_\perp$ towards the center of Fig.24 can come to a stop, and the impact parameter steplength stays finite and constant. About the s-dependence of σ_{total} very little can be said as long as this argument has not been made quantitative yet: if the series of the $T^{(n)}$ converges, the cross section (Fig.25a) must flatten out at high energies in order to satisfy the Froissart bound.

It is important to mention that the same type of analysis has also to be carried out for the abelian case of QED. At the level of the LLA for the Pomeron channel, the leading singularity of the tower diagrams [38),39)] in QED is also a fixed cut to the right of j= 1 and has very much the same characteristics as in the nonabelian case. Important differences between the two cases are expected to come in when the effects of the higher approximations $T^{(n)}$ are included (it seems that the sum of all those diagrams which are described in Ref.38 and Ref.4o for QED represents the analogue of the T that we have discussed for the nonabelian case.The eikonal graphs of Ref.39 and even the "operator eikonal" expansion of Ref.33 only form a subset of the more general class of diagrams in Refs. 38 and 4o and do not satisfy full s-channel unitarity).

Let me now describe the other approach towards analyzing the structure of T that I mentioned at the beginning of this section. It starts from the reggeon calculus representation of T and then uses the phase structure of reggeon field theory (RFT) which has been investigated during the last years. As it is well-known[3], RFT lives in two space and one time dimension (impact parameter and rapidity), and it has a nonrelativistic energy-momentum relation: $E = \Delta + \alpha' k_\perp^2$ (E=1-j, j = angular momentum; $\Delta = 1 - \alpha(0)$; $\alpha(0)$ and α' are intercept and slope, respectively, of the trajectory function). This it is quite different from relavistic quantum field theory, and since, moreover, the triple interaction vertex (at least for the Pomeron case) is purely imaginary, it is clear that the phase structure of RFT, as a function of the "mass" Δ , is not the same as in usual quantum field theory models. It will, therefore, be useful to first review what we know about the phases of RFT. As I have said in the beginning, RFT is designed to satisfy t-channel unitarity (to be more precise: partial wave unitarity),and, there is no a priori restriction on the parameters such as Δ and α' : as long as no connection was made between RFT and a specific underlying theory, it was, therefore, the stategy to vary the RFT parameters and to see for which values a realistic strong interaction theory emerges.

Fig. 26 shows the two phases of RFT: the intercept of the output singularity, i.e. the power of s of the elastic forward scattering amplitude, has been plotted as a function of the negative bare mass: $-\Delta_0 = \alpha(0) - 1$ In the subcritical phase to the left of the dotted line the total cross section is falling. There is no problem with s-channel unitarity, but from the physical point of view this phase has little interest, since in nature σ_{total} is far from being falling. When $-\Delta_0$ approaches the critical value slightly to the right of zero (i.e. $\alpha_{0c}(0)$ is slightly above one), the total cross section becomes less and less falling until the power of s reaches zero: at the critical point $\sigma_{total} \sim (\ln s)^{-\gamma}$, $-\gamma \sim 0.2$. At this critical point a phase transition occurs: particle production shows long range correlations, and for the elastic scattering amplitude a scaling law with two anomalous dimensions holds. Consistency of this solution with s-channel unitarity is highly nontrivial: it has been checked quite extensively (including the decoupling problems of the Pomeron), and all tests have been passed successfully. Thus this critical RFT is an excellent candidate for strong interaction theory at high energies. Those energies, however, for which asymptopia of strong interactions is expected to set in, lie above presently available energy ranges, and it remains to explain how the finite energy tail of asymptopia connects up with critical RFT. In the supercritical phase to the right of the dotted line in Fig.26 (the bare mass is now negative) two solutions have been suggested (and there is still disagreement on which of them is the correct one): the first one has been obtained by Amati et al[41]) and leads to a total cross section which saturates the Froissart bound $\sigma_{total} \sim (\ln s)^2$. For such a behavior of the total cross section unitarity in the s and t-channel presents certain problems, and a complete check is still missing. The most important physical implication of this solution lies in the fact that the rise of the total cross section as observed at ISR energies, does not require any special value for Δ_o: as long as $\Delta_o < \Delta_{o\,critical}$, the behavior of σ_{total} has the same s-dependence. The other solution to RFT in the supercritical phase has been presented by A.White[42]). It leads to a falling total cross section, thus making the phase picture in Fig. 26 quite symmetric with respect to the critical point. While this solution has no problems with unitarity, its physical implications are very strong: the <u>only</u> possibility for having a nonfalling total cross section is <u>critical RFT</u>, and this requires a very <u>special</u> reason why the bare Pomeron intercept takes just the critical value. White[43]) also gives an explanation for this: he argues that criticality of RFT can be explained within QCD as being equivalent to confinement. I shall now try to explain this argument, which, of course, relies upon the validity of the <u>second</u> solution to supercritical RFT. However, I should emphasize once more that several people consider the <u>first</u> solution to be the correct one.

The basic idea is this: one <u>reformulates</u> the reggeon calculus (which has been derived in the previous section and, as elementary reggeon, only contains the quantum number carrying vector particle but no Pomeron), in terms of a <u>new</u> RFT which now contains, in addition to the vector particle reggeon field, also a Pomeron field (in terms of the vector particle, the Pomeron is a bound state of an even number of vector particles). Then one investigates the structure of this RFT as a function of the parameters of the Yang-Mills theory, in particular the mass M of the vector particle. For the Yang-Mills theory at $M^2 \neq o$ (one now considers generalizations of the SU(2) Higg's model: the gauge group could be SU(3), and the pattern of generating masses for the vector particles may be more complex), it is argued that the normal $i\epsilon$-prescription should be replaced by a principal value regularization: by assumption, this is the way to reach, in the limit $M \to o$, the pure Yang-Mills case in the confining phase. The RFT obtained from such a modified Yang-Mills is found to be in the supercritical phase with a falling cross section, as long as $M^2 \neq o$, and it becomes critical at $M^2 = o$. As a result, the nonfalling cross section, being a very special feature of strong interaction physics, can be explained only in a massless confining vector theory, where confinement, by assumption, is reached in the zero mass limit of massive Yang-Mills theory with a modified $i\epsilon$ -prescription.

In order to explain this argument in somewhat more detail, it will be necessary to say a few more words about the solution to supercritical RFT, as it has been obtained before any connection to an underlying theory was made. Let us start with a RFT that contains, as the only field, the Pomeron in the subcritical phase, and decrease the mass from positive values to negative ones. Beyond the critical point the effective potential (for simplicity, our RFT contains only a triple interaction) has its stable minimum no longer at the origin, and by redefining the field variables one has to expand around other field configurations (note that in contrast to, say, the simplest Higgs model one cannot simply perform a shift of the field variables by a constant (i.e. time independent) amount: a detailed description of the "generalized shifting" procedure can be found in Ref.42). As a result, new interaction terms (Fig. 27a) and new diagrams (Fig.27b) appear, involving creation and annihilation of Pomeron pairs out of the vacuum,and the mass of the Pomeron propagator is positive again. The "new" elements in Fig. 27a, b lead to additions to the triple Pomeron interaction (Fig. 27c), giving rise to a nontrivial momentum dependence. In fact, this momentum dependence is singular: the reggeon line inside the vertex of Fig.27c carries a factor $[\alpha_0' t_1 - |\Delta_0 - \Delta_{0c}|]^{-1}$ which for positive t_1 produces a pole. This singularity is of the same form as if the upper reggeon in Fig. 27c would be an odd-signature massive vector particle of mass $|\Delta_0 - \Delta_{0c}|$, accompanied by its signature factor $[\cos \frac{\pi}{2}\alpha(t_1)]^{-1}$: this suggests that the supercritical phase has a more complex reggeon content than the subcritical phase we started with. A more detailed investigation (which, via cut reggeon field theory, takes into account the s-channel unitarity content of RFT) shows, in fact, that a consistent interpretation of this solution of supercritical RFT requires the presence of several massive reggeizing vector particles in addition to the Pomeron: in Fig. 27b, for example, the two-reggeon intermediate state receives contributions from both the two-Pomeron cut and the two vector particle cut. At the critical point $\Delta_0 = \Delta_{0c}$ these vector particles become massless (together with the Pomeron), but they completely decouple from the Pomeron because the vertices in Fig.27a are proportional to $\Delta_0 - \Delta_{0c}$.

In the next logical step of the argument one wants to identify these massive vector particles with massive gluons that exist in an unconfining phase of Yang-Mills theories (QCD). For this it is necessary to show how the (massive) reggeon calculus of the previous section (the SU(2) Higgs model now being generalized to other gauge groups and Higg's patterns) can be mapped into such a supercritical RFT. Let me show, as an example, that with an appropriate definition of the Pomeron certain elements of the reggeon calculus have, in fact, the same structure as the "singular" RFT vertex of Fig. 27c. One of the simplest elements of the reggeon calculus, the $2\to 2$ reggeon vertex, consists of several contributions one of which is illustrated in Fig. 28. Its momentum dependence comes from the exchange of an elementary gluon between the two reggeized gluons. Each reggeon line in Fig. 28a carries its signature factor which, in the small g approximation is simply a propagator $[t-M^2]^{-1}$. The singularity structure of the two-reggeon state to the left of the interaction vertex is easily analyzed: besides the two-reggeon cut, there is the reggeon particle singularity which for the normal $i\epsilon$-prescription sits on an unphysical angular momentum sheet, and the two-particle cut. Now it becomes crucial to modify the $i\epsilon$- prescription such that the reggeon particle singularity appears on the physical sheet (simultaneously the two-particle cut disappears on the unphysical sheet): in the limit $M\to o$ it becomes a pole degenerate with the Regge pole of the vector particle, but it still has the quantum numbers of a bound state of two gluons and can be identified as the Pomeron singularity. Taking this singularity on the lhs in Fig.28a and drawing a single Pomeron line for this bound state of a reggeizing gluon and an elementary gluon, we arrive at Fig.28b which (always in the limit $M^2\to o$) is of the same form as Fig. 27c. This shows that a certain part of the reggeon calculus has, after changing the $i\epsilon$- prescription, the same structure in angular momentum and transverse momentum as supercritical RFT. In the same way more complicated parts of the reggeon calculus can be identified with higher order elements of RFT in the supercritical phase. It is, however, clear that this way of dividing the reggeon calculus of massive Yang-Mills theories into several pieces each of which goes into different elements of the RFT raises counting problems which still remain to be solved: before this can be done it will be necessary to complete the calculation of the most general element of the reggeon calculus which has not been found yet .

Finally, the limit $M^2\to o$ is taken and by assumption, massive Yang-Mills theory with the modified infrared regularization reaches QCD in the confining phase. At the same time, the masses of the RFT elements, being of the order M^2, approach zero, and the singular elements à la Fig.28b disappear: from the analysis of the supercritical phase of RFT it then follows that the RFT has become critical with the nonfalling cross section $\sigma_{total} \sim [\ln s]^{-\gamma}$ [44].

IV. Summary: the Regge limit in QCD

In these lectures I have reviewed the present status of the high energy (Regge) limit of nonabelian gauge theories, distinguishing between what has been achieved already, what sort of strategies and approaches seem to emerge, and what remains to be done in the future. Since (almost) all existing calculations start from perturbation theory of spontaneously broken gauge theories, hoping that at the end the limit, where the Higgs sector decouples, can be taken and reaches pure Yang-Mills theory, I have first tried to illustrate how good perturbation theory can be for this Regge limit: there is hope that perturbation theory is a valid starting point, since the Regge limit can be studied very close to the perturbative regime of QCD. But selection and summation of terms in the perturbution expansion must be much more complicated, because the Regge limit is also sensitive to features that have to do with confinement. It is, therefore, necessary to keep a certain control, throughout all calculations, of how reliable the perturbative approach is, and this can be done by keeping an eye on the hadron radius $\langle b^2 \rangle$.

After dividing gauge models into two classes - those where all vector particles reggeize and those where some of them don't - I have spent some time on describing, for the first type, how unitarity in both the s and t-channel can be used to classify those terms in the perturbation expansion which (at least) have to be summed up for obtaining a valid high energy description. The result (for the massive, i.e. spontaneously broken, Yang-Mills case) comes in form of a complete reggeon calculus, thus generalizing that property of the theory which at a lower level had manifested itself in the reggeization of the vector particles. The elements of this reggeon calculus are computable, but an expression for the general interaction vertex has still to be found. The zero-mass limit seems to exist, provided the external couplings are taken to be a model for hadronic bound states (e.g. $q\bar{q}$).

For the summation of all these contributions two different approaches seem to emerge. The first one, being more geometrical, investigates the distribution in impact parameter space of the wee partons. A diffusion picture then emerges which is quite different from the random walk picture in multiperipheral models: diffusion, as a function of the number of steps, takes place in the variable $\ln k_\perp^2$ rather than b. It is argued that, after summing <u>all</u> contributions required by unitarity, the hadronic radius $\langle b^2 \rangle$ <u>may</u> stay finite when the mass of the vector particles is taken to zero, but a new technique has to be developped in order to put this on a firm ground. Such a technique would also allow to study the abelian case (QED), where the summation of diagrams is still incomplete. The second approach makes use of the phase structure of reggeon field theory, and is based upon one of the two competing solutions that have been advocated for the supercritical phase. By assuming that QCD in the confining phase can be reached from spontaneously broken gauge theories in the zero mass limit, but only after the $i\epsilon$-prescription of the massive case has been altered, it is argued that such a massive case corresponds to supercritical reggeon field theory with a falling cross section, whereas in the zero mass limit the reggeon field theory becomes critical with the nonfalling cross section $\sigma_{total} \sim (\ln s)^{-\gamma}$.

Acknowledgement:

For very helpful discussions I am indebted to Profs. V.N.Gribov, L.N.Lipatov and A.R. White.

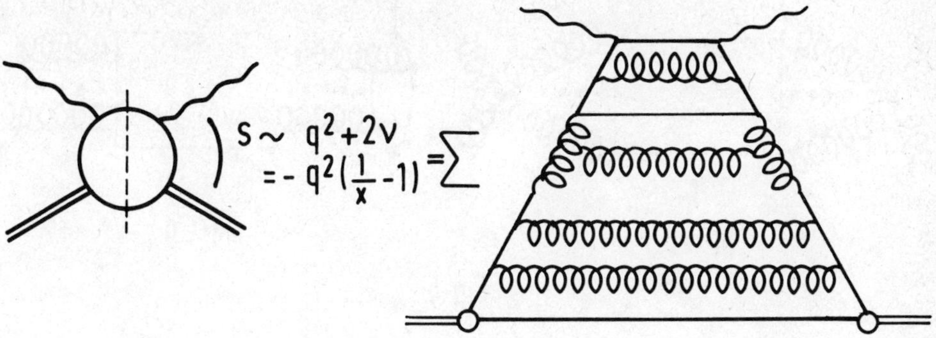

Fig.1

Fig.1 : Hadronic part of the deep inelastic leptoproduction process in QCD

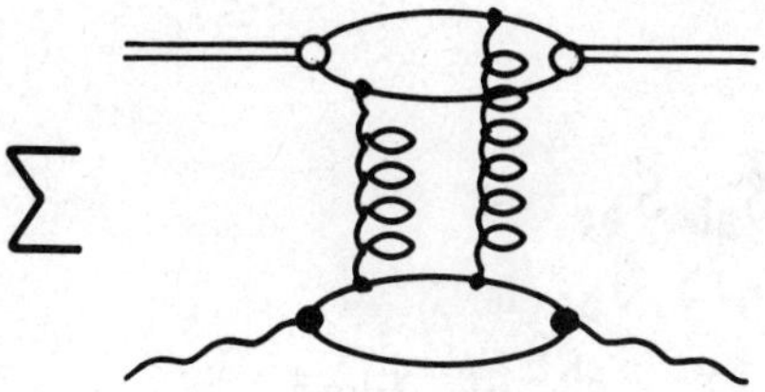

Fig. 2

Fig.2 : Simplest model for elastic photon-hadron scattering in QCD: the sum goes over all possibilities of coupling two gluons to the quark lines

Fig.3 : Reggeization of an elementary particle in field theory: the exchange on the lhs is elementary, on the rhs the particle reggeizes

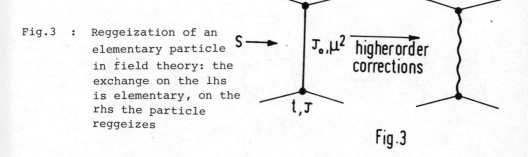

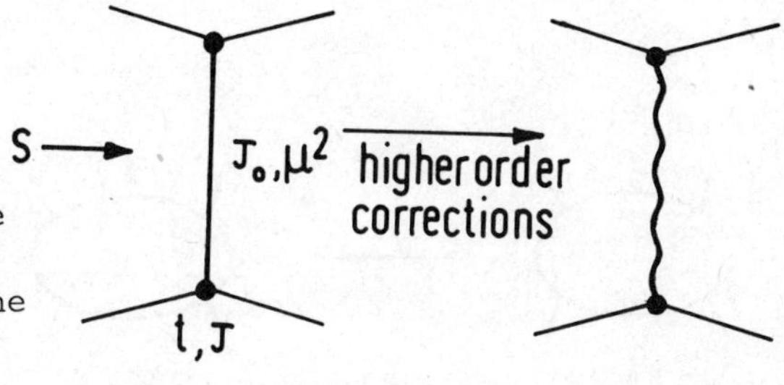

Fig.3

$$M^2 \neq 0 \qquad\qquad M^2 = 0$$

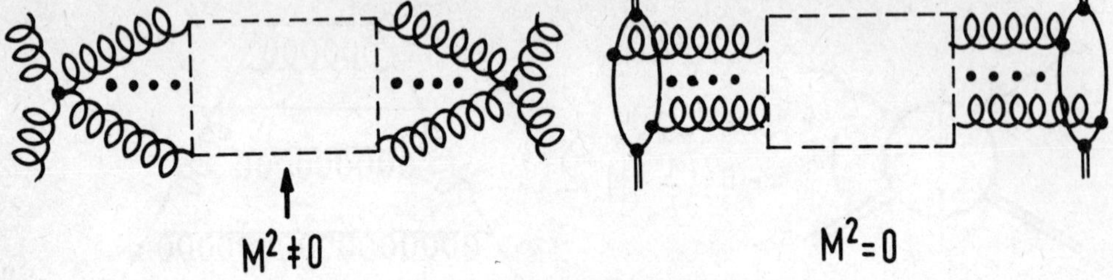

Fig.4

Fig.4 : Model for the Pomeron in QCD: the lhs denotes the unitary high energy expression for vector-vector scattering in the massive Higgs model; on the rhs the external particles are replaced by $q\bar{q}$ bound states, and the gluon mass is taken to zero

$$S, S_{ab}, S_{bc} \longrightarrow \infty$$
$$S_{ab}/S, S_{bc}/S \longrightarrow 0$$
$$t_1, t_2, \frac{S_{ab}S_{bc}}{S} \text{ fixed}$$

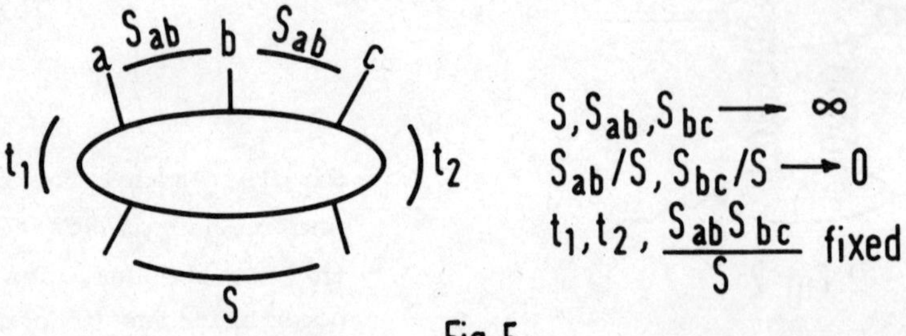

Fig.5

Fig.5 : Kinematics of the 2->3 process in the double Regge limit

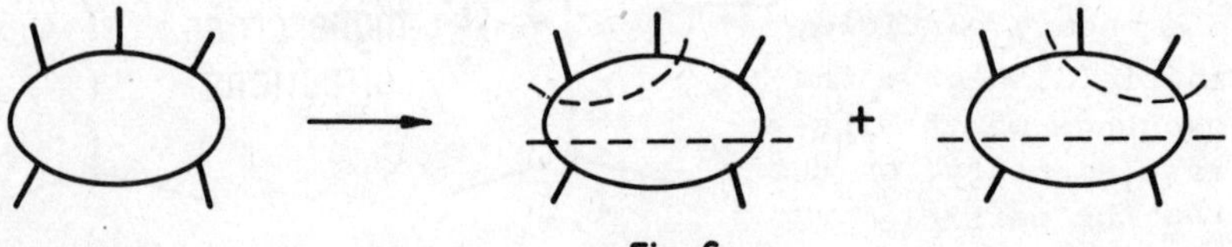

Fig.6

Fig.6 : Analytic decomposition of the 2->3 amplitude in the double Regge limit

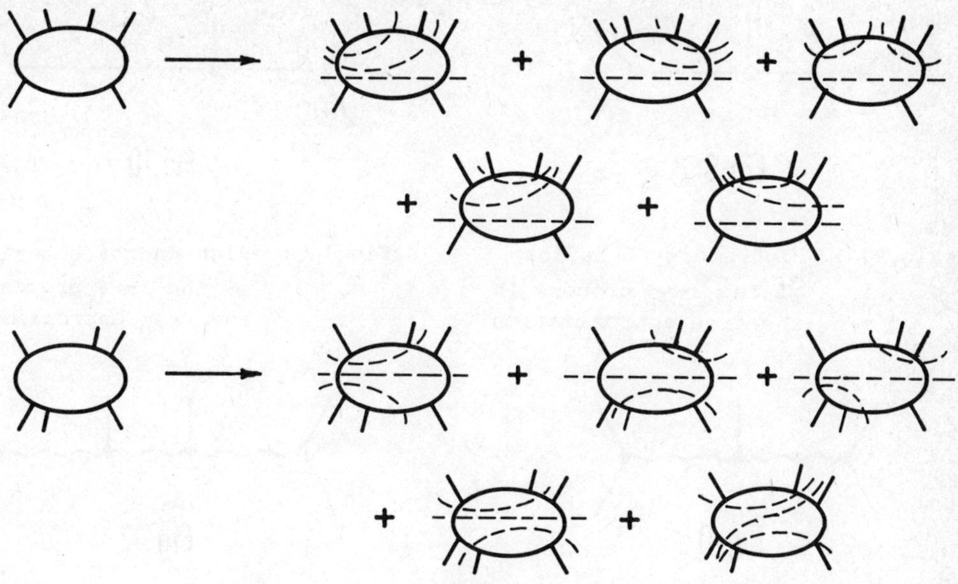

Fig.7

Fig.7 : Analytic decomposition of the 2 -> 4 and 3 -> 3 amplitudes

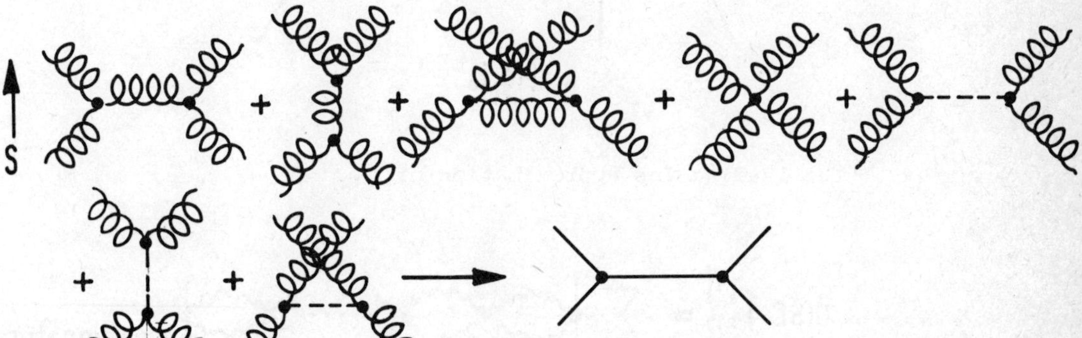

Fig.8

Fig.8 : Seven Feynmann diagrams for vector-vector scattering in lowest order perturbation theory and their high energy behavior

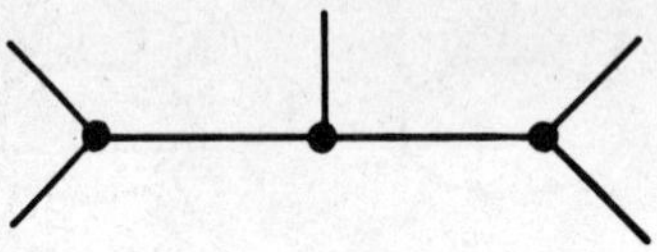

Fig.9

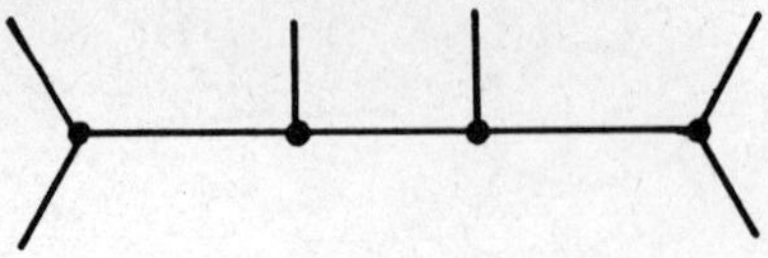

Fig.10

Fig.9 : High energy behavior of the 2-> 3 process in the tree approximation

Fig.1o : High energy behavior of the 2-> 4 process in the tree approximation

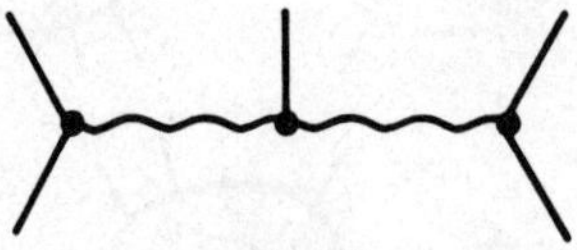

Fig.11

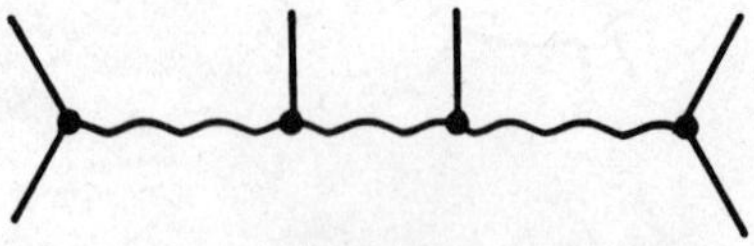

Fig.12

Fig.11 : The leading-logarithm approximation of $T_{2\to 3}$: the wavy lines denote the exchange of a reggeized vector particle

Fig.12 : The leading-lns approximation of $T_{2\to 4}$

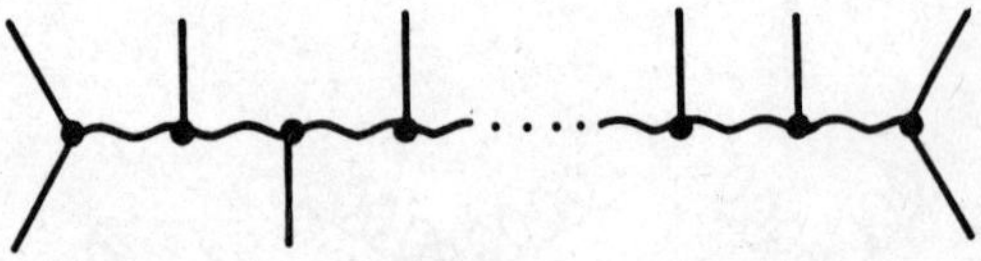

Fig.13

Fig.13 : The leading-lns approximation of $T_{n\to m}$

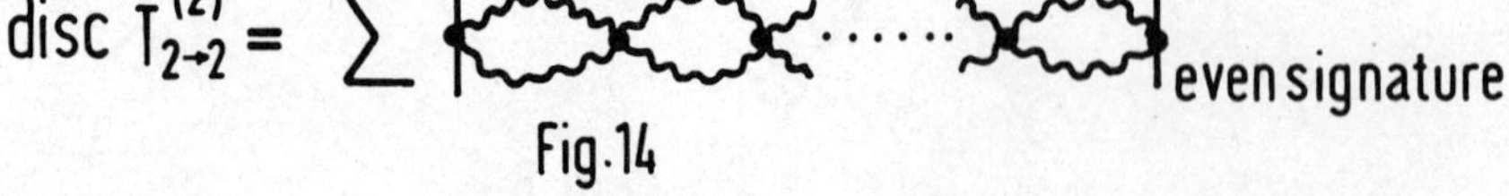

Fig.14

Fig.14 : The leading-lns approximation for even signature amplitudes $T_{2\to 2}$, as defined by its discontinuity

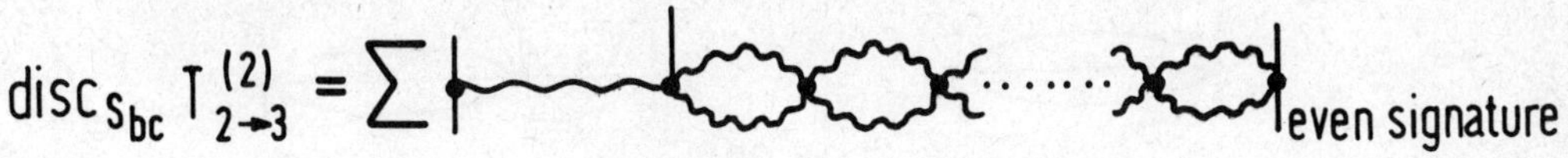

Fig.15

Fig.15 : The leading-lns approximation for the 2->3 amplitude with signatures $(\tau_1, \tau_2) = (-,+)$

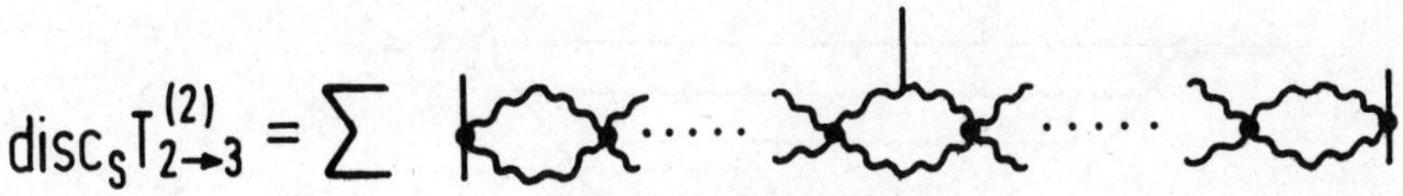

Fig.16

Fig.16 : The leading-lns approximation for the 2->3 amplitude with signatures $(\tau_1, \tau_2) = (+ , +)$

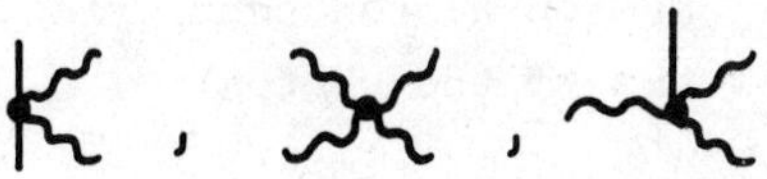

Fig.17

Fig.18

Fig.17 : Elements of the reggeon calculus for $T^{(2)}$

Fig.18 : Reggeon diagrams for $T^{(3)}_{2\to 2}$ with odd signature

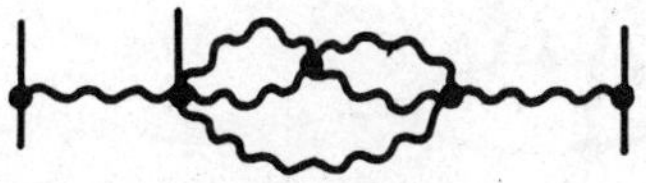

Fig.19

Fig.19 : Reggeon diagrams for $T^{(3)}_{2\to 3}$ with signature (- , -).

Fig.20

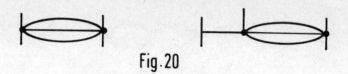

Fig.2o : Lowest order perturbation theory for Figs.18 and 19

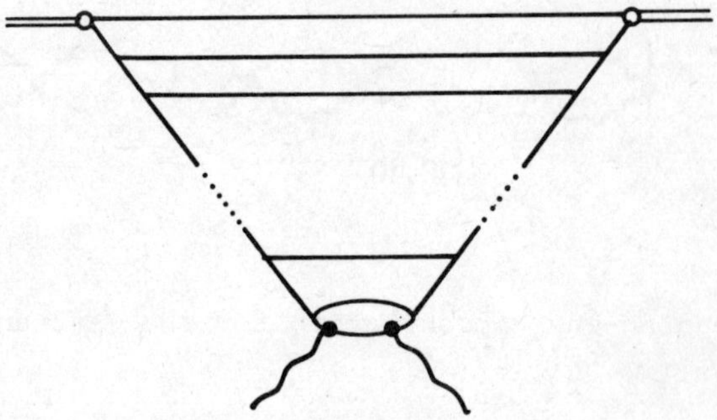

Fig. 21

Fig.21 : Space time picture for the elastic photon-hadron scattering process in the Regge limit (rest frame of the photon)

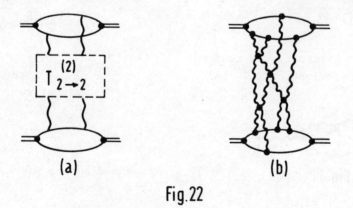

Fig.22

Fig.22 : Elastic scattering of two $q\bar{q}$-systems in QCD: (a) the zero mass limit of $T^{(2)}_{2\to 2}$; (b) Parts of $T^{(4)}_{2\to 2}$ for which the zero mass limit can be shown to exist

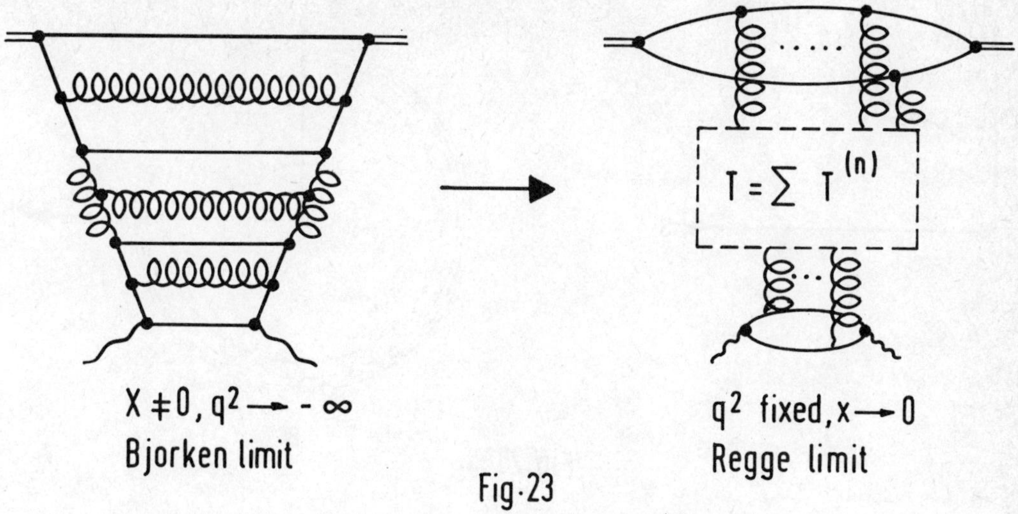

Fig.23 : Elastic photon hadron scattering in QCD: on the lhs in the Bjorken limit, on the rhs in the Regge limit

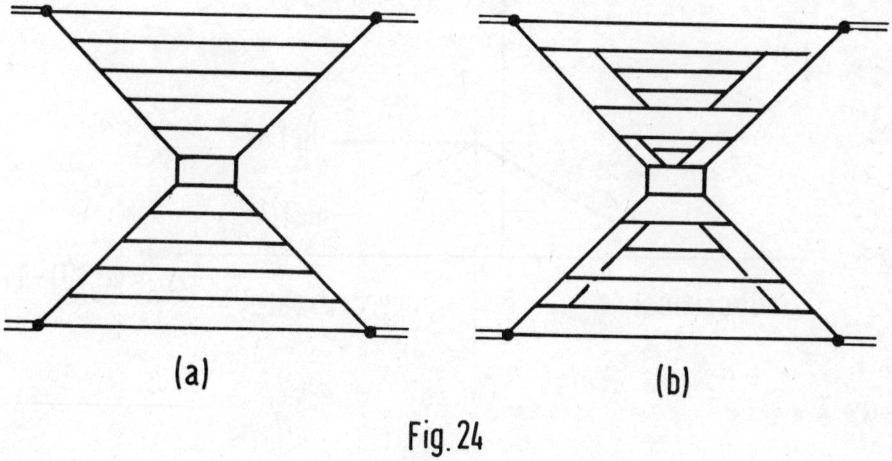

Fig.24 : Space-time picture for the elastic photon-hadron scattering process in the Regge limit (CM-system): (a) $T^{(2)}_{2\to 2}$; (b) $T^{(4)}_{2\to 2}$

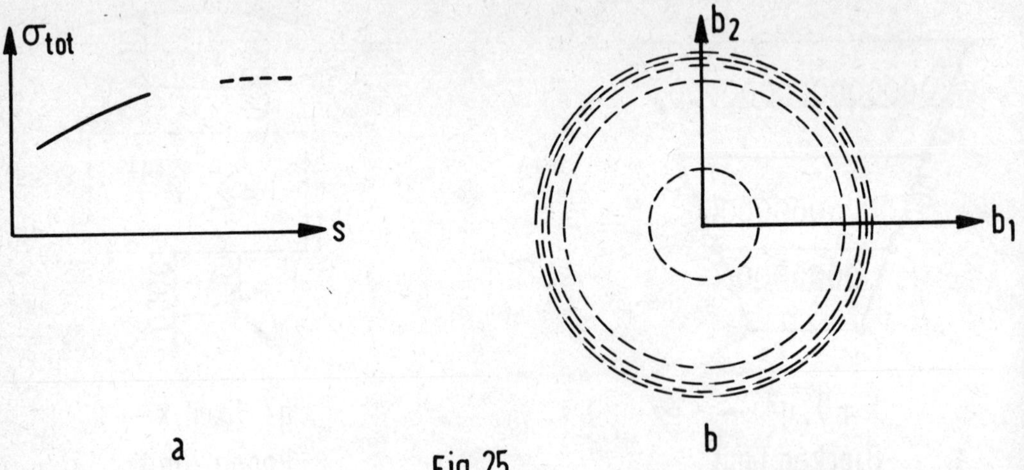

Fig.25

Fig.25 : (a) The total cross section as obtained in $T^{(2)}_{2\to 2}$; (b) the hadron extension in b-space for $T^{(2)}_{2\to 2}$ (massive case)

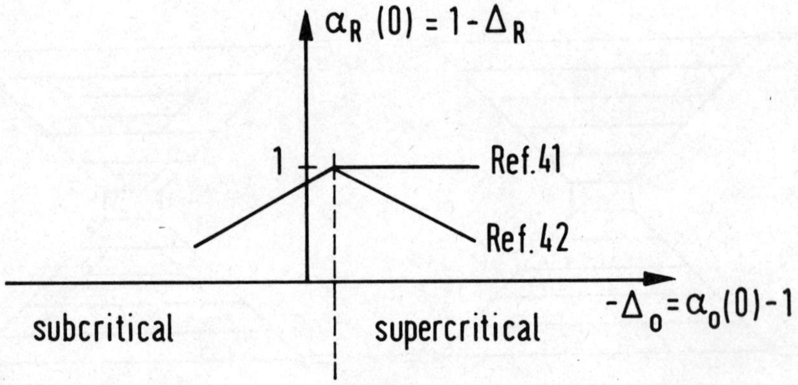

Fig.26

Fig.26 : Phase structure of reggeon field theory: the intercept of the renormalized Pomeron singularity is plotted against the bare negative mass. The two curves to the right of the critical point (dashed line) indicate the two solutions described in Refs. 41 and 42.

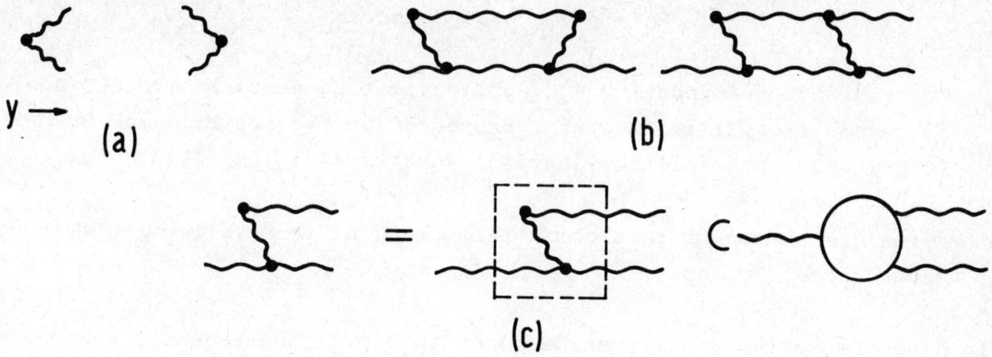

Fig. 27

Fig.27 : Elements of supercritical RFT according to Ref. 42: (a) Pomeron creation and annihilation; (b) new diagrams which appear only in this phase of RFT; (c) interpretation of (b): new additions to the triple -Pomeron vertex

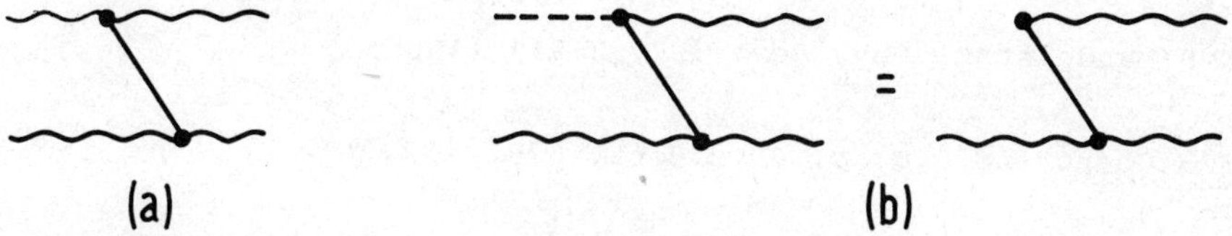

Fig. 28

Fig.28 : A part of the 2- 2 reggeon vertex(a), as obtained in massive Yang-Mills theory, is identified as the "singular" vertex (b) of RFT (Fig.27c)

References and Footnotes

1) For reviews of foundation and applications of perturbative QCD see, for example: J.Ellis, Lectures presented at the Les Houches Summer School 1976; H.D.Politzer, Physics Reports 14 , 129 (1974)

2) For a discussion of this point I am grateful to Dr.J.Kwieczinsky from Cracov, Poland.

3) A comprehensive review can be found in H.D.I. Abarbanel, J.B.Bronzan, R.L.Sugar, and A.R.White, Physics Reports 21c, 121 (1975)

4) M.Moshe, Physics Reports 37c, 257 (1978) and references therein

5) M.Gell-Mann and M.L.Goldberger, Phys.Rev.Letters 9, 275 (1962); M.Gell-Mann M.L.Goldberger, F.E.Low, and F.Zachariasen, Phys.Letters 4, 265 (1963); M.Gell-Mann, M.Goldberger, F.E.Low, E.Marx and F. Zachariasen, Phys.Rev. 133, B 145 (1964); M.Gell-Mann,M.L.Goldberger F.E.Low, V.Singh, and F.Zachariasen, Phys.Rev. 133, B 949 (1964)

6) S.Mandelstam, Phys.Rev. 137, B 949 (1965)

7) H.Cheng and C.C.Lo, Phys.Lett. 57B, 177 (1975)

8) M.T.Grisaru, Phys.Rev. D 16, 1962 (1977); P.H. Dondi and H.R.Rubinstein, Phys.Rev. D 18, 4819 (1978)

9) K.Bardakci and M.B.Halpern, Phys.Rev. D6, 696 (1972)

1o) L.F.Li, Phys.Rev. D9, 1723 (1974)

11) M.T.Grisaru, H.J.Schnitzer, and H.-S. Tsao, Phys.Ref. D8, 4498 (1973

12) M.T.Grisaru, H.J.Schnitzer, and H.-S.Tsao, Phys.Rev. D9, 2864 (1974)

13) M.T.Grisaru and H.J.Schnitzer, Brandeis Preprint 1979

14) L.Lukaszuk and L.Szymanowski, Preprint of Institute for Nuclear Research,Warsaw, 1979

15) H.Georgi and S.L.Glashow, Phys.Rev.Lett. 32, 438 (1974)

16) M.C.Bergere and C.de Calan, Saclay preprint DPh -T/79-7

17) H.P.Stapp, in Les Houches Lectures 1975 (North-Holland, Amsterdam) p.159; A.R.White, ibid.p.427

18) V.N.Gribov, JETP 26, 414 (1968)

19) R.C.Brower, C.E.Detar and J.Weis,Physics Reports 14c, 257 (1974)

2o) J.Bartels, Phys.Rev. D11, 2977 and 2989 (1975)

21) J.Bartels, Nucl.Phys. B 151, 293 (1979)

22) L.N.Lipatov, Yadernaya Fiz. 23, 642 (1976)

23) E.A.Kuraev, L.N.Lipatov, V.S.Fadin, JETP 71, 84o (1976)

24) Ya.Ya. Balitsky, L.N.Lipatov, and V.S.Fadin in "Materials of the 14 th Winter School of Leningrad Institut of Nuclear Research 1979", p.1o9

25) E.A.Kuraev, L.N.Lipatov, and V.S.Fadin, JETP 72, 377 (1977)

26) J.Bartels, in preparation

27) V.N.Gribov in "Materials of the 8 th Winter School of Leningrad Institute of Nuclear Research 1973", p.5.

28) S.-J.Chang and S.-K. Ma, Phys.Rev. 188, 2385 (1969)

29) R.K.Ellis, H.Georgi, M.Machacek, H.D.Politzer, and G.G.Ross, CALT 68-684

3o) H.T.Nieh and Y.P.Yao, Phys.Rev. D 13, 1o82 (1976); B.M.McCoy and T.T.Wu, Phys.Rev. D 12, 2357 (1976) and Rhys.Rev. D13, 1o76 (1976); L.Tyburski, Phys.Rev. D 13, 11o7 (1976);

31) C.Y.Lo and H.Cheng, Phys.Rev. D 13, 1131 (1976) and Phys.Rev. D 15, 2959 (1977)

32) J.A.Dickinson, Phys.Rev. D 16, 1863 (1977)

33) H.Cheng, J.Dickinson, C.Y.Lo, K.Olausen and P.S.Yeung, Phys.Letters 76 B, 129 (1978)

34) H.Cheng, J.A.Dickinson, C.Y.Lo, and K.Olausen, Preprint 1977 and Stony Brook I TP-SB 79-7

35) P.Carruthers and F.Zachariasen, Physics Letters 62 B, 338 (1976)

36) J.B.Bronzan and R.L.Sugar, Phys.Rev. D 17, 585 (1978)

37) J.Bartels, unpublished

38) V.N.Gribov, L.N.Lipatov, and G.V.Frolov, Yad.Fiz 12, 994 (1971) Sov.Journ. of Nucl.Phys. 12, 543 (71)

39) H.Cheng and T.T.Wu, Phys.Rev. D1, 2775 (197o) and Phys.Lett.24, 1456 (197o)

4o) S.-J.Chang and P.M.Fishbane, Phys.Rev. D 2, 11o4 (197o)

41) For a review of this solution see M.Le Bellac in "19 th International conference on High Energy Physics, Tokyo 1978", p.153 and references therein.

42) A.R.White, Ref.TH 2592-CERN

43) A.R.White, Ref.TH 2629-CERN

44) It should be emphasized that this argument is not strictly based on the reggeon calculus which has been derived in the previous section: there it was characterized as the $g \to o$ limit of the unitary S-matrix, and this approximation does not include renormalization of the parameters g, M^2 etc. In order to use the concept of asymptotic freedom of g^2 $(k_\perp^2)$ for large values of transverse momentum, as it is done in Ref.43, it is necessary to go beyond this approximation and include more nonleading terms. Whether this can be done in a consistent way, i.e. without destroying the subtle constraints of unitarity order by order in g^2, remains to be seen. It may also be that some of these new contributions are nonperturbative, i.e.they cannot be expanded in powers of g^2 at all.

Selected Issues from

Lecture Notes in Mathematics

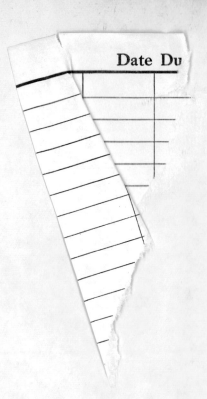